Applied Survival Analysis

Applied Survival Analysis

Regression Modeling of Time to Event Data

DAVID W. HOSMER, Jr.

Department of Biostatistics and Epidemiology
University of Massachusetts
Amherst, Massachusetts

STANLEY LEMESHOW

Department of Biostatistics and Epidemiology
University of Massachusetts
Amherst, Massachusetts

A Wiley-Interscience Publication
JOHN WILEY & SONS, INC.
New York • Chichester • Weinheim • Brisbane • Singapore • Toronto

Copyright © 1999 by John Wiley & Sons, Inc.

All rights reserved. Published simultaneously in Canada.

Library of Congress Cataloging in Publication Data:

Hosmer, David W.
    Applied survival analysis : regression modeling of time to event
data / David W. Hosmer, Jr., Stanley Lemeshow
      p. cm. — (Wiley series in probability and statistics)
    Includes bibliographical references and indexes.
    ISBN 0-471-15410-5 (cloth : alk. paper)
    1. Medicine—Research—Statistical methods. 2. Medical sciences—
Statistical methods—Computer programs. 3. Regression analysis—
Data processing. 4. Prognosis—Statistical methods. 5. Logistic
distribution. I. Lemeshow, Stanley. II. Title. III. Series.
R853.S7H67 1998
610'.7'27—dc21   98-27511

Printed in the United States of America

10 9 8 7 6 5 4 3

To Trina, Wylie, Tri,
and the memory of my parents
D. W. H.

To Elaine, Jenny, Adina, Steven,
my mother, Jack, and the memory of my father & Marisha
S. L.

Contents

Preface

The study of events involving an element of time has a long and important history in statistical research and practice. Examples chronicling the mortality experience of human populations date from the 1700s [see Hald (1990)]. Recent advances in methods and statistical software have placed a seemingly bewildering array of techniques at the fingertips of the data analyst. It is difficult to find either a subject matter or a statistical journal that does not have at least one paper devoted to use or development of these methods.

In spite of the importance and widespread use of these methods there is a paucity of material providing an introduction to the analysis of time to event data. A course dealing with this subject tends to be more advanced and often is the third or fourth methods course taken by a student. As such, the student typically has a strong background in linear regression methods and usually some experience with logistic regression. Yet most texts fail to capitalize on this statistical and experiential background. The approach is either highly mathematical or does not emphasize regression model building. The goal of this book is to provide a focused text on regression modeling for the time to event data typically encountered in health related studies. For this text we assume the reader has had a course in linear regression at the level of Kleinbaum, Kupper, Muller and Nizam (1998) and one in logistic regression at the level of Hosmer and Lemeshow (1989). Emphasis is placed on the modeling of data and the interpretation of the results. Crucial to this is an understanding of the nature of the "incomplete" or "censored" data encountered. Understanding the censoring mechanism is important as it may influence model selection and interpretation. Yet, once understood and accounted for, censoring is often just another technical detail handled by the computer software allowing emphasis to return to model building, assessment of model fit and assumptions and interpretation of the results.

The increase in the use of statistical methods for time to event data is directly related to their incorporation into major and minor (specialized) statistical software packages. To a large extent there are no major differences in the capabilities of the various software packages. When a particular approach is available in a limited number of packages it will be noted in this text. In general, analyses have been performed in STATA [Stata Corp. (1997)]. This easy to use package combines reasonably good graphics and excellent analysis routines, is fast, is compatible across Macintosh, Windows and UNIX platforms and interacts well with Microsoft Word 6.0. Other major statistical packages employed at various points during the preparation of this text include BMDP [BMDP Statistical Software (1992)], SAS [SAS Institute Inc. (1989)] and S-PLUS [S-Plus Statistical Sciences (1993)].

This text was prepared in camera ready format using Microsoft Word 6.0.1 on a Power Macintosh platform. Mathematical equations and symbols were built using Math Type 3.5 [Math Type: Mathematical Equation Editor (1997)]. When necessary, graphics were enhanced and modified using MacDraw.

Early on in the preparation of the text we made a decision that data sets used in the text would be made available to readers via the World Wide Web rather than on a diskette distributed with the text. The ftp site at John Wiley & Sons, Inc. for the data in this text is ftp://ftp.wiley.com/public/sci_tech_med/survival. In addition, the data may also be found, by permission of John Wiley & Sons Inc., in the archive of statistical data sets maintained at the University of Massachusetts at Internet address http://www-unix.oit.umass.edu/~statdata in the survival analysis section. Another advantage to having a text web site is that it provides a convenient medium for conveying to readers text changes after publication. In particular, as errata become known to us they will be added to an errata section of the text's web site at John Wiley & Sons, Inc. Another use that we envision for the web is the addition, over time, of new data sets to the statistical data set archive at the University of Massachusetts.

As in any project with the scope and magnitude of this text, there are many who have contributed directly or indirectly to its content and style and we feel quite fortunate to be able to acknowledge the contributions of others. One of us (DWH) would like to express special thanks to a friend and colleague, Petter Laake, Head of the Section of Medical Statistics at the University of Oslo, for arranging for a Senior Scientist Visiting Fellowship from the Research Council of Norway that supported a sabbatical leave visit to the Section in Oslo during the winter of 1997. We would like to thank Odd Aalen for reading and commenting on several sections of the text. His advice was most helpful in preparing the material on frailty and additive models in Chapter 9. While in Oslo, and afterwards, Ørnulf Borgan was especially helpful in clarifying some of the details of the counting process approach and graciously shared some, at that time, unpublished research of his and his student, J. K. Grønnesby. Thoughtful and careful commentary by outside reviewers, in particular Daniel Commenges, of the UFR de Santé Publique at the University of Bordeaux II, improved the content and quality of the text.

We are grateful to colleagues in our Department who have contributed to the development of this book. These include Drs. Jane McCusker, Anne Stoddard and Carol Bigelow for the use and insights into the data from the Project IMPACT Study and Janelle Klar and Elizabeth K. Donohoe for their extraordinarily careful reading of the manuscript and editorial suggestions.

DAVID W. HOSMER, JR.
STANLEY LEMESHOW

Amherst, Massachusetts
August, 1998

CHAPTER 1

Introduction to Regression Modeling of Survival Data

1.1 INTRODUCTION

Regression modeling of the relationship between an outcome variable and independent predictor variable(s) is commonly employed in virtually all fields. The popularity of this approach is due to the fact that biologically plausible models may be easily fit, evaluated and interpreted. Statistically, the specification of a model requires choosing both systematic and error components. The choice of the systematic component involves an assessment of the relationship between an "average" of the outcome variable and the independent variable(s). This may be guided by an exploratory analysis of the current data and/or past experience. The choice of an error component involves specifying the statistical distribution of what remains to be explained after the model is fit (i.e., the residuals).

In an applied setting, the task of model selection is, to a large extent, based on the goals of the analysis and on the measurement scale of the outcome variable. For example, a clinician may wish to model the relationship between a measure of nutritional status (e.g., caloric intake) and various demographic and physical characteristics of the child such as gender, socio-economic status, height and weight, among children between the ages of two and six seen in the clinics of a large health maintenance organization (HMO). A good place to start would be to use a model with a linear systematic component and normally distributed errors, the usual linear regression model. Suppose instead that the clinician decides to convert the nutrition data into a dichotomous variable that indicated whether the child's diet met specified intake criteria (1 =

1

yes and 0 = no). If we assume the goal of this analysis is to estimate the "effect" of the various factors via an odds-ratio, then the logistic regression model would be a good choice. The logistic regression model has a systematic component that is linear in the log-odds and has binomial/Bernoulli distributed errors. There are many issues involved in the fitting, refinement, evaluation and interpretation of each of these models. However, the clinician would follow the same basic modeling paradigm in each scenario.

This basic modeling paradigm is commonly used in texts taking a data-based approach to either linear or logistic regression [e.g., Kleinbaum, Kupper, Muller and Nizam (1998) and Hosmer and Lemeshow (1989)]. We use it in this text to motivate our discussion of the similarities and differences between the linear (and the logistic) regression model and regression models appropriate for survival data. In this spirit we begin with an example.

Example

A large HMO wishes to evaluate the survival time of its HIV+ members using a follow-up study. Subjects were enrolled in the study from January 1, 1989 to December 31, 1991. The study ended on December 31, 1995. After a confirmed diagnosis of HIV, members were followed until death due to AIDS or AIDS-related complications, until the end of the study or until the subject was lost to follow-up. We assume that there were no deaths due to other causes (e.g., auto accident). The primary outcome variable of interest is survival time after a confirmed diagnosis of HIV. Since subjects entered the study at different times over a 3-year period, the maximum possible follow-up time is different for each study participant. Possible predictors of survival time were collected at enrollment into the study. Data listed in Table 1.1 for 100 subjects are: TIME: the follow-up time is the number of months between the entry date (ENT DATE) and the end date (END DATE), AGE: the age of the subject at the start of follow-up (in years), DRUG: history of prior IV drug use (1 = Yes, 0 = No), and CENSOR: vital status at the end of the study (1 = Death due to AIDS, 0 = Lost to follow-up or alive).[1] Of many possible covariates, age and prior drug use

---

[1] Although it may seem odd that if the subject's time to failure is *not* censored the subject receives a "1" for this variable, this is the convention followed in the literature and will be followed throughout this text as well.

were chosen for their potential clinical relevance as well as for statistical purposes to illustrate techniques for continuous and nominal scale predictor variables.

One of the most important differences between the outcome variables modeled via linear and logistic regression analyses and the time variable in the current example is the fact that we may only observe the survival time partially. The variable TIME listed in Table 1.1 actually records two different things. For those subjects who died, it is the outcome variable of interest, the actual survival time. However, for subjects who were alive at the end of the study, or for subjects who were lost, TIME indicates the length of follow-up (which is a partial or incomplete observation of survival time). These incomplete observations are referred to as being *censored*. For example, subject 1 died from AIDS 5 months after being seen in the HMO clinic (CENSOR = 1) while subject 2 was not known to have died from AIDS at the conclusion of the study and had been followed for 6 months (CENSOR = 0). It is possible for a subject to have entered the study 6 months before the end or he/she could have entered the study much earlier, eventually becoming lost to follow-up as a result of moving, failing to return to the clinic or some other reason. For the time being we do not differentiate between these possibilities and consider only the two states: dead (as a result of AIDS) and not known to be dead.

The main goal for a statistical analysis of these data is to fit a model that will yield biologically plausible and interpretable estimates of the effect of age and drug use on survival time, for HIV+ patients. Before beginning any statistical modeling, we should perform a thorough univariate analysis of the data to obtain a clear sense of the distributional characteristics of our outcome variable as well as all possible predictor variables. The fact that some of our observations of the outcome variable, survival time, are incomplete is a problem for conventional univariate statistics such as the mean, standard deviation, median, etc. If we ignore the censoring and treat the censored observations as if they were measurements of survival time, then the resulting sample statistics are not estimators of the respective parameters of the survival time distribution. They are estimators of parameters of a combination of the survival time distribution and a second distribution that depends on survival time as well as statistical assumptions about the censoring mechanism. For example, the average of TIME for subjects 1 and 2 in Table 1.1 is 5.5 months. The number 5.5 months is not an estimate of the mean length of survival. We can say the mean survival is estimated to be *at least* 5.5 months. But how can we appropriately use the fact that the survival time

Table 1.1   Study Entry and Ending Dates, Survival Time (Time), Age, History of IV Drug Use (Drug) and Vital Status (Censor) at Conclusion of Study

ID	Ent Date	End Date	Time	Age	Drug	Censor	ID	Ent Date	EndDate	Time	Ag	Drug	Censor
1	15May90	14Oct90	5	46	0	1	51	11Nov89	10Feb91	15	33	0	1
2	19Sep89	20Mar90	6	35	1	0	52	1Oct90	31Oct90	1	31	0	1
3	21Apr91	20Dec91	8	30	1	1	53	20Mar90	18Jan91	10	33	0	1
4	3Jan91	4Apr91	3	30	1	1	54	30Jul90	29Aug90	1	50	1	1
5	18Sep89	19Jul91	22	36	0	1	55	17Jul89	14Feb90	7	36	1	1
6	18Mar91	17Apr91	1	32	1	0	56	10Nov90	9Feb91	3	30	1	1
7	11Nov89	11Jun90	7	36	1	1	57	5Mar89	4Jun89	3	42	1	1
8	25Nov89	25Aug90	9	31	1	1	58	2Mar91	1May91	2	32	1	1
9	11Feb91	13May91	3	48	0	1	59	11Sep89	11May92	32	34	0	1
10	11Aug89	11Aug90	12	47	0	1	60	12Sep89	12Dec89	3	38	1	1
11	11Apr90	10Jun90	2	28	1	0	61	8Apr90	6Feb91	10	33	0	0
12	11May91	10May92	12	34	0	1	62	20Apr89	20Mar90	11	39	1	1
13	17Jan91	16Feb89	1	44	1	1	63	31Jan91	2May91	3	39	1	1
14	16Feb91	17May92	15	32	1	1	64	15Sep89	15Apr90	7	33	1	1
15	9Apr91	6Feb94	34	36	0	1	65	7Dec91	7May92	5	34	1	1
16	9Mar91	8Apr91	1	36	0	1	66	4Mar90	1Oct92	31	34	0	1
17	3Aug90	2Dec90	4	54	0	1	67	20Apr89	19Sep89	5	46	1	1
18	10Jun90	8Jan92	19	35	0	0	68	16Jun89	15Apr94	58	22	0	1
19	12Jun91	11Sep91	3	44	1	0	69	1Oct90	31Oct90	1	44	1	1
20	7Jan91	8Mar91	2	38	0	1	70	1Feb91	3May91	3	37	0	0
21	29Aug89	28Oct89	2	40	0	0	71	13May89	10Dec92	43	25	0	1
22	29May89	27Nov89	6	34	1	1	72	9Aug90	8Sep90	1	38	0	1
23	16Nov90	14Nov95	60	25	0	0	73	18Dec91	17Jun92	6	32	0	1
24	9May90	8Apr91	11	32	0	1	74	23Aug90	21Jan95	53	34	0	1
25	10Sep91	9Nov91	2	42	1	0	75	19Jan91	19Mar92	14	29	0	1
26	26Dec91	26May92	5	47	0	1	76	26Aug91	25Dec91	4	36	1	1
27	29May91	27Sep91	4	30	0	0	77	16May91	13Nov95	54	21	0	1
28	1May90	31May90	1	47	1	1	78	20Mar89	19Apr89	1	26	1	1
29	24Mar91	22Apr92	13	41	0	1	79	5Oct91	4Nov91	1	32	1	1
30	18Jul89	17Oct89	3	40	1	1	80	21May91	19Jan92	8	42	0	1
31	16Sep90	15Nov90	2	43	0	1	81	10Jun91	9Nov91	5	40	1	1
32	22Jun89	22Jul89	1	41	0	1	82	31Aug89	30Sep89	1	37	1	1
33	27Apr90	25Oct92	30	30	0	1	83	28Dec91	27Jan92	1	47	0	1
34	16May90	14Dec90	7	37	0	1	84	29Sep90	28Nov90	2	32	1	1
35	19Feb89	20Jun89	4	42	1	1	85	20Nov91	19Jun92	7	41	1	0
36	17Feb90	18Oct90	8	31	1	1	86	2Jul89	1Aug89	1	46	1	0
37	6Aug91	5Jan92	5	39	1	1	87	11Oct91	10Aug92	10	26	1	1
38	10Aug89	10Jun90	10	32	0	1	88	11Oct90	10Oct92	24	30	0	0
39	27Dec90	25Feb91	2	51	0	1	89	5Dec90	5Jul91	7	32	1	1
40	26Apr89	24Jan90	9	36	0	1	90	8Sep89	8Sep90	12	31	1	0
41	4Dec90	3Dec93	36	43	0	1	91	10Apr90	9Aug90	4	35	0	1
42	28Apr91	28Jul91	3	39	0	1	92	11Dec90	9Sep95	57	36	0	1
43	9Jul91	7Apr92	9	33	0	1	93	15Dec90	14Jan91	1	41	1	1
44	31Dec89	1Apr90	3	45	1	1	94	13Jan89	13Jan90	12	36	1	0
45	20Dec89	18Nov92	35	33	0	1	95	22Aug91	21Mar92	7	35	1	1
46	22Jun91	20Feb92	8	28	0	1	96	2Aug91	1Sep91	1	34	1	1
47	11Apr90	11Mar91	11	31	0	1	97	22May91	21Oct91	5	28	0	1
48	22May90	19Jan95	56	20	1	0	98	2Apr90	1Apr95	60	29	0	0
49	11Nov91	10Jan92	2	44	0	0	99	1May91	30Jun91	2	35	1	0
50	18Jan91	19Apr91	3	39	1	1	100	11May89	10Jun89	1	34	1	1

for subject 1 is *exactly* 5 months while that of subject 2 is *at least* 6 months? We return to the univariate descriptive statistics problem shortly.

Suppose for the moment that we have performed the univariate analysis and wish to explore possibilities for an appropriate regression model. In linear regression modeling the first step is usually to examine a scatterplot of the outcome variable versus all continuous variables to see if the "cloud" of data points supports the use of a straight-line model. We also assess if there appears to be anything unusual in the scatter about a potential model. For example, is the linear model plausible except for one or two points? The fact that we have censored data presents a problem for the interpretation of a scatterplot with survival time data. If we were to ignore the censoring in survival time, then we would have an extension of the problem we noted with use of the arithmetic mean as an estimator of the "true" mean. The values obtained from any "line" fit to the cloud of points would not estimate the "mean" at that point. We would only know that the "mean" is *at least* as large as the point on the "line."

Regardless of this "at least" problem, a scatterplot is still a useful and informative descriptive tool with censored survival time data. However, to interpret the plot correctly we must keep track of the different types of observations by using different plotting symbols for the values assigned to the censoring variable. Figure 1.1 presents the scatterplot of TIME versus AGE for the data in Table 1.1, where different plotting symbols are used for the two levels of CENSOR. We formalize the statistical assumptions about the censoring later in Chapter 1, but for the moment we assume that it is independent of the values of survival time and all covariate variables.

Under the independence assumption the censored and non-censored points should be mixed in the plot with the mix dictated by the study design. Any trend in the plot is controlled by the nature and strength of the association between the covariate and survival time. For example, if age has a strong negative association with survival time, then observed survival times should be shorter for older subjects than for younger ones. If all subjects were followed for the *same fixed length of time*, then we would expect to find proportionally more censored observations among younger subjects than older ones. However, if subjects enter the study *uniformly over the study period* and independently of their age, then we would expect an equal proportion of censored observations at all ages. The example data are assumed to be from a study of this type. We see in Figure 1.1 that the censored and non-censored observations are mixed at about a 4 to 1 ratio at all ages.

In the linear regression model the basic shape of the scatterplot is controlled by the nature and strength of the relationship between the outcome and covariate variables and the fact that the errors follow a normal distribution (a relatively short-tailed symmetric distribution). For example, if the relationship is systematically linear and strongly positive, then the cloud of points should be a tight ellipse oriented from lower left to upper right. If the relationship is weakly linear and positive, then the cloud will be more circular in shape with a left to right orientation. If the relationship is quadratic with a strong association, then the cloud may look like a banana. With survival data the shape of the plot is also controlled by the nature of the systematic relationship between "time" and the covariate, but the distribution of the errors is typically skewed to the right. The shape of the plot in Figure 1.1 is controlled by the strong association in these data between age and survival time, the fact that survival time is skewed to the right and the constraint that subjects can be followed for at most 84 months. The cloud of points in Figure 1.1 is densest for short survival times and slowly

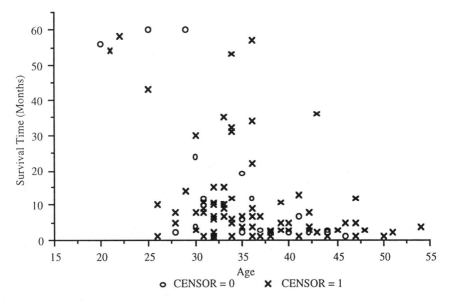

Figure 1.1 Scatterplot of survival time versus age for 100 subjects in the HMO-HIV+ study. The plotting symbols represent values of CENSOR.

trickles out to longer times with the plot truncated at the maximum length of follow-up.

In order to illustrate the shape of the plot when the covariate is strongly positively related to survival time, we reverse the order of age by creating a new variable IAGE = 1000/AGE. The scatterplot of TIME versus the created variable is shown Figure 1.2. In this case we see that the plot has the same shape but in the other direction.

We are still faced with the task of how to use the scatterplot to postulate a model for the systematic component and the issue of identifying an appropriate distribution for the errors. In linear regression when a choice for the parametric model is neither clearly indicated by the scatterplot nor provided by past experience or by some underlying biologic or clinical theory, we can use a technique called "scatterplot smoothing" to yield a non-parametric estimate of the systematic component. Cleveland (1993) discusses scatterplot smoothing and several of the methods are available in the STATA and S-Plus software packages as well as others. A scatterplot smoothing of a plot such as the one in Figure 1.1 could be difficult to interpret since censored and non-censored times have been treated equally. That is, the presence of the censored observations in the smoothing process could, in some examples, make it difficult to visualize the systematic component of the survival times.

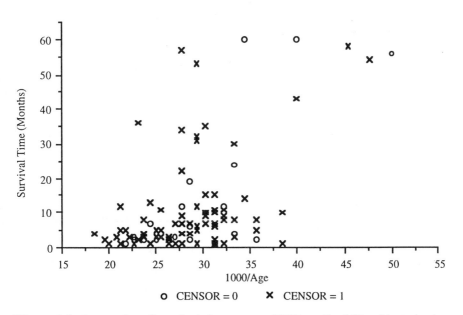

Figure 1.2 Scatterplot of survival time versus 1000/age for 100 subjects in the HMO-HIV+ study. The plotting symbols represent values of CENSOR.

The scatterplot in Figure 1.1 can be used to illustrate other fundamental differences between an analysis of censored survival time and a
normal errors linear regression. The dependent variable, TIME, must
take on positive values. Thus any model we choose for the systematic
component of the model must yield fitted values which are strictly positive. This discourages use of a strictly linear model, as fitted values
could be negative, especially for subjects with short survival times. If we
look at Figure 1.1 and try to draw a smooth curve (systematic component) which, by eye, best fits the points, it would begin in the top left
corner and drop sharply, curving to the lower right. Curves of this basic
shape can often be described by a function with the basic form $t = e^{-x}$.

We noted that the distribution of survival times in Figure 1.1 appears
to be skewed to the right. The simplest statistical distribution with this
characteristic is the exponential distribution. The combination of an
exponential systematic component and exponentially distributed errors
suggests, as a beginning point, a regression model which is called the
exponential regression model. If we assume that we have a single independent variable, x, then this model may be expressed as follows:

$$T = e^{\beta_0 + \beta_1 x} \times \varepsilon, \tag{1.1}$$

where T denotes survival time and ε follows the exponential distribution
with parameter equal to one and is denoted $E(1)$ in this text.[2]   The
model in (1.1) has the desired properties of yielding positive values
from a "curved" systematic component with a skewed error distribution. Note that this model is not linear in its parameters. However, it
may be "linearized" by taking the natural log.   (In this text
$\log \equiv \log_e \equiv \ln$.) This yields the following model:

$$Y = \beta_0 + \beta_1 x + \theta, \tag{1.2}$$

where $Y = \ln(T)$ and $\theta = \ln(\varepsilon)$. The model in (1.2) looks like the equation for the usual normal errors linear regression model except that the
distribution of the errors, θ, is not normal. Instead, the errors follow an
"extreme minimum value" distribution. This distribution is not encountered often outside of applications in survival analysis but plays a
central role in models of life-length and is often referred to as the
Gumbel distribution. The mean of this distribution is 0 and its shape

---

[2] The $E(1)$ density function is $f(\varepsilon) = e^{-\varepsilon}$ and the survivorship function is
$S(\varepsilon) = e^{-\varepsilon}$.

parameter is 1 (denoted $G(0,1)$ in this text[3]). The details of this distribution are presented in Lawless (1982). [Other texts such as Evans, Hastings and Peacock (1993) present the distribution of $-\theta = -\ln(\varepsilon)$, the "extreme maximum value" distribution.] The extreme minimum value distribution is derived by considering the statistical distribution of the minimum value from a simple random sample of observations. As the size of the sample increases, the distribution of the minimum value may be shown, after appropriate scaling, to be $G(0,1)$. The notion of a survival time being the minimum of many other times is an appealing, but somewhat simplistic, way to conceptualize survival time. For example, if the survival time of a complex object, such as a computer, depends on the continued survival of each of a large number of components whose failures are independent, then survival of the computer terminates when the first component fails (i.e., the minimum value of many independent, identically distributed, observations of time). The same analogy could be used to characterize the death of a human being.

The use of the distribution $G(0,1)$ in (1.2) is somewhat like using the standard normal distribution in linear regression. The standard normal distribution is denoted $N(0,1)$ in this text. From practical experience we know that, in linear regression, the errors rarely if ever have variance equal to one. The usual assumption is that the variance is neither a function of the outcome variable nor of the independent variables. It is assumed to be constant and equal to the parameter σ^2. This distribution is denoted $N(0,\sigma^2)$. An additional parameter may be introduced into (1.2) by multiplying θ by σ to yield the model

$$y = \beta_0 + \beta_1 x + \sigma \times \theta. \tag{1.3}$$

The distribution of $\sigma \times \theta$ is denoted as $G(0,\sigma)$.

The problem we face now is not only how to fit models like those in (1.1)–(1.3) but how to fit them when some of the observations of the outcome variable are censored. In linear regression with normal errors, *least squares* is the method discussed in regression texts such as Kleinbaum, Kupper, Muller and Nizam (1998) and used by most (probably all) computer software packages. This approach yields estimators with a number of desirable statistical properties. They are normally distributed with variances and covariances whose estimates are available in the output from the regression programs in all software packages. This allows

---

[3] The density function of the $G(0,1)$ is $f(\theta) = e^{[\theta - \exp(\theta)]}$ and the survivorship function is $S(\theta) = e^{-\exp[\theta]}$.

put from the regression programs in all software packages. This allows for the t-distribution, with appropriately chosen degrees-of-freedom, to be used to form confidence intervals and to test hypotheses about individual parameters. The F-distribution, with appropriate degrees-of-freedom, may be used to assess overall model significance. Least squares is an estimation method with its own statistical properties, but it may also be viewed, with normally distributed errors, as a special case of an estimation method called *Maximum Likelihood Estimation* (MLE). We use MLE with an adaptation for censored data to fit the models in (1.1)–(1.3). This allows us to appeal to the well-developed theory for maximum likelihood estimators to test hypotheses and form confidence intervals for individual parameters and to assess overall model significance with the same ease and simplicity of computation as in linear regression.

The simplest way to conceptualize our data is to assume that continued observation of a subject is controlled by two completely independent time processes. The first is the actual survival time associated with the disease of interest. For example, in the HMO-HIV+ study it would be the length of survival after diagnosis as HIV+. The second is the length of time until a subject is lost to follow-up. Again in the HMO-HIV+ study this would be the length of time until the subject moved, died from another cause such as an auto accident, etc. We assume both of these are under observation and that the recorded time represents time to the event that occurred first. Two variables are used to characterize a subject's time, the actual observed time, T, and a censoring indicator variable, C. In this text we use $c = 1$ to denote that the observed value of T measures the actual survival time of the subject (i.e., death from the "disease" of interest was the reason follow-up ended on the subject). We use $c = 0$ to denote that follow-up ended on the subject for reasons other than death from the disease of interest. Actual observed values of these variables and a covariate for a subject are denoted by lower case letters in the triplet (t, c, x) where x denotes the value of a covariate of interest. For example, the triplet for subject 1 in Table 1.1 with AGE as the covariate is (5, 1, 46), where x = age at the time of enrollment into the study. This triplet states that subject 1 was observed for $t = 5$ months when the subject died from AIDS or AIDS-related causes ($c = 1$) and was $x = 46$ years old at the time of enrollment into the study. The triplet for subject 2 is (6, 0, 35). This triplet states that subject 2 was observed for 6 months before being lost for some reason unrelated to being HIV+ ($c = 0$) and was 35 years old at the time of enrollment into the study.

The first step in maximum likelihood estimation is to create the specific likelihood function to be maximized. In simplest terms, the likelihood function is an expression that yields a quantity similar to the probability of the observed data under the model. First, we create a fairly general likelihood function, then we apply the method to the models in (1.1)–(1.3). Suppose that the distribution of survival time for a subject with covariate x and the disease of interest can be described by the cumulative distribution function $F(t, \beta, x)$. For example, the value of the function $F(5, \beta, 46)$ gives the proportion of 46-year-old subjects expected to die from AIDS or AIDS-related causes in less than 5 months. The quantity β denotes the parameters of the distribution, which we need to estimate. For example, when we use the models in (1.1)–(1.3) the unknown parameters are $\beta = (\beta_0, \beta_1)$. The *survivorship function* is obtained from the cumulative distribution and is defined as $S(t, \beta, x) = 1 - F(t, \beta, x)$. The value of the function $S(5, \beta, 46)$ gives the proportion of 46 year olds expected to live at least 5 months. To create the likelihood function, we also need a function that we think of, for the moment, as giving the "probability" that the survival time is exactly t. This function is derived mathematically from the distribution function and is called the density function. We denote the density function corresponding to $F(t, \beta, x)$ as $f(t, \beta, x)$. For example, the value of the function $f(5, \beta, 46)$ gives the "probability" that a subject 46 years old survives exactly 5 months.[4]

We construct the actual likelihood function by considering the contribution of the triplets $(t, 1, x)$ and $(t, 0, x)$ separately. In the case of the triplet $(t, 1, x)$ we know that the survival time was exactly t. Thus the contribution to the likelihood for this triplet is the "probability" that a subject with covariate value x dies from the disease of interest at time t units. This is given by the value of density function $f(t, \beta, x)$. For the triplet $(t, 0, x)$ we know that the survival time was at least t. Thus the contribution to the likelihood function of this triplet is the *probability* that a subject with covariate value x survives at least t time units. This probability is given by the survivorship function $S(t, \beta, x)$. Under the assumption of independent observations, the full likelihood function is obtained by multiplying the respective contributions of the observed triplets, a value of $f(t, \beta, x)$ for a noncensored observation and a value

---

[4] Readers having had some mathematical statistics know that the density function does not yield a probability but a probability per-unit of time over a small interval of time, $f(t, \beta, x) = \lim_{\Delta t \to 0} \{ F(t + \Delta t, \beta, x) - F(t, \beta, x) / \Delta t \}$.

of $S(t, \beta, x)$ for censored observations. In general, a concise way to denote the contribution of each triplet to the likelihood is the expression

$$[f(t, \beta, x)]^c \times [S(t, \beta, x)]^{1-c}, \qquad (1.4)$$

where $c = 0$ or 1.

We denote the observed data for a sample of n independent observations as (t_i, c_i, x_i) for $i = 1, 2, \ldots, n$. Since the observations are assumed to be independent, the likelihood function is the product of the expression in (1.4) over the entire sample and is

$$l(\beta) = \prod_{i=1}^{n} \left\{ [f(t_i, \beta, x_i)]^{c_i} \times [S(t_i, \beta, x_i)]^{1-c_i} \right\}. \qquad (1.5)$$

To obtain the maximized likelihood with respect to the parameters of interest, β, we maximize the log-likelihood function,

$$L(\beta) = \sum_{i=1}^{n} \left\{ c_i \ln[f(t_i, \beta, x_i)] + (1 - c_i) \ln[S(t_i, \beta, x_i)] \right\}. \qquad (1.6)$$

Since the log function is monotone, the maximum of (1.5) and (1.6) occur at the same value of β; however, maximizing (1.6) is computationally simpler than maximizing (1.5). The procedure to obtain the values of the MLE involves taking derivatives of $L(\beta)$ with respect to β, the unknown parameters, setting these equations equal to zero, and solving for β.

Before becoming completely involved in maximum likelihood estimation, let us consider the implications and assumptions of our model. There are several key points to be made. We have assumed that we are in "constant contact" with our subjects and thus are able to record the exact time of survival or follow-up. In essence we have treated time as a continuous variable. Scenarios where time is observed less precisely are considered in Chapter 7. We have accounted for the partial information on survival time contained in the censored observations. That is, we have explicitly used the fact that we know survival is at least as large as the recorded follow-up time via the inclusion in the likelihood of the term $S(t, \beta, x)$ for all censored observations. Another key point is that the reasons for observing a censored observation are assumed to be completely unrelated to the disease process of interest. In the example, we assume that being lost to follow-up is unrelated to the progression of disease in an HIV+ subject. We exclude the possibility that subjects have moved to another location which they perceive to offer better care for an HIV+ individual.

After a careful examination of the scatterplot in Figure 1.1, we arrived at the conclusion that the exponential regression model in (1.1) might be a good starting point to model these data. We also noted that the model in (1.1) could be linearized to the model shown in (1.2) and further generalized by the inclusion of a shape parameter in (1.3). We now apply MLE to each of these models in turn to show that (1.1) and (1.2) are equivalent, with (1.1) yielding fitted values for time and (1.2) for log-time. Comparison of (1.1) and (1.2) to (1.3) requires discussion of the role of the extra shape parameter in the analysis.

Suppose we wish to use a software package to fit the exponential regression model in (1.1) to the data displayed in Figure 1.1. We would find that many packages (e.g., BMDP, EGRET, SAS and STATA) fit, as a default, the model in (1.2). Once this model has been fit, we can convert it by exponentiation to estimate the model in (1.1). The equations to be solved to obtain the MLE of $\boldsymbol{\beta}$ are identical for the models in (1.1) and (1.2). Thus we show in detail the application of MLE to the log-linearized model in (1.2).

The model in (1.2) states that the values of log(survival time) come from a distribution of the form $\beta_0 + \beta_1 x + G(0,1)$. This is the extreme minimum value distribution with mean equal to $\beta_0 + \beta_1 x$ and is denoted $G(\beta_0 + \beta_1 x, 1)$. Another way to describe the model is to subtract the part involving the unknown parameters (the systematic component) from both sides of the equation in (1.2) and note that since this difference, $y - (\beta_0 + \beta_1 x)$, is equal to θ, it is distributed $G(0,1)$. Thus we may obtain the contributions to the likelihood function by substituting the expression $y - (\beta_0 + \beta_1 x)$ into the equations defining the survivorship and density function for $G(0,1)$ as follows:

$$S(y, \boldsymbol{\beta}, x) = e^{-\exp\left[y - (\beta_0 + \beta_1 x)\right]} \tag{1.7}$$

and

$$f(y, \boldsymbol{\beta}, x) = e^{\left\{y - (\beta_0 + \beta_1 x) - \exp\left[y - (\beta_0 + \beta_1 x)\right]\right\}}. \tag{1.8}$$

Substituting the expressions in (1.7) and (1.8) into (1.6) yields the following log-likelihood:

$$L(\beta) = \sum_{i=1}^{n} c_i \ln\left(e^{\left\{y_i-(\beta_0+\beta_1 x_i)-\exp[y_i-(\beta_0+\beta_1 x_i)]\right\}}\right) + (1-c_i)\ln\left(e^{-\exp[y_i-(\beta_0+\beta_1 x_i)]}\right)$$

$$= \sum_{i=1}^{n} c_i[y_i-(\beta_0+\beta_1 x_i)]-e^{[y_i-(\beta_0+\beta_1 x_i)]}. \qquad (1.9)$$

In order to obtain the MLE of β, we must take the derivatives of the log-likelihood in (1.9) with respect to β_0 and β_1, set the two resulting expressions equal to zero and solve them for β_0 and β_1. The two equations to be solved are

$$\sum_{i=1}^{n}\left(c_i - e^{[y_i-(\beta_0+\beta_1 x_i)]}\right) = 0 \qquad (1.10)$$

and

$$\sum_{i=1}^{n} x_i\left(c_i - e^{[y_i-(\beta_0+\beta_1 x_i)]}\right) = 0. \qquad (1.11)$$

The equations in (1.10) and (1.11) are nonlinear in β_0 and β_1 and must be solved using an iterative method. It is not important to understand the details of how these equations are solved at this point since any software package we choose to use will have such a method. We used the exponential regression command in STATA, "ereg," to fit this model to the data in Table 1.1 using $x = $ AGE.

Table 1.2 presents the parameter estimates in the column labeled "Coeff." and estimates of the standard error of the estimated parameters in the column labeled "Std. Err." The standard error estimates are obtained from theoretical results of maximum likelihood estimation. The column labeled "z" is the ratio of the estimated coefficient to its estimated standard error and is the Wald statistic for the respective parameter. Under the usual assumptions for maximum likelihood es-

Table 1.2 Estimated Parameters, Standard Errors, z-Scores, Two-Tailed p-Values and 95 Percent Confidence Intervals for the Log-Time Exponential Regression Model Fit to the Data in Table 1.1

Variable	Coeff.	Std. Err.	z	$P>|z|$	95% Conf. Int.
Age	-0.094	0.0158	-5.96	0.00	-0.124, -0.063
Constant	5.859	0.5853	10.01	0.00	4.711, 7.006

timation, the Wald statistic follows the standard normal distribution under the hypothesis that the true parameter value is zero. The last two columns provide a two-tailed *p*-value and the endpoints of a 95 percent confidence interval computed under these assumptions.

The output in Table 1.2 shows that the maximum likelihood estimates of the two parameters are

$$\hat{\beta}_0 = 5.859 \text{ and } \hat{\beta}_1 = -0.094.$$

In this text the " ^ " is used to indicate that a particular quantity is the maximum likelihood estimate. We can use the estimates in Table 1.2 in the same manner as is used in linear regression to obtain an equation which provides predicted (i.e., fitted) values of the outcome variable, log-time. The resulting equation is $\hat{y} = 5.859 - 0.094\,\text{AGE}$. This equation may be converted to one providing fitted values for time by exponentiation, namely

$$\hat{t} = e^{5.859-0.094\,\text{AGE}}.$$

This conversion is similar to that used to convert parameter estimates in logistic regression to estimates of odds ratios. In order to see the results of fitting this model, we add it to the scatterplot that was shown in Figure 1.1. The new scatterplot with the fitted values is presented in Figure 1.3.

Recall that the objective of the analysis was to postulate and then fit a model which would yield positive fitted values and display the curvature observed in Figure 1.1 for the systematic component. Examining the plot in Figure 1.3, we can see that the fitted model has both of these properties. The curve does not go through the middle of the data in the sense that 26 data points lie above the curve and 74 below it. Intuitively, since censored observations represent lower bounds on unobserved survival times, one would expect the curve to be shifted upward. The actual location of the fitted curve on the graph depends on the value of $\hat{\beta}_0$, whose value depends on the percentage of censored observations. It suffices for the moment to note that if 80 percent of the data had been censored, the curve could have fallen above all of the points on the graph. On the whole we find, at least visually, that the model seems to provide an adequate descriptor of the trend in the data.

One possible approach to improving the fitted model would be to see whether the addition of the shape parameter, σ, in (1.3) contributes significantly to the model. The model in (1.3) is a log-Weibull distri-

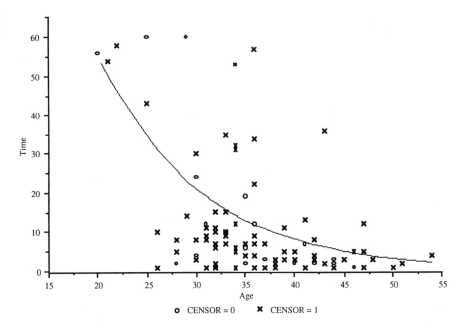

Figure 1.3 Scatterplot of survival time versus age for 100 subjects in the HMO-HIV+ study. The values of censor are the plotting symbol. The smooth curve is the fitted values, $\hat{t} = \exp(5.859 - 0.094\,\mathrm{AGE})$, from the exponential regression model in Table 1.2.

bution (see Chapter 8). We note, without showing the actual output, that the shape parameter is not significant. (For those interested, this was done by fitting the model using STATA's Weibull Regression command, "weibull.") Thus we conclude that, of the two models considered, model (1.2) describes the data as well as the more complicated model (1.3).

If we were to continue to use the linear regression modeling paradigm to motivate our approach to the analysis of these data, the next step would be to check the scale of age in the systematic component, making sure that the data support a linear model. If not, a suitable transformation must be identified. Once we felt we had done the best possible job of building the systematic component, we would use appropriately formulated regression diagnostics to search for overly influential and/or poorly fit points. This would be followed by an examination of the distribution of the estimated residuals to see if our assumptions about the error component hold. Once convinced that our model was the best fitting model possible, we would provide a clinical interpretation of the estimated model parameters. This important series of tasks is not addressed at this point, but it provides the approach for much of what follows in this text.

In summary, the HMO-HIV+ example has served to highlight the similarities and, more importantly, the differences we must address when trying to apply the linear regression modeling paradigm to the analysis of survival time data. The fact that we observed "time" places restrictions on the types of models that can be used. Any model must yield positive fitted values and its error component will be more likely to have a skewed distribution (e.g., exponential-like) than a symmetric one such as the normal. In addition, the presence of incompletely observed or censored values of "time" necessitates modifications to the standard maximum likelihood approach to estimation. It is this latter point that tends to make the analysis of survival data more complicated than a typical linear or logistic regression analysis. Thus we present a more detailed discussion of typically encountered censoring mechanisms.

1.2 TYPICAL CENSORING MECHANISMS

It may seem somewhat obvious, but we cannot discuss a censored observation until we have carefully defined an uncensored observation. This point may seem trivial, but in applied settings confusion about censoring may not be due to the incomplete nature of the observations but rather may be the result of an unclear definition of survival time. The observation of survival time, life-length, or whatever other term may be used has two components which must be unambiguously defined: a beginning point where $t = 0$ and a reason or cause for the observation of time to end. For example, in a randomized clinical trial, observation of survival time may begin on the day a subject is randomized to receive one of the treatment protocols. In an occupational exposure study, it may be the day a subject began work at a particular plant. In the HMO-HIV+ study discussed above, it was when a subject met the clinical criteria for being diagnosed as HIV+ and entered the study. In some applications it may not be obvious what the best $t = 0$ point should be. For example, in the HIV+ study, the best $t = 0$ point might be infection date; another choice might be the date of diagnosis; and a third, the criteria used in the example, might be diagnosis *and* enrollment in the study. Observation may end at the time when a subject literally "dies" from the disease of interest, or it may end upon the occurrence of some other non-fatal, well-defined, condition such as meeting clinical criteria for remission of a cancer. The survival time is the distance on the time scale between these two points.

In practice, a value of time may be obtained by calculating the number of days (or months, or years, etc.) between two calendar dates. Table 1.1 presents the entry date and end date for the subjects in the HMO-HIV+ study. Most statistical software packages have functions which allow the user to manipulate calendar dates in a manner similar to other numeric variables. They do this by creating a numeric value for each calendar date, which is defined as the number of days from some predetermined reference date. For example, the reference date used by BMDP, SAS and STATA is January 1, 1960. Subject 1 entered the study on May 15, 1990 which is 11,092 days after the reference date, and died October 14, 1990 which is 11,244 days after the reference date. The interval between these two dates is $11,244 - 11,092 = 152$ days. The number of days is converted into the number of months by dividing by $30.4375 (= 365.25/12)$. Thus the survival time in months for subject 1 is $4.994 (= 152/30.4375)$. It is common, when reporting results in tabular form, to round to the nearest whole number as shown in Table 1.1 (i.e., 5 months). The level of precision used for survival time will depend on the particular application. Clock time may be combined with calendar date to obtain survival time in units of fractions of days.

Two mechanisms that can lead to incomplete observation of time are censoring and truncation. A censored observation is one whose value is incomplete due to random factors for each subject. A truncated observation is one which is incomplete due to a selection process inherent in the study design. The most commonly encountered form of a censored observation is one in which observation begins at the defined time $t = 0$ and terminates before the outcome of interest is observed. Since the incomplete nature of the observation occurs in the right tail of the time axis, such observations are said to be *right censored*. For example, in the HMO-HIV+ study, a subject could move out of town, could die in an auto accident or the study could end before death from the disease of interest could be observed. In a study where right censoring is the only type of censoring possible, observation on subjects may begin at the same time or at varying times. For example, in a test of computer life length we may begin with all computers started at exactly the same time. In a randomized clinical trial or observational study, such as the HMO-HIV+ study, patients may enter the study over a several year enrollment period. As we see from the data reported in Table 1.1, subject 2 entered the study on September 19, 1989 while subject 4 entered on January 3,

1991. In this type of study, each subject's calendar beginning point is assumed to define the $t = 0$ point.

For obvious practical reasons all studies have a point at which observation ends on all subjects; therefore subjects entering at different times will have variable lengths of maximum follow-up time. In the HMO-HIV+ study, subjects were enrolled between January 1, 1989 and December 31, 1991, with follow-up ending December 31, 1995. Thus, the longest any subject could have been followed was 7 years. For example, subject 5 entered the study on September 18, 1989. Thus the longest this subject could have been followed was 6 years and 3.5 months. However, this subject was not followed for the maximum length of time as the subject died of AIDS or AIDS-related causes on July 19, 1991, yielding a survival time of 22 months. Incomplete observation of survival time due to the end of the study is also a right-censored observation.

A typical pattern of entry into a follow-up study is shown in Figure 1.4. This is a hypothetical 2-year-long study in which patients are enrolled during the first year. We see that subject 1 entered the study on January 1, 1990 and died on March 1, 1991. Subject 2 entered the study on February 1, 1990 and was lost to follow-up on February 1,

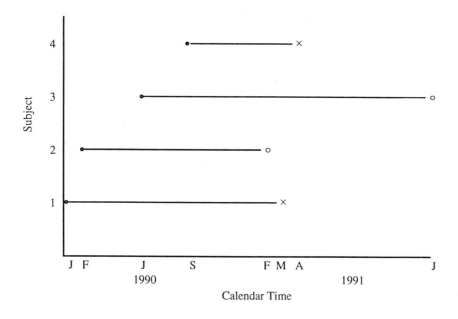

Figure 1.4 Line plot in calendar time for four subjects in a hypothetical follow-up study.

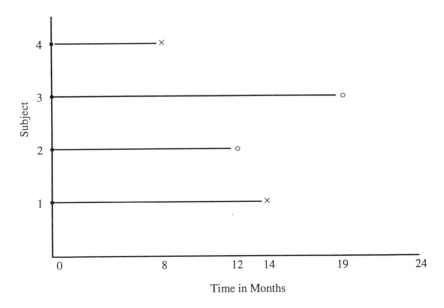

Figure 1.5 Line plot in the time scale for four subjects in a hypothetical follow-up study.

1991. Subject 3 entered the study on June 1, 1990 and was still alive on December 31, 1991, the end of the study. Subject 4 entered the study on September 1, 1990 and died on April 1, 1991. Subjects 2 and 3 have right-censored observations of survival time. These data are plotted on the actual time scale in months in Figure 1.5. Note that each subject's time has been plotted as if he or she were enrolled at exactly the same calendar time and were followed until his or her respective end point.

In some studies, there may be a clear definition of the beginning time point; but subjects may not come under actual observation until after this point has passed. For example, in modeling age at menarche, suppose we define the zero value of time as 8 years. Suppose a subject enters the study at age 10, still not having experienced menarche. We know that this subject was "at risk" for experiencing menarche since age 8 but, due to the study design, was not enrolled in the study until age 10. This subject would not enter the analysis until time 10. This type of incomplete observation of time is called *left truncation* or *delayed entry*.

Another censoring mechanism that can occur in practice is *left censoring*. An observation is left censored if the event of interest has already occurred when observation begins. For example, in the study of age at menarche, if a subject enrolls in the study at age 10, and has already experienced menarche, this subect's time is left censored.

A less common form of incomplete observation occurs when the entire study population has experienced the event of interest before the study begins. An example would be a study of risk factors for time to diagnosis of colorectal cancer among subjects in a cancer registry with this diagnosis. In this study, being in the cancer registry represents a selection process assuring that time to the event is known for each subject. This selection process must be taken into account in the analysis. This type of incomplete observation of time is called *right truncation*.

In some practical settings one may not be able to observe time continuously. For example, in a study of educational interventions to prevent IV drug use, the protocol may specify that subjects, after completion of their "treatment," will be contacted every 3 months for a period of 2 years. In this study, the outcome might be time to first relapse to IV drug use. Since subjects are contacted every 3 months, time is only accurately measured to multiples of 3 months. Given the discrete nature of the observed time variable, it would be inappropriate to use a statistical model which assumed that the observed values of time were continuous. Thus, if a subject reports at the 12-month follow-up that she has returned to drug use, we know only that her time is between 9 and 12 months. Data of this type are said to be *interval censored*.

We consider mechanisms and analysis of right-censored data throughout this text since this is the most commonly occurring form of censoring. Modifications of the methods of analysis appropriate for right-censored data to other censoring mechanisms is discussed in Chapter 7.

Prior to the development of a regression model for the relationship between age and survival time among the subjects in the HMO-HIV+ study, we mentioned that the first step in any analysis of survival time, or for that matter any set of data, should be a thorough univariate analysis. In the absence of censoring, this would use the techniques covered in an introductory course on statistical methods. The exact combination of statistics used would depend on the application. It might include graphical descriptors such histograms, box and whisker plots, cumulative percent distribution polygons or other methods. It would also include a table of descriptive statistics containing point estimates and confidence intervals for the mean, median, standard deviation and various percentiles of the distribution of each continuous variable. The presence of censored data in the sample complicates the calculations but not the fundamental goal of univariate analysis. In the next chapter we present the methods for univariate analysis in the presence of right-censored data.

1.3 EXAMPLE DATA SETS

In addition to the data from the hypothetical HMO-HIV+ study introduced in this chapter, data from two additional studies will be used throughout the text to illustrate methods and provide data for the end of chapter exercises. The data from all three studies may be obtained from the John Wiley & Sons (ftp://ftp.wiley.com/public/sci_tech_med/survival) web site. They may also be obtained from the web site of statistical data sets at the University of Massachusetts/Amherst in the section on survival data (http://www-unix.oit.umass.edu/~statdata).

Our colleagues, Drs. Jane McCusker, Carol Bigelow, and Anne Stoddard, have provided us with a subset of data from the University of Massachusetts Aids Research Unit (UMARU) IMPACT Study (UIS). This was a 5-year (1989–1994) collaborative research project (Benjamin F. Lewis, P.I., National Institute on Drug Abuse Grant #R18-DA06151) comprised of two concurrent randomized trials of residential treatment for drug abuse. The purpose of the study was to compare treatment programs of different planned durations designed to reduce drug abuse and to prevent high-risk HIV behavior. The UIS sought to determine whether alternative residential treatment approaches are variable in effectiveness and whether efficacy depends on planned program duration.

We refer to the two treatment program sites as A and B in this text. The trial at site A randomized 444 participants and was a comparison of 3- and 6-month modified therapeutic communities which incorporated elements of health education and relapse prevention. Clients in the relapse prevention/health education program (site A) were taught to recognize "high-risk" situations that are triggers to relapse and were taught the skills to enable them to cope with these situations without using drugs. In the trial at site B, 184 clients were randomized to receive either a 6- or 12-month therapeutic community program involving a highly structured life-style in a communal living setting. Our colleagues have published a number of papers reporting the results of this study, see McCusker et. al. (1995, 1997a, 1997b).

As is shown in the coming chapters, the data from the UIS provide a rich setting for illustrating methods for survival time analysis. The small subset of variables from the main study we use in this text is described

Table 1.3 Description of Variables in the UMARU IMPACT Study (UIS), 628 Subjects

Variable	Description	Codes/Values
ID	Identification Code	1–628
AGE	Age at Enrollment	Years
BECKTOTA	Beck Depression Score at Admission	0.000–54.000
HERCOC	Heroin/Cocaine Use During 3 Months Prior to Admission	1 = Heroin & Cocaine 2 = Heroin Only 3 = Cocaine Only 4 = Neither Heroin nor Cocaine
IVHX	IV Drug Use History at Admission	1 = Never 2 = Previous 3 = Recent
NDRUGTX	Number of Prior Drug Treatments	0–40
RACE	Subject's Race	0 = White 1 = Other
TREAT	Treatment Randomization Assignment	0 = Short 1 = Long
SITE	Treatment Site	0 = A 1 = B
LOT	Length of Treatment (Exit Date – Admission Date)	Days
TIME	Time to Return to Drug Use (Measured from Admission)	Days
CENSOR	Returned to Drug Use	1 = Returned to Drug Use 0 = Otherwise

in Table 1.3. Since the analyses we report in this text are based on this small subset of variables, the results reported here should not be thought of as being in any way comparable to results of the main study. In addition we have taken the liberty in this text of simplifying the study design by representing the planned duration as short versus long. Thus, short versus long represents 3 months versus 6 months planned duration at site A, and 6 months versus 12 months planned duration at site B. The time variable considered in this text is defined as the number of days from admission to one of the two sites to self-reported return to drug use. The censoring variable is coded 1 for return to drug or lost to follow-up and 0 otherwise. The study team felt that a subject who was lost to follow-up was likely to have returned to drug use. The original data have been modified in such a way as to preserve subject confidentiality.

Another data set has been provided by our colleague Dr. Robert Goldberg of the Department of Cardiology at the University of Massachusetts Medical School. The data come from The Worcester Heart Attack Study (WHAS). The main goal of this study is to describe trends over time in the incidence and survival rates following hospital admission for acute myocardial infarction (AMI). Data have been collected

Table 1.4   Description of the Variables Obtained from the Worcester Heart Attack Study (WHAS), 481 Subjects

Variable	Description	Codes / Values
ID	Identification Code	1–481
AGE	Age at Hospital Admission	Years
SEX	Gender	0 = Male, 1 = Female
CPK	Peak Cardiac Enzymes	International Units (IU/100)
SHO	Cardiogenic Shock Complications	0 = No, 1 = Yes
CHF	Left Heart Failure Complications	0 = No, 1 = Yes
MIORD	MI Order	0 = First, 1 = Recurrent
MITYPE	MI Type	1 = Q-wave, 2 = Not Q-wave 3 = Indeterminate
YEAR	Cohort Year	1 = 1975, 2 = 1978, 3 = 1981, 4 = 1984, 5 = 1986, 6 = 1988
YRGRP	Grouped Cohort Year	1 = 1975 & 1978 2 = 1981 & 1984 3 = 1986 & 1988
LENSTAY	Length of Hospital Stay	Days between Hospital Discharge and Hospital Admission
DSTAT	Discharge Status from Hospital	0 = Alive 1 = Dead
LENFOL	Total Length of Follow-up	Days between Date of Last Follow-up and Hospital Admission Date
FSTAT	Status as of Last Follow-up	0 = Alive 1 = Dead

during ten 1-year periods beginning in 1975 on all AMI patients admitted to hospitals in the Worcester, Massachusetts, metropolitan area. The main data set has information on more than 8,000 admissions. The data in this text were obtained by taking a 10 percent random sample within 6 of the cohort years. In addition only a small subset of variables is included in our data set, and subjects with any missing data were dropped from the sampled data set. Dr. Goldberg and his colleagues have published more than 30 papers reporting the results of various analyses from the WHAS. The reader interested in learning more about the WHAS and its findings should see Goldberg et. al. (1986, 1988, 1989, 1991, 1993) and Chiriboga et al. (1994). A complete list of WHAS papers may be obtained by contacting the authors of this text.

Table 1.4 describes the subset of variables used along with their codes and values. One should not infer that results reported and/or obtained in exercises in this text are comparable in any way to analyses of the complete data from the WHAS.

Various survival time variables can be created from the hospital admission date, the hospital discharge date and the date of the last follow-up. Two times have been calculated from these dates and are included in the data set, length of hospital stay (hospital admission to discharge) and total length of follow-up (hospital admission to last follow-up). Each has its own censoring variable denoting whether the subject had died or was alive at hospital discharge or last follow-up, respectively. As noted, the data set we use in this text contains a few key patient demographic characteristics and variables describing the nature of the AMI. One should be aware of the fact that the values of the variable peak cardiac enzymes are unadjusted to the respective hospital norm. The principle rationale for inclusion of this covariate is to provide a continuous covariate that may be predictive of survival and require some sort of non-linear transformation when included in the regression models discussed in this text.

EXERCISES

1. Using the data from the Worcester Heart Attack Study:

(a) Graph length of follow-up versus age using the censoring variable at follow-up as the plotting symbol for each of the pooled cohorts defined by YRGRP. Are the plots basically the same or do they differ in shape in an important way? Is it possible to tell from the shape of the plot if age is a predictor of survival time?

(b) What key characteristics of the data plotted in problem 1(a) should be kept in mind when choosing a possible regression model?

(c) By eye, draw on each of the three scatterplots from problem 1(a) what you feel is the best regression function for a survival time regression model.

(d) Obtain a cross tabulation of YRGRP and the censoring variable FSTAT and compute the percent dead and the percent censored in each of the three groups. What effect do you think the difference in the percent censored should have on the location of the lines drawn in problem 1(e)?

(e) Fit the exponential regression model to the data in each of the three scatterplots and add the fitted values to the plot (e.g., see Figure 1.3). How do the regression fitted values compare to the ones drawn in problem 1(c)? Is the response to problem 1(d) correct?

2. What key characteristics about the observations of total length of follow-up must be kept in mind when considering the computation of simple univariate descriptive statistics?

3. Repeat problems 1 and 2 using time to return to drug use and age in the UIS and grouping by study site.

CHAPTER 2

Descriptive Methods for Survival Data

2.1 INTRODUCTION

In any applied setting, a statistical analysis should begin with a thoughtful and thorough univariate description of the data. The fundamental building block of this analysis is an estimate of the cumulative distribution function. Typically, not much attention is paid to this fact in an introductory course on statistical methods, where directly computed estimators of measures of central tendency and variability are more easily explained and understood. However, routine application of standard formulas for estimators of the sample mean, variance, median, etc., will not yield estimates of the desired parameters when the data include censored observations. In this situation, we must obtain an estimate of the cumulative distribution function in order to obtain values of statistics which do estimate the parameters of interest.

In the HMO-HIV+ study described in Chapter 1, we assume that the recorded data are continuous and are subject to right censoring only. Remember that time itself is always continuous, but our inability to measure it precisely is an issue that we must deal with. We introduced the cumulative distribution function in Chapter 1 along with its complement, the survivorship function. Simply stated, the cumulative distribution function is the probability that a subject selected at random will have a survival time less than some stated value, t. This is denoted as $F(t) = \Pr(T < t)$. The survivorship function is the probability of observing a survival time greater than or equal to some stated value t, denoted $S(t) = \Pr(T \geq t)$. In most applied settings we are more interested in describing how long the study subjects live, than how quickly they die. Thus estimation (and inference) focuses on the survivorship function.

2.2 ESTIMATION OF THE SURVIVORSHIP FUNCTION

The Kaplan-Meier estimator of the survivorship function [Kaplan and Meier (1958)], also called *the product limit* estimator, is the estimator used by most software packages. This estimator incorporates information from all of the observations available, both uncensored and censored, by considering survival to any point in time as a series of steps defined by the observed survival and censored times. It is analogous to considering a toddler who must take five steps to walk from a chair to a table. This journey of five steps must begin with one successful step. The second step can only be taken if the first was successful. The third step can be taken only if the second (and also the first) was successful. Finally the fifth step is possible only if the previous four were completed successfully. In an analysis of survival time, we estimate the conditional probabilities of "successful steps" and then multiply them together to obtain an estimate of the overall survivorship function.

To illustrate these ideas in the context of survival analysis, we describe estimation of the survivorship function in detail using data for the first five subjects in the HMO-HIV+ study in Table 1.1, as shown in Table 2.1.

The "steps" are intervals defined by a rank ordering of the survival times. Each interval begins at an observed time and ends just before the next ordered time and is indexed by the rank order of the time point defining its beginning. Subject 4's survival time of 3 months is the shortest and is used to define the interval $I_0 = \{t : 0 \le t < 3\} = [0,3)$. The expression in curly brackets, { }, defines a collection or set of values that includes all times beginning with and including 0 and up to, but not including, 3. This is more concisely denoted using the mathematical notation of a square bracket to mean the value is included, a parenthesis to mean the value is not included, and the comma to mean all values in between. We use both notations in this text. The second rank-ordered

Table 2.1   Survival Times and Vital Status (Censor) for Five Subjects from the HMO-HIV+ Study

Subject	Time	Censor
1	5	1
2	6	0
3	8	1
4	3	1
5	22	1

time is subject 1's survival time of 5 months. This survival time, in conjunction with the ordered survival time of subject 4, defines interval $I_1 = \{t : 3 \leq t < 5\} = [3,5)$. The next ordered time is subject 2's censored time of 6 months and, in conjunction with subject 1's value of 5 months, defines interval $I_2 = \{t : 5 \leq t < 6\} = [5,6)$. The next interval uses subject 3's value of 8 months and the previous value of 6 months and defines $I_3 = \{t : 6 \leq t < 8\} = [6,8)$. Subject 5's value of 22 months and subject 3's value of 8 months are used to define the next to last interval $I_4 = \{t : 8 \leq t < 22\} = [8,22)$. The last interval is defined as $I_5 = \{t : t \geq 22\} = [22,\infty)$.

All subjects were alive at time $t = 0$ and remained so until subject 4 died at 3 months. Thus, the estimate of the probability of surviving through interval I_0 is 1.0; thus, the estimate of the survivorship function is

$$\hat{S}(t) = 1.0$$

at each t in I_0. Just before time 3 months, five subjects were alive, and at 3 months one subject died. In order to describe the value of the estimator at 3 months, consider a small interval beginning just before 3 months and ending at 3 months. We designate such an interval as $(3 - \delta, 3]$. The estimated conditional probability of dying in this small interval is 1/5 and the probability of surviving through it is $1 - 1/5 = 4/5$. At any specified time point, the number of subjects alive is called the number at risk of dying or simply the number at risk. At time 3 months this number is denoted as n_1, the 1 referring to the fact that 3 months is the first observed time. The number of deaths observed at 3 months was 1 but, with a larger sample, more than one could have been observed. To allow for this, we denote the number of deaths observed as d_1. In this more general notation, the estimated probability of dying in the small interval around 3 is d_1/n_1 and the estimated probability of surviving is $(n_1 - d_1)/n_1$. The probability that a subject survives to 3 months is estimated as the probability of surviving through interval I_0 times the conditional probability of surviving through the small interval around 3. Throughout the discussion of the Kaplan-Meier estimator, the word "conditional" refers to the fact that the probability applies to those who survived to the point or interval under consideration. Since we observed the death at exactly 3 months, this estimated probability would be the same no matter how small a value of δ we use to define the interval around 3 months. Thus, we consider the estimate of the survival probability to be at exactly 3 months. The value of this estimate is

$$\hat{S}(3) = 1.0 \times (4/5) = 0.8.$$

We now consider estimation of the survivorship function at each time point in the remainder of interval I_1. No other failure times (deaths) were observed, hence the estimated conditional probability of survival through small intervals about every time point in the interval is 1.0. Cumulative multiplication of these times the estimated survivorship function leaves it unchanged from its value at 3 months.

The next observed failure time is 5 months. The number at risk is $n_2 = 4$ and the number of deaths is $d_2 = 1$. The estimated conditional probability of surviving through a similarly defined small interval at 5 months, $(5 - \delta, 5]$, is $(4-1)/4 = 0.75$. By the same argument used at 3 months, the estimate of the survivorship function at 5 months is the product of the respective estimated conditional probabilities,

$$\hat{S}(5) = 1.0 \times (4/5) \times (3/4) = 0.6.$$

No other failure times were observed in I_2, thus the estimate remains at 0.6 through the interval.

The number at risk at the next observed time, 6 months, is $n_3 = 3$ and the number of deaths is zero since subject 2 was lost to follow-up at 6 months. The estimated conditional probability of survival through a small interval at 6 months is $(3-0)/3 = 1.0$. Again, the estimated survivorship function is obtained by successive multiplication of the estimated conditional probabilities and is

$$\hat{S}(6) = 1.0 \times (4/5) \times (3/4) \times (3/3) = 0.6.$$

No failure times were observed in I_3 and the estimate remains the same until the next observed failure time.

The number at risk 8 months after the beginning of the study is $n_4 = 2$ and the number of deaths is $d_4 = 1$. The estimated conditional probability of survival through a small interval at 8 months is $(2-1)/2 = 0.5$. Hence, by the same argument used at 3, 5 and 6 months, the estimated survivorship function at 8 months after the beginning of the study is

$$\hat{S}(8) = 1.0 \times (4/5) \times (3/4) \times (3/3) \times (1/2) = 0.3.$$

No other failure times were observed in I_4, thus the estimated survivorship function remains constant and equal to 0.3 throughout the interval.

The last observed failure time was 22 months. There was a single subject at risk and this subject died, hence $n_5 = 1$ and $d_5 = 1$. The estimated conditional probability of surviving through a small interval at 22 months is $(1-1)/1 = 0.0$. The estimated survivorship function at 22 months is

$$\hat{S}(22) = 1.0 \times (4/5) \times (3/4) \times (3/3) \times (1/2) \times (0/1) = 0.0.$$

No subjects were alive after 22 months; thus the estimated survivorship function is equal to zero after that point.

Through this example, we have demonstrated the essential features of the Kaplan-Meier estimator of the survivorship function. The estimator at any point in time is obtained by multiplying a sequence of conditional survival probability estimators. Each conditional probability estimator is obtained from the observed number at risk of dying and the observed number of deaths and is equal to " $(n-d)/n$." This estimator allows each subject to contribute information to the calculations as long as they are known to be alive. Subjects who die contribute to the number at risk up until their time of death, at which point they also contribute to the number of deaths. Subjects who are censored contribute to the number at risk until they are lost to follow-up.

The estimate obtained from the data in Table 2.1 is presented in tabular form in Table 2.2. Computer software packages often present an abbreviated version of this table containing only the observed failure times and estimates of the survivorship function at these times with the implicit understanding that it is constant between failure times.

A graph is an effective way to display an estimate of a survivorship function. The graph shown in Figure 2.1 is obtained from the survivorship function in Table 2.2. The graph shows the decreasing step function defined by the estimated survivorship function. It drops at the values of the observed failure times and is constant between observed failure times. An embellishment provided by some software packages, but rarely presented in published articles, is an indicator on the graph where censored observations occurred. The censored time of 6 months appears as a small × in the figure.

Table 2.2   **Estimated Survivorship Function Computed from the Survival Times for the Five Subjects from the HMO-HIV+ Study Shown in Table 2.1**

Interval	$\hat{S}(t)$
$0 \leq t < 3$	1.0
$3 \leq t < 5$	0.8
$5 \leq t < 6$	0.6
$6 \leq t < 8$	0.6
$8 \leq t < 22$	0.3
$t \geq 22$	0.0

In our example, no two subjects shared an observation time, and the longest observed time was a failure. Simple modifications to the method described above are required when either of these conditions is not met. Consider a case where a failure and a censored observation have the same recorded value. We assume that, since the censored observation was known to be alive when last seen, its survival time is longer than the recorded time. Thus a censored subject contributes to the number at risk at the recorded time but is not among those at risk immediately after that time. Along the same lines, suppose we have multi-

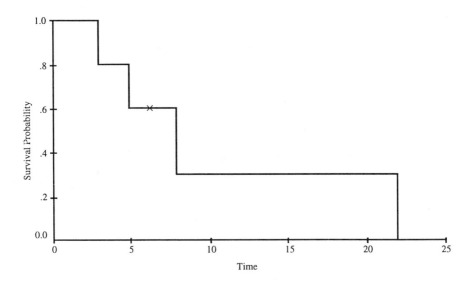

Figure 2.1 Graph of the estimated survivorship function from Table 2.2.

ple failures, $d > 1$, at some time t. It is unlikely that each subject died at the exact same time t; however, we were unable to record the data with any more accuracy. One way to break these ties artificially would be to order the d tied failure times randomly by subtracting a tiny random value from each. For example, if we had observed three values at 8 months we could subtract from each failure time the value of a uniformly distributed random variable on the interval (0, 0.01). This would artificially order the times, yet not change their respective positions relative to the rest of the observed failure times. We would estimate the survivorship function with $d = 1$ at each of the randomly ordered times. The resulting estimate of the survivorship function at the last of the d times turns out to be identical to that obtained using $(n - d)/n$ as the estimate of the conditional probability of survival for all d considered simultaneously. Thus, it is unnecessary to make adjustments for ties when estimating the survivorship function. However, if there are extensive numbers of tied failure times, then a discrete time model may be a more appropriate model choice (see Chapter 7).

If the last observed time corresponds to a censored observation, then the estimate of the survivorship function does not go to zero. Its smallest value is that estimated at the last observed survival time. In this case the estimate is considered to be undefined beyond the last observed time. If both censored and non-censored values occur at the longest observed time, then the protocol of assuming that censoring takes place after failures dictates that $(n - d)/n$ is used to estimate the conditional survival probability at this time. The estimated survivorship function does not go to zero and is undefined after this point. When these types of ties occur, software packages, which provide a tabular listing of the observed survival times and estimated survivorship function, list the censored observations after the survival time, with the value of the estimated survivorship function at the survival time. Simple examples demonstrating each of these situations are obtained by adding additional subjects to the five shown in Table 2.1.

In order to use the Kaplan–Meier estimator in other contexts, we need a more general formulation. Assume we have a sample of n independent observations denoted (t_i, c_i), $i = 1, 2, \ldots, n$ of the underlying survival time variable T and the censoring indicator variable C.[1] Assume that among the n observations there are $m \leq n$ recorded times of failure.

---

[1] Unless stated otherwise we assume recorded values of time are continuous and subject only to right censoring.

We denote the rank-ordered survival times as $t_{(1)} < t_{(2)} < \cdots < t_{(m)}$. In this text, when quantities are placed in rank order we use the same variable notation but place subscripts in parentheses. Let the number at risk of dying at $t_{(i)}$ be denoted n_i and the observed number of deaths be denoted d_i. The Kaplan–Meier estimator of the survivorship function at time t is obtained from the equation

$$\hat{S}(t) = \prod_{t_{(i)} \leq t} \frac{n_i - d_i}{n_i} \tag{2.1}$$

with the convention that

$$\hat{S}(t) = 1 \text{ if } t < t_{(1)}.$$

This formulation differs slightly from that described using the data in Table 2.1 in that intervals defined by censored observations are not considered. We saw in the example that conditional survival probabilities are equal to one at censored observations and that the estimate of the survivorship function is unchanged from the value at the previous survival time. Thus the general formula in (2.1) uses only the points at which the value of the estimator changes.

Figure 2.2 presents the graph of the Kaplan–Meier estimate of the

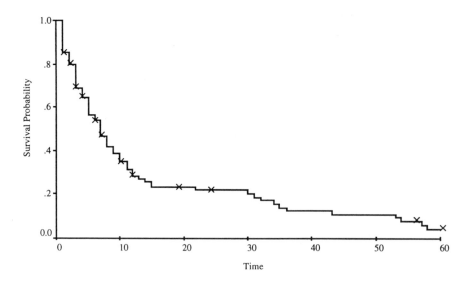

Figure 2.2 Kaplan–Meier estimate of the survivorship function for the HMO-HIV+ study.

survivorship function in (2.1), using all subjects in the HMO-HIV+ study. The construction of the estimate in this case demonstrates conventions for handling tied survival times as well as tied survival and censored times. The data, along with calculations for the beginning and end of the survivorship function, are presented in Table 2.3. The columns in Table 2.3 present the time interval, the number at risk of dying (n), the number of deaths (d), the number of subjects lost to follow-up (c), the estimate of the conditional survival probability $[(n-d)/n]$ and the estimate of the survivorship function $[\hat{s}(t)]$. All quantities are evaluated at the time point defined by the end of the previous interval and the beginning of the current interval.

The first observed survival time is 1 month; thus the value of the estimated survivorship function at each point in the interval [0,1) is 1.0. At 1 month there were 100 subjects at risk. Of these, 15 died and 2 were lost to follow-up (censored), yielding an estimate of the conditional survival probability of $0.85 = (100-15)/100$. The estimate of the survivorship function at 1 month is $0.85 = 1.0 \times 0.85$. The estimate remains at this value at each point in the interval [1,2). At the next observed survival time, 2 months, there were only 83 subjects at risk since 15 died and 2 were lost to follow-up one month before. At 2 months, 5 subjects died and 5 more were lost to follow-up; thus the estimate of the conditional survival probability is $(83-5)/83 = 0.9398$. The estimate of the survivorship function is obtained as the product of the value of the survivorship function just prior to 2 months and the conditional survival probability at 2 months and is $0.85 \times 0.9398 = 0.7988$. The estimate remains at this value throughout the interval [2,4). At the next observed survival time, 4 months, there were 73 subjects at risk, since 5 died and 5 were censored at 2 months. At 4 months, 10 subjects died and 2 were censored. The estimate of the conditional survival probability is

Table 2.3 Partial Calculations of the Kaplan–Meier Estimate Shown in Figure 1.2

Interval	n	d	c	$(n-d)/n$	$\hat{s}$
[0,1)	100	0	0	1.0	1.0
[1,2)	100	15	2	0.85	0.85
[2,4)	83	5	5	0.9398	0.7988
[4,5)	73	10	2	0.8630	0.6894
⋮	⋮	⋮	⋮	⋮	⋮
[58,60)	3	1	0	0.6667	0.0389
[60,60]	2	0	2	1.0	0.0389

$(73-10)/73 = 0.8630$ and the estimate of the survivorship function is $0.7988 \times 0.8630 = 0.6894$. The estimate remains at this value until the next observed survival time, 5 months, at which time 61 subjects are at risk. This process continues, sequentially, considering each observed survival time, until the last observed survival time, which was 58 months. At that time 3 subjects were at risk, 1 died and none were censored. The estimate of the conditional survival probability is $(3-1)/3 = 0.6667$. The estimate of the survivorship function is $0.0584 \times 0.6667 = 0.0389$, where 0.0584 is the value just prior to 58 months. The largest observed time is 60 months, when 2 subjects remained at risk and both were censored. Thus, the estimate of the conditional survival probability is $(2-0)/2 = 1.0$ and the estimate of the survivorship function remains at the value 0.0389. The function is undefined beyond 60 months, which is denoted in Table 2.3 by recording the last interval as [60,60].

When we have a large study whose mortality experience is presented in calendar time units (such as quarterly, semi-annually, etc.), the life-table estimator of the survivorship function may be used as an alternative to the Kaplan–Meier estimator. The life-table estimator has been used for more than 100 years to describe human mortality experience and is among the earliest examples of the application of statistical methods. It will not play a large role in the analysis of survival data in this text, but we present it because of its historical importance and the fact that it is a grouped-data analog of the Kaplan–Meier estimator. More detail on the various types of life-table estimators may be found in Lee (1992).

In some applied settings the data may be quite extensive with sample sizes in the many hundreds of subjects. In these situations it can be quite cumbersome to tabulate or graph the Kaplan–Meier estimator of the survivorship function. In a sense, the problem faced is similar to one addressed in a first course on statistical methods: how best to reduce the volume of data but not the statistical information that can be gleaned from it. To this end the histogram is usually introduced as an estimator of the density function and the resulting cumulative percent distribution polygon as an estimator of the cumulative distribution function. This process could be reversed. That is, we might first derive the estimator of the cumulative distribution and, afterwards, compute the histogram as a function of the cumulative distribution. When the data contain censored observations, using the second approach and deriving an estimator of the survivorship function (instead of the cumulative distribution function) is the more feasible tactic. The first step is to define the intervals that will be used to group the data. The goal in the choice of intervals is the same as for the construction of a histogram—the intervals should be biologically meaningful, yield an adequate description of the data and, if convenient, be of equal width. There are no mechanized rules for construction of the histogram, to guide in the choice of number of intervals. However, the meaningful unit will likely be some multiple of a year.

Once a set of intervals has been chosen, the construction of the estimator follows the basic idea used for the Kaplan–Meier estimator. Suppose we decide to use 6-month intervals. A typical interval will be of the form $[t, t+6)$. As before, let n denote the number of subjects at risk of dying at time t. These subjects are often described as the number who enter the interval alive. As we follow these subjects across the interval, d subjects have survival times and c subjects have censored times in this interval. Thus, not all subjects were at risk of dying for the entire interval. A modification typically employed is to reduce the size of the risk set by one-half of those censored in the interval. The rationale behind this adjustment is that if we assume the censored observations were uniformly distributed over the interval, then the average size of the risk set in the interval is $n - (c/2)$. This average risk set size is used to calculate the estimate of the conditional probability of survival through the interval as $(n - (c/2) - d)/(n - (c/2))$. These estimates of the conditional probabilities are multiplied to obtain the life-table estimator of the survivorship function.

The life-table estimator of the survivorship function for the HMO-HIV+ data using 6-month intervals is shown in Table 2.4. The estimated value of the survivorship function in the first interval is

$$0.5684 = (100 - (10/2) - 41)/(100 - (10/2)).$$

The value in the second interval is computed as

$$0.3171 = 0.5684 \times (49 - (3/2) - 21)/(49 - (3/2)).$$

The remaining values are calculated in a similar fashion.

When we graph the estimate, we have to decide how to represent the actual values. Consider the first interval $[0, 6)$, where the value of the estimated survivorship function is reported in Table 2.4 as 0.5684. If, as in Figure 2.3, we were to represent the graph as a step function, then this interval would be represented by a horizontal straight line of height

Table 2.4   Life-Table Estimator of the Survivorship Function for the HMO-HIV+ Study

Interval	Enter	Die	Censored	$\hat{S}$
[0 , 6)	100	41	10	0.5684
[6 , 12)	49	21	3	0.3171
[12 , 18)	25	6	2	0.2378
[18 , 24)	17	1	1	0.2234
[24 , 30)	15	0	1	0.2234
[30 , 36)	14	5	0	0.1436
[36 , 42)	9	1	0	0.1277
[42 , 48)	8	1	0	0.1117
[48 , 54)	7	1	0	0.0958
[54 , 60)	6	3	1	0.0435
[60 , 66)	2	0	2	0.0435

1 until 6 months when it would drop to 0.5684. Other intervals would be represented in a similar manner. An alternative representation, used by some software packages, is a polygon connecting the value of the estimator drawn at the end of the interval. The first interval would be

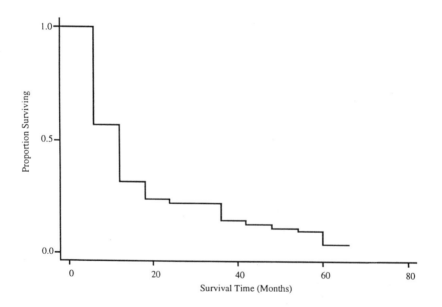

Figure 2.3 Step function representation of life-table estimate of the survivorship function for the HMO-HIV+ study in Table 2.4.

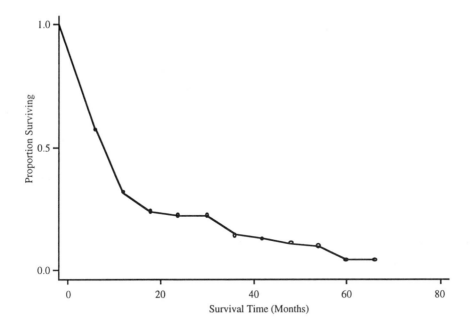

Figure 2.4 Polygon representation of life-table estimate of the survivorship function for the HMO-HIV+ study in Table 2.4.

represented by a point of height 0.5684 plotted at 6 months, the second by a point of height 0.3171 at 12 months, the third by a point of height 0.2378 at 18 months, and so on. These points are then connected by straight lines. The rationale for using the polygon is to better represent the assumed underlying continuous distribution of survival time. Some, but not all, programs will plot a point equal to 1.0 at time zero since, by definition, that is the value of the survivorship function at zero. This point is then connected to the point representing the first interval. The polygonal representation of the life-table estimator from Table 2.4 is shown in Figure 2.4.

Because the graph in Figure 2.4 has been drawn as a polygon, it looks smoother than the step function of the Kaplan–Meier estimator. The life-table estimate in Figure 2.3 in this example does a reasonable job of estimating the survivorship function. Since it is a grouped-data statistic, it is not as precise an estimate as the Kaplan–Meier estimator, which uses the individual values. Later in this chapter we discuss estimation of percentiles of the survival time distribution and these use the Kaplan–Meier estimator.

2.3 USING THE ESTIMATED SURVIVORSHIP FUNCTION

In Section 2.2 we described in detail how to calculate the Kaplan–Meier and life-table estimators of the survivorship function with little if any discussion of how to interpret the resulting estimate or how it may be used to derive point estimates of quantiles of the distribution. One of the biggest challenges in survival analysis is becoming accustomed to using the survivorship function as a descriptive statistic. This function describes the complement of what we typically describe in a set of data. The change from thinking about the percentage of observations less than a value to thinking about the percentage greater than that value, like many things, becomes easier with practice.

The survivorship function estimate shown in Figure 2.2 descends sharply at first and then tails off gradually, reaching its minimum value of 0.04 at 60 months. The initial steep descent shows that there were many subjects who died shortly after enrollment in the study. The relatively long right tail is a result of the few subjects who had long survival times. The minimum value of the survivorship function is not zero since the largest observed time was a censored observation. The shape of the curve depends on the observed survival times and the proportion of observations that are censored. If many subjects in the HMO-HIV+ study had long survival times with the same pattern of censored observations, then the curve would descend slowly at first and then more rapidly until the minimum is reached. If the survival times were more evenly distributed over the 60 months, then the curve would descend gradually to its minimum value. The pattern of enrollment in a follow up study can influence the shape of the curve. A study with a 2-year enrollment period and 5 years overall length with many late entries is likely to have more censored observations and thus a different looking estimated survivorship function than the same study with many early entries. Many factors influence the shape of the survivorship function, and thus it is difficult to make accurate statements about what a "typical" survivorship function will look like.

In most, if not all, applied settings we will need a confidence interval estimate for the survivorship function as well as point and confidence interval estimates of various quantiles of the survival time distribution. We begin by discussing confidence interval estimation of the survivorship function.

Several different approaches may be taken when deriving an estimator for the variance of the Kaplan–Meier estimator. We derive it from a technique which is referred to as the *delta method* and is based on a first-order Taylor series expansion. This method is presented in general terms in Appendix 1. The Kaplan–Meier estimator at any time t may be viewed as a product of proportions. Rather than derive a variance estimator of this product, we derive one for its log since the variance of a sum is simpler to calculate than variance of a product. The log of the Kaplan–Meier estimator is

$$\ln\left(\hat{S}(t)\right) = \sum_{t_{(i)} \le t} \ln\left(\frac{n_i - d_i}{n_i}\right)$$
$$= \sum_{t_{(i)} \le t} \ln\left(\hat{p}_i\right),$$

where

$$\hat{p}_i = \left(n_i - d_i\right)/n_i .$$

If we consider the observations in the risk set at time $t_{(i)}$ to be independent Bernoulli observations with constant probability, then $\hat{p}_i$ is an estimator of this probability and an estimator of its variance is $\left(\hat{p}_i(1 - \hat{p}_i)\right)/n_i$. As shown in Appendix 1, the variance of the log of variable X is approximately:

$$\text{Var}[\ln(X)] \cong \frac{1}{\mu_X^2}\sigma_X^2, \qquad (2.2)$$

where the mean and variance of X are denoted μ_X and σ_X^2, respectively. An estimator for the variance is obtained by replacing μ_X and σ_X^2 in (2.2) with estimators of their respective values. Applying this result to $\ln\left(\hat{p}_i\right)$ yields the estimator

$$\widehat{\text{Var}}[\ln(\hat{p}_i)] \cong \frac{1}{\hat{p}_i^2}\frac{\hat{p}_i(1 - \hat{p}_i)}{n_i}$$
$$\cong \frac{d_i}{n_i(n_i - d_i)} .$$

If we assume that observations at each time are independent, then the estimator of the variance of the log of the survivorship function is

$$\widehat{Var}\left[\ln\left(\hat{S}(t)\right)\right] = \sum_{t_{(i)} \le t} \widehat{Var}\left[\ln(\hat{p}_i)\right]$$

$$= \sum_{t_{(i)} \le t} \frac{d_i}{n_i(n_i - d_i)} . \tag{2.3}$$

An estimator of the variance of the survivorship function is obtained by another application of the delta method shown in Appendix 1. This time an approximation is applied to find the variance of an exponentiated variable and is

$$Var(e^X) \cong \left(e^{\mu_X}\right)^2 \sigma_X^2. \tag{2.4}$$

Using the fact that $\hat{S}(t) = e^{\ln\left(\hat{S}(t)\right)}$, we let X stand for $\ln\left(\hat{S}(t)\right)$, σ_X^2 stand for the variance estimator in (2.3) and approximate μ_X by $\ln\left(\hat{S}(t)\right)$ in expression (2.4). Then we obtain Greenwood's formula [Greenwood (1926)] for the variance of the survivorship function:

$$\widehat{Var}\left(\hat{S}(t)\right) = \left(\hat{S}(t)\right)^2 \sum_{t_{(i)} \le t} \frac{d_i}{n_i(n_i - d_i)} . \tag{2.5}$$

The method shown to derive the estimator in (2.5) is, in some sense, the "traditional" approach in that it may be found in most texts on survival analysis published prior to 1990. In contrast, the texts by Fleming and Harrington (1991) and Andersen, Borgan, Gill and Keiding (1993) consolidate a large number of results derived from applications of theory based on counting processes and martingales. This theory is well beyond the scope of this text, but we mention it here as it has allowed development of many useful tools and techniques for the study of survival time. The current thrust in the development of software is based on the counting process paradigm as its methods and tools may be used to analyze, in a relatively uncomplicated manner, some rather complex problems. The estimator in (2.5) may also be obtained from the counting process approach.

The counting process approach to the analysis of survival time plays a central role in many of the methods discussed in this text. A brief presentation of the central ideas behind the counting process formulation of survival analysis is given in Appendix 2. We will use results from this theory to provide justification for estimators, confidence interval estimators and hypothesis testing methods.

After obtaining the estimated survivorship function, we may wish to obtain pointwise confidence interval estimates. The counting process theory has been used to prove that the Kaplan–Meier estimator and functions of it are asymptotically normally distributed [Andersen, Borgan, Gill and Keiding (1993, Chapter IV) or Fleming and Harrington (1991, Chapter 6)]. Thus, we may obtain pointwise confidence interval estimates for functions of the survivorship function by adding and subtracting the product of the estimated standard error times a quantile of the standard normal distribution. We could apply this theory directly to the Kaplan–Meier estimator using the variance estimator in (2.5). However, this approach could easily lead to confidence interval endpoints that are less than zero or greater than one. In addition, the assumption of normality implicit in the use of the procedure may not hold for the small to moderate sample sizes often seen in typical problems. To address these problems, Kalbfleisch and Prentice (1980, page 15) suggest that confidence interval estimation should be based on the function

$$\ln\left[-\ln\left(\hat{S}(t)\right)\right],$$

called the *log-log survivorship function*. One advantage of this function over the survivorship function is that its possible range is from minus to plus infinity. The expression for the variance of the log-log survivorship function is obtained from a second application of the delta method for a log transformed variable shown in (2.2). The estimator of the variance of the log-log survivorship function is

$$\widehat{\mathrm{Var}}\left\{\ln\left[-\ln\left(\hat{S}(t)\right)\right]\right\} = \frac{1}{\left[\ln\left(\hat{S}(t)\right)\right]^2} \sum_{t_{(i)} \le t} \frac{d_i}{n_i(n_i - d_i)}. \qquad (2.6)$$

The endpoints of a $100(1-\alpha)$ percent confidence interval for the log-log survivorship function are given by the expression

$$\ln\left[-\ln\left(\hat{S}(t)\right)\right] \pm z_{1-\alpha/2}\widehat{\mathrm{SE}}\left\{\ln\left[-\ln\left(\hat{S}(t)\right)\right]\right\}, \qquad (2.7)$$

where $z_{1-\alpha/2}$ is the upper $\alpha/2$ percentile of the standard normal distribution and $\widehat{\mathrm{SE}}(\cdot)$ represents the estimated standard error of the argument, which in this case is the positive square root of (2.6). If we denote the lower and upper endpoints of this confidence interval as $\hat{c}_l$ and $\hat{c}_u$, it follows that the lower and upper endpoints of the confidence interval for the survivorship function are

$$\exp\left[-\exp(\hat{c}_u)\right] \text{ and } \exp\left[-\exp(\hat{c}_l)\right], \qquad (2.8)$$

respectively. That is, the lower endpoint from (2.7) yields the upper endpoint in (2.8). These are the endpoints reported by most, if not all, software packages for each observed value of survival time. The confidence interval is valid only for values of time over which the Kaplan–Meier estimator is defined, which is basically the observed range of survival times. Borgan and Leistøl (1990) studied this confidence interval and found that it performed well for sample sizes as small as 25 with up to 50 percent right-censored observations.

Figure 2.5 presents the Kaplan–Meier estimator of the survivorship function for the HMO-HIV+ study and the upper and lower pointwise 95 percent confidence bands computed using (2.8). The endpoints of the pointwise confidence intervals are connected to form a "confidence band." (Recall that any time one has a collection of individual 95 percent confidence interval estimates, the probability that they all contain their respective parameters is much less than 95 percent.) An alternative presentation used by some software packages connects the endpoints of the confidence intervals with vertical lines. This is useful for small data sets, but for large data sets the resulting graph becomes cluttered with too many lines, and we lose the visual conciseness seen in Figure 2.5. This figure demonstrates some of the properties of the log-log-based confidence interval estimator. The intervals are skewed for large and, though harder to see in Figure 2.5, small values of the estimated survivorship function and are fairly symmetric around 0.5. The direction of skewness is opposite for the two tails, toward zero for values of the estimated survivorship function near one and toward one for values near zero. In all cases, the endpoints lie between zero and one. In Figure 2.5, the confidence intervals further support the observation of a survivorship function describing many early deaths with a few deaths near the maximum of 5 years of follow-up.

Simultaneous confidence bands for the entire survivorship function are not as readily available as the pointwise estimates, since they require percentiles for statistical distributions not typically computed by software packages. The band proposed by Hall and Wellner (1980) is discussed in some detail in Andersen, Borgan, Gill and Keiding (1993) and

Fleming and Harrington (1991). It is also discussed in Marubini and Valsecchi (1995). A table of percentiles obtained from Hall and Wellner (1980) is provided in Appendix 3. Given the tabled percentiles, confidence bands based on the estimated survivorship function itself, or its log-log transformation, are not difficult to calculate. Borgan and Leistøl (1990) show that the performance of the Hall and Wellner confidence bands is comparable for both functions and is adequate for samples as small as 25 with up to 50 percent censoring. To maintain consistency with the pointwise intervals calculated in (2.8), which are based on the log-log transformation, we present the Hall and Wellner bands for the transformed function. Hall and Wellner, as well as Borgan and Leistøl, recommend that these confidence bands be restricted to values of time smaller than or equal to the largest observed survival time, e.g., the largest non-censored value of time denoted $t_{(m)}$. The endpoints of the $100(1-\alpha)$ percent confidence bands in the interval $[0, t_{(m)}]$ for the log-log transformation are

$$\ln\left[-\ln\left(\hat{S}(t)\right)\right] \pm H_{\hat{a},\alpha} \frac{\left(1 + n\hat{\sigma}^2(t)\right)}{\sqrt{n}\left|\ln\left(\hat{S}(t)\right)\right|}, \tag{2.9}$$

where

$$\hat{\sigma}^2(t) = \sum_{t_{(i)} \le t} \frac{d_i}{n_i(n_i - d_i)},$$

the estimator of the variance of the log of the Kaplan–Meier estimator from (2.3), and $H_{\hat{a},\alpha}$ is a percentile from Appendix 3, where

$$\hat{a} = n\hat{\sigma}^2(t_{(m)}) \Big/ \left[1 + n\hat{\sigma}^2(t_{(m)})\right].$$

If we denote the lower and upper endpoints of this confidence band as $\hat{b}_l$ and $\hat{b}_u$, then the lower and upper endpoints of the confidence band for the survivorship function are

$$\exp\left[-\exp\left(\hat{b}_u\right)\right] \text{ and } \exp\left[-\exp\left(\hat{b}_l\right)\right]. \tag{2.10}$$

To obtain the bands for the survivorship function from the HMO-HIV+ study, we note that the largest observed survival time is 58 months and $\hat{\sigma}^2(58) = 0.423$. Most software packages will provide either the values of the estimated variance of the log of the Kaplan–Meier estimator

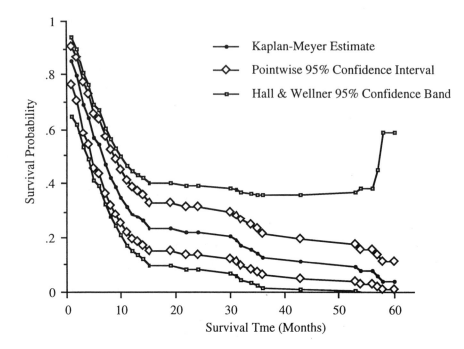

Figure 2.5 Kaplan–Meier estimate, pointwise 95% confidence intervals, and Hall and Wellner 95% confidence bands for the survivorship function for the HMO–HIV+ study.

or those of the Greenwood estimator of the variance of the survivorship function. The values of $\hat{\sigma}^2(t)$ are easily obtained by dividing the Greenwood estimator by the square of the Kaplan–Meier estimator. To obtain the percentile from Appendix 3 we compute

$$\hat{a} = (100 \times 0.423)/(1 + 100 \times 0.423) = 0.98$$

and note that, since both $H_{0.9,0.95}$ and $H_{1.0,0.95}$ equal 1.358, linear interpolation of tabled values is not necessary and we use 1.358. In cases when $\hat{a} < 0.9$, linear interpolation between two tabled values may be required to obtain the most accurate value. To obtain the confidence bands, we compute the endpoints in (2.9) and (2.10) for each observed value of time. We can ignore the censoring since the estimated survivorship function and its variance are constant between observed failure times. These endpoints may be plotted, along with the estimated survivorship function, restricting the plot to the interval [0,58]. This plot is also shown in Figure 2.5. The increased width of the confidence bands relative to the pointwise confidence intervals is seen in this figure. The increased width is needed to assure that the probability is 95 percent that each of the individual 95 percent confidence interval estimates simultaneously covers its respective parameter. In particular, we note the lack of precision in the band for times near the maximum of 58 months. The bands do support the observation of many early deaths and a few at or near the maximum follow-up time of 60 months.

The estimated survivorship function and its confidence intervals and/or bands provide a useful descriptive measure of the overall pattern of survival times. However, it is often useful to supplement the presentation with point and interval estimates of key quantiles. The estimated survivorship function may be used to estimate quantiles of the survival time distribution in the same way that the estimated cumulative distribution of, say, height or weight may be used to estimate quantiles of its distribution. This may be done graphically and the graphical procedure can be codified into a formula for analytic calculations based on the tabular form of the estimate.

The quantiles most frequently reported by software packages are the three quartile boundaries of the survival time distribution. To obtain graphical estimates, begin on the percent survival (y) axis at the quartile of interest and draw a horizontal line until it first touches the estimated survivorship function. A vertical line is drawn down to the time axis to obtain the estimated quartile. In order for the estimate to be finite, the horizontal line must hit the survivorship function. Thus, the minimum possible estimated quantile which has a finite value is the observed minimum of the survivorship function, and only quantiles within the observed range of the estimated survivorship function may be estimated. For example, if the range was from 1.0 to 0.38 then we could estimate the 75th and 50th percentiles but not the 25th percentile. Graphically determined estimates of the three quartile boundaries, denoted $\hat{t}_{75}$, $\hat{t}_{50}$ and $\hat{t}_{25}$, based on the Kaplan–Meier estimate of the survivorship function for the data in Table 2.1 are shown in Figure 2.6.

The graphical method is easy to use, but it is not especially precise. The method may be described in a formula, from which a more accurate numerical value may be determined from a tabular presentation of the estimated survivorship function. We illustrate the method by estimating the median or second quartile, $\hat{t}_{50}$, and we then generalize it into a formula that may be used for any quantile. By referring to Table 2.2, and Figure 2.6 we see that the horizontal line hits the survivorship funct-

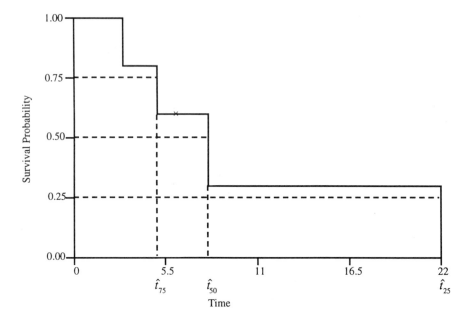

Figure 2.6 Kaplan–Meier estimate of the survivorship function for the data in Table 2.1 and graphically determined estimates of the quartiles.

ion at the riser connecting steps ending, respectively (looking right to left), at 8 and 5 months. The vertical line hits at exactly 8 months. Thus the estimated median survival time in this example is $\hat{t}_{50} = 8$. A formula to describe this estimator is

$$\hat{t}_{50} = \min\{t : \hat{S}(t) \le 0.50\}.$$

The formula says to proceed as if you are walking up a set of stairs from the right to the riser where the horizontal line hits. The estimate is the time value defining the left-most point of the step you're standing on. If we assume that the riser is attached at the top and bottom, then the description also works when the horizontal line hits one of the steps. The estimate is, again, the value of time defining the left-most point of the step. In general, the estimate of the pth percentile is

$$\hat{t}_p = \min\{t : \hat{S}(t) \le (p/100)\}.$$

The estimates of the other quartiles from Table 2.2 are $\hat{t}_{75} = 5$ and $\hat{t}_{25} = 22$.

For the full data set for the HMO-HIV+ study, the estimates of the three quartiles are $\hat{t}_{75} = 3$, $\hat{t}_{50} = 7$ and $\hat{t}_{25} = 15$. The interpretation of these values is that we estimate that 75 percent will live at least three months, half are estimated to live at least 7 months, and only 25 percent are estimated to live at least 15 months.

A confidence interval estimate for the quantiles can add further understanding about possible values for the parameter being estimated. Approximate confidence intervals may be obtained by appealing to the theory that, for large samples, the quantile estimator is normally distributed with mean equal to the quantile being estimated. An estimator of the variance of this distribution may be obtained from an application of the delta method, as outlined in Collet (1994) and discussed in greater detail from the counting process approach in Andersen, Borgan, Gill and Keiding (1993). The suggested estimator for the variance of the estimator of the pth percentile is

$$\widehat{\text{Var}}\left(\hat{t}_p\right) = \frac{\widehat{\text{Var}}\left(\hat{S}(\hat{t}_p)\right)}{\left[\hat{f}(\hat{t}_p)\right]^2}. \tag{2.11}$$

The numerator of (2.11) is Greenwood's estimator and the denominator is an estimator of the density function of the distribution of survival time. The estimator of the density function used by many software packages is

$$\hat{f}(\hat{t}_p) = \frac{\hat{S}(\hat{u}_p) - \hat{S}(\hat{l}_p)}{\hat{l}_p - \hat{u}_p}. \tag{2.12}$$

The values $\hat{u}_p$ and $\hat{l}_p$ are chosen such that $\hat{u}_p < \hat{t}_p < \hat{l}_p$ and most often are obtained from the equations shown below:

$$\hat{u}_p = \max\left\{t : \hat{S}(t) \geq (p/100) + 0.05\right\} \text{ and } \hat{l}_p = \min\left\{t : \hat{S}(t) \leq (p/100) - 0.05\right\}. \tag{2.13}$$

While values other than 0.05 could have been used in (2.13), 0.05 seems to work well in practice and is used by a number of statistical packages. The endpoints of a $100(1 - \alpha)$ percent confidence interval are

$$\hat{t}_p \pm z_{1-\alpha/2} \hat{SE}(\hat{t}_p), \tag{2.14}$$

where $\hat{SE}(\hat{t}_p) = \sqrt{\hat{Var}(\hat{t}_p)}$.

Evaluation of (2.11) through (2.14) is most easily illustrated with an example. In the HMO-HIV+ study, the estimated median survival time is $\hat{t}_{50} = 7$ months. The value of $\hat{u}_{50}$ is the largest value of time, t, such that $\hat{S}(t) \geq 0.55$. After sorting on survival time and listing the values of the Kaplan–Meier estimator, we find that $\hat{S}(5) = 0.56$ and $\hat{S}(6) = 0.54$, hence $\hat{u}_{50} = 5$. The value of $\hat{l}_{50}$ is the smallest value of t, time, such that $\hat{S}(t) \leq 0.45$. From the same listing we find that $\hat{S}(7) = 0.47$ and $\hat{S}(8) = 0.42$, hence $\hat{l}_{50} = 8$. Thus the estimate of the density function in (2.12) is

$$\hat{f}(\hat{t}_{50}) = \frac{\hat{S}(5) - \hat{S}(8)}{8 - 5} = \frac{0.56 - 0.42}{8 - 5} = 0.0467.$$

The value of Greenwood's estimator at $t = 7$ months is

$$\hat{Var}\left(\hat{S}(7)\right) = 0.002672$$

and evaluation of (2.11) yields

$$\hat{Var}\left(\hat{t}_{50}\right) = \frac{0.00267}{[0.0467]^2} = 1.224.$$

The end points of the 95 percent confidence interval for median survival time are

$$7 \pm 1.96 \times \sqrt{1.224} = (4.8, \ 9.2).$$

Table 2.5   Estimated Quartiles, Estimated Standard Errors and 95% Confidence Intervals for Survival Time in the HMO-HIV+ Study

Quantile	Estimate	Std. Err.	95% CIE
75	3	0.59	1.8, 4.2
50	7	1.11	4.8, 9.2
25	15	7.45	1.4, 29.6

Table 2.5 presents the estimated survival times for the quartiles, their estimated standard errors, and 95 percent confidence intervals. The results in Table 2.5 further quantify our previous observation of many early deaths with a few at nearly the maximum of follow-up. We note that the confidence interval is quite wide for the 25th percentile. After 15 months only 17 subjects remained at risk. The lack of precision in the confidence interval estimate for this percentile is due to the smaller number of subjects at risk. In general, the right tail of the survivorship function is estimated with considerably less precision than the left tail.

The confidence interval estimator in (2.14) requires that we compute an estimator of the density function at the estimator of the quantile, and the endpoints depend on the assumption that the distribution of the estimated quantile is normal. The sensitivity of the confidence interval to the choice of estimator of the density and the assumption of normality has not been studied. Brookmeyer and Crowley (1982) proposed an alternative method which does not require estimation of the density function [this is discussed in general terms in Andersen, Borgan, Gill and Keiding (1993)]. In this method, the confidence interval for a quantile consists of the values t such that

$$\frac{\left|\hat{S}(t) - p/100\right|}{\widehat{\text{SE}}\left(\hat{S}(t)\right)} \leq z_{1-\alpha/2}.$$

The expression on the left side is a test statistic for the hypothesis $H_o: S(t) = p/100$. The confidence interval is the set of values of t for which we would fail to reject the hypothesis. In other words, it is the set of observed survival times for which the confidence interval estimates for the survivorship function contain the quantile. This interval may be determined graphically in a manner similar to Figure 2.6 by drawing a horizontal line from $p/100$ to where it intersects the step functions defining the upper and lower pointwise confidence intervals. The endpoints of the confidence interval are found by drawing vertical lines down to the time axis. If the software package provides the capability to list the endpoints of the confidence intervals for the estimated survivorship function, then the upper and lower endpoints can be precisely determined. Alternative test statistics based on transformations of the estimated survivorship function, such as the log-log transformation, could be used equally well. Brookmeyer and Crowley recommend that this interval be used when there are no tied survival times. The data from the HMO-HIV+ study contain many tied survival times and thus it would be inappropriate to use the Brookmeyer–Crowley limits in a definitive analysis. However, these data may be used to illustrate the calculations for the median survival time.

Table 2.6 lists the values of the estimated survivorship function and the endpoints of 95 percent confidence intervals determined by the log-log transformation for survival times around the median value of 7 months.

In Table 2.6, we see that the confidence interval estimate at 4 months does not contain 0.5, while at 5 months it does contain 0.5. Thus the lower endpoint of the Brookmeyer–Crowley interval is 5 months. We see that the confidence interval at 9 months does not contain 0.5, while the interval at 8 months does contain it. Hence, the upper limit is 8 months. Brookmeyer–Crowley limits could be determined in a similar manner for other quantiles, though those for the median are most often calculated and reported by software packages. The Brookmeyer–Crowley confidence interval for the median of (5, 8) is comparable to the interval (4.8, 9.2) from Table 2.5, which was based on the large sample distribution of the estimator of the median.

In the analysis of survival time, the sample mean is not as important a measure of central tendency as it is in other settings. (The exception is in fully parametric modeling of survival times when the estimator of the mean, or a function of it, provides an estimator of a parameter vital to the analysis and interpretation of the data. We discuss parametric modeling in Chapter 8.) This is due to the fact that censored survival time data are most often skewed to the right and, in these situations, the median usually provides a more intuitive measure of central tendency. For the sake of completeness, we describe how the estimator of the mean

Table 2.6 Listing of Observed Survival Times, the Estimated Survivorship Function and Individual 95% Confidence Limits for Values of Time near the Estimated Median Survival Time of 7 months for the HMO-HIV+ Data

Time	Estimate	95% CIE
4	0.64	0.54, 0.66
5	0.56	0.46, 0.66
6	0.54	0.43, 0.64
7	0.47	0.36, 0.57
8	0.42	0.32, 0.52
9	0.39	0.28, 0.49

and the estimator of its variance are calculated and illustrate their use with examples from the HMO-HIV+ study.

Computational questions arise if the largest observation is censored, in which case one has two choices: (1) Use only the observed survival times (in which case the estimator is biased downwards) or (2) use all observations (in which case one "pretends" that the largest observation was actually a survival time, but the estimator is interpreted conditionally on the observed range). There is no uniform agreement on which is the best approach. For example, SAS (PROC LIFETEST) uses the former approach while BMDP (1L) uses the latter approach. In the absence of censoring, both approaches yield the usual arithmetic mean.

The estimator used for the mean is obtained from a mathematical result which states that, for a positive continuous random variable, the mean is equal to the area under the survivorship function. From mathematical methods of calculus this may be represented as the integral of the survivorship function over the range, that is,

$$\mu = \int_0^\infty S(u)\, du.$$

If we restrict the variable to the interval $[0, t^*]$, then the mean of the variable in this interval is

$$\mu(t^*) = \int_0^{t^*} S(u)\, du.$$

The estimator is obtained by using the Kaplan–Meier estimator of the survivorship function. The reason for restricting the range over which the mean is calculated is that the Kaplan–Meier estimator is undefined beyond the largest value of time. The value of t^* used depends on which of the two previously described approaches is chosen. Recall that the observed ordered survival times are denoted $t_{(i)}$, $i = 1, \ldots, m$. We denote the largest observed value of time in the sample as $t_{(n)}$. The two approaches to calculating the estimator of the mean correspond to defining $t^* = t_{(m)}$, that is, using the interval $[0, t_{(m)}]$, or defining $t^* = t_{(n)}$, i.e., using the interval $[0, t_{(n)}]$. In situations where the largest observed value of time is an observed failure time, the two approaches yield identical estimators.

The value of the estimator is the area under the step function defined by the Kaplan–Meier estimator and the particular interval chosen.

To illustrate the calculation, consider the data in Table 2.1 for which the estimated survivorship function is presented in Table 2.2 and is graphed in Figure 2.1. In this example, the largest observed value of time is 22 months and it represents a survival time. Thus, the value of the estimated mean is the area under the step function shown in Figure 2.1. This area is the sum of the areas of four rectangles defined by the heights of the four steps and the four observed survival times. The actual calculation is performed as follows (refer to Table 2.2):

$$\hat{\mu}(22) = 1.0 \times [3-0] + 0.8 \times [5-3] + 0.6 \times [8-5] + 0.3 \times [22-8]$$
$$= 10.6.$$

This is the value which would be reported by both BMDP and SAS.

For sake of illustration, suppose that the value recorded at 22 months was a censored observation. If we use the interval $[0,22]$ (BMDP's method), we would report the estimated mean as $\hat{\mu}(22) = 10.6$. If we use the interval $[0,8]$ (SAS's method), then we would report the estimated mean as $\hat{\mu}(8) = 6.4$. This is the area of the first three rectangles in Figure 2.1. In this example, the two estimates of the mean are quite different since the largest observation, 22 months, is much larger than the largest observed survival time, 8 months.

The equation defining the estimator based on the observed range of survival times only is

$$\hat{\mu}(t_{(m)}) = \sum_{i=1}^{m} \hat{S}(t_{(i-1)})\left(t_{(i)} - t_{(i-1)}\right), \qquad (2.15)$$

where $\hat{S}(t_{(0)}) = 1.0$ and $t_{(0)} = 0.0$. The equation defining the estimator for the entire observed range of data is

$$\hat{\mu}(t_{(n)}) = \hat{\mu}(t_{(m)}) + (1 - c_{(n)})\hat{S}(t_{(m)})(t_{(n)} - t_{(m)}), \qquad (2.16)$$

where $c_{(n)}$ denotes the censoring status, $(0, 1)$, of this observation. Each term in the summation in (2.15) denotes the calculation of the area of one of the rectangles defined by the Kaplan–Meier estimator and two observed times. Note that the estimators in (2.15) and (2.16) are identical when the largest observation and the largest observed survival time are equal.

We recommend that the estimator based on the entire observed range of the data (2.16) be used since the one based on the observed range of survival times (2.15) does not use the information on survival available in times larger than the largest survival time. We note that if those observations that are long and censored had actually been observed survival times, then the estimated mean survival time would have been increased substantially. However, there may be situations (e.g., when there is considerable uncertainty in measuring the longest censored time [$t_{(n)}$ in (2.16)], when the estimator based on survival times only is preferred.

The estimator of the variance of the sample mean is neither particularly intuitive nor easy to motivate, so we just provide it and demonstrate the calculation. In the case of no censored data, it reduces to the usual "sample variance divided by the sample size" estimator. Andersen, Borgan, Gill and Keiding (1993) present a mathematical derivation of the estimator of the mean and its variance, as well as results which show that the standard normal distribution may be used to form a confidence interval estimator. The equation defining the estimator of the variance of the sample mean computed using (2.15) is as follows:

$$\widehat{Var}(\hat{\mu}(t_{(m)})) = \frac{n_d}{n_d - 1} \sum_{i=1}^{m-1} \frac{A_i^2 d_i}{n_i(n_i - d_i)}, \tag{2.17}$$

where $n_d = \sum_{i=1}^{m} d_i$ denotes the total number of subjects with an observed survival time and

$$A_i = \sum_{j=i}^{m-1} \hat{S}(t_{(j)})(t_{(j+1)} - t_{(j)}).$$

The estimator of the variance using (2.16) is obtained by "pretending" that the largest observed time is an observed survival time for purposes of the summation in (2.17), but n_d is not changed. An example will help distinguish between the two cases. The data in Table 2.1 yielded an estimated mean $\hat{\mu}(22) = 10.6$. Evaluation of the estimator in (2.17) yields

$$\widehat{Var}[\hat{\mu}(22)] = \frac{4}{4-1}\left[\frac{7.6^2}{5(5-1)} + \frac{6.0^2}{4(4-1)} + \frac{4.2^2}{2(2-1)}\right]$$

$$= 19.61$$

where

$$7.6 = A_1 = 0.8(5-3) + 0.6(8-5) + 0.3(22-8) ,$$
$$6.0 = A_2 = 0.6(8-5) + 0.3(22-8)$$

and

$$4.2 = A_3 = 0.3(22-8) .$$

Assume for the moment that the largest value, 22 months, is a censored observation and that we use (2.16) to estimate the mean. Then the estimate of the variance is

$$\widehat{\text{Var}}[\hat{\mu}(22)] = \frac{3}{3-1}\left[\frac{7.6^2}{5(5-1)} + \frac{6.0^2}{4(4-1)} + \frac{4.2^2}{2(2-1)}\right]$$
$$= 22.06 .$$

If we restrict estimation of the mean to observed survival times and estimate the mean using (2.15), then the estimate of the variance obtained by evaluating (2.17) is

$$\widehat{\text{Var}}[\hat{\mu}(8)] = \frac{3}{3-1}\left[\frac{3.4^2}{5(5-1)} + \frac{1.8^2}{4(4-1)}\right]$$
$$= 1.27 ,$$

where

$$3.4 = A_1 = 0.8(5-3) + 0.6(8-5)$$

and

$$1.8 = A_2 = 0.6(8-5) .$$

Approximate confidence intervals are obtained using percentiles from the standard normal distribution. Using the data in Table 2.1, the endpoints of a 95 percent confidence interval are $10.6 \pm 1.96\sqrt{19.61}$. This is shown only for purposes of illustration since the sample size is only five with four survival times and any asymptotic theory will not hold. In practice, the estimated mean and its estimated standard error would typically be included in the table containing the estimates of the key quantiles and their estimated standard errors.

For the whole HMO-HIV+ study the estimate of the mean using all of the observed times is $\hat{\mu}(60) = 14.67$ and the estimated variance from (2.17) is 3.93, yielding a 95 percent confidence interval of (10.78,

18.56). We note that, in this example, the largest survival time was 58 months and $\hat{\mu}(58) = 14.59$. Thus, the means from the two approaches are not too different. The right skewness evident in the plot of the survivorship function shown in Figure 2.2 is further quantified by the difference between the estimate of the median (7 months) and the estimate of the mean (approximately 15 months). In these data, as is the case with most analyses of survival time, the median is the better measure of central tendency.

2.4 COMPARISON OF SURVIVORSHIP FUNCTIONS

After providing a description of the overall survival experience in the study, we usually turn our attention to a comparison of the survivorship experience in key subgroups in the data. These groups might be defined by treatment arms in a clinical trial or by other key factors thought to be related to survival. The goals in this analysis are identical to those of the two sample t-test, the nonparametric rank sum test and the one-way analysis of variance. Namely, we wish to quantify differences between groups through point and interval estimates of key measures. Standard statistical procedures, such as those named above, may be used without modification when there are no censored observations.

Since survival data are typically right skewed, we would likely use rank-based non-parametric tests followed by estimates and confidence intervals of medians (and possibly other quantiles) within groups. Modifications of these procedures are required when censored observations are present in the data. These tests are described and illustrated with the HMO-HIV+ study data beginning with methods for comparing two groups.

When comparing groups of subjects, it is always a good idea to begin with a graphical display of the data in each group. In studies of survival time, we should graph the Kaplan–Meier estimator of the survivorship function for each of the groups. In the HMO-HIV+ study, a variable thought to be related to the survival experience of the subjects was a history of IV drug use, coded 0 = No and 1 = Yes. Figure 2.7 presents the graphs of the estimated survivorship functions for these two groups of subjects.

Both groups show a similar pattern of survival: a rapidly descending survivorship function with a long right tail. This is the result of a number of early deaths and a few subjects with survival near the maximum follow-up time. Since the estimated survivorship functions do not go to zero, we know that the largest observation in each group was a censored value. The figure also shows a separation of the functions for the two groups. The estimated survivorship function for the non-IV drug users lies completely above that for the IV drug users. In general, the pattern of one survivorship function lying above another means the group de-fined by the upper curve lived longer, or had a more favorable survival experience, than the group defined by the lower curve. In other words, at any point in time the proportion of subjects estimated to be alive is greater for one group (represented by the upper curve) than the other (represented by the lower curve). Estimates of the within-group statis-tics such as the median are computed using the methods described in Section 2.3. The statistical question is whether the observed difference seen in Figure 2.7 is significant.

A number of statistical tests have been proposed to answer this question, and most software packages provide results from at least two of these tests. However, comparison of the results obtained by differ-ent packages can become confusing due to small but annoying differ-ences in terminology and methods used to calculate the tests. The original developers [Mantel (1966), Peto and Peto (1972), Gehan (1965), Breslow (1970), Prentice (1978)] of these tests sought ways to extend tests used with non-censored data to the censored data setting.

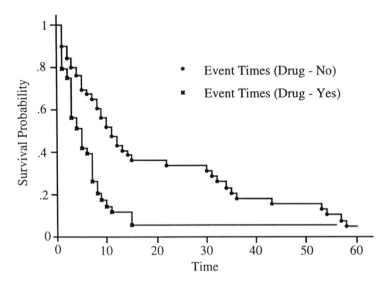

Figure 2.7 Estimated survivorship functions for subjects with and without a history of IV drug use.

The derivation and algebraic representation of the tests can, at times, seem complex and confusing. Lawless (1982) presents a concise summary of the traditional approach to the development of these tests, based on the theory of nonparametric tests, using exponentially ordered scores. However, in recent years, these tests have been reexamined from the counting process point of view and have been shown to be special cases of a more general class of counting process based tests. These results are summarized in Andersen, Borgan, Gill and Keiding (1993).

The calculation of each test is based on a contingency table of group by status at each observed survival time, as shown in Table 2.7. In this table, the number at risk at observed survival time $t_{(i)}$ is denoted by n_{0i} in Group 0 and by n_{1i} in group 1; the number of observed deaths in each of the these two groups is denoted by d_{0i} and d_{1i}, respectively; the total number at risk is denoted by n_i; and the total number of deaths is denoted by d_i. The contribution to the test statistic at each time is obtained by calculating the expected number of deaths in group 1 or 0, assuming that the survivorship function is the same in each of the two groups. This yields the usual *row total times column total divided by grand total* estimator. For example, using group 1, the estimator is

$$\hat{e}_{1i} = \frac{n_{1i}d_i}{n_i}. \tag{2.18}$$

Most software packages base their estimator of the variance of d_{1i} on the hypergeometric distribution, defined as follows:

$$\hat{v}_{1i} = \frac{n_{1i}n_{0i}d_i(n_i - d_i)}{n_i^2(n_i - 1)}. \tag{2.19}$$

Table 2.7 Table Used for Test of Equality of the Survivorship Function in Two Groups at Observed Survival Time $t_{(i)}$

Event/Group	1	0	Total
Die	d_{1i}	d_{0i}	d_i
Not Die	$n_{1i} - d_{1i}$	$n_{0i} - d_{0i}$	$n_i - d_i$
At Risk	n_{1i}	n_{0i}	n_i

The contribution to the test statistic depends on which of the various tests is used, but each may be expressed in the form of a ratio of weighted sums over the observed survival times. These tests may be defined in general as follows:

$$Q = \frac{\left[\sum_{i=1}^{m} w_i (d_{1i} - \hat{e}_{1i}) \right]^2}{\sum_{i=1}^{m} w_i^2 \hat{v}_{1i}}. \tag{2.20}$$

Under the null hypothesis that the two survivorship functions are the same, and assuming that the censoring experience is independent of group, and that the total number of observed events and the sum of the expected number of events is large, then the significance level for Q may be obtained using the chi-square distribution with one degree-of-freedom [i.e., $p = \Pr(\chi^2(1) \geq Q)$]. Exact methods of inference for use with small samples have been implemented in the software package StatXact 3 (1995) but will not be discussed in this text.

The most frequently used test is based on weights equal to one, $w_i = 1$. In this case, the test mimics the well-known Mantel–Haenszel test of the hypothesis that the stratum specific odds-ratio is equal to one [see Mantel (1966) for further details]. However, this test is most often called the log-rank test, due to Peto and Peto (1972). The test is related to a test proposed by Savage (1956) for noncensored data, and BMDP calls it the generalized Savage test.

Gehan (1965) and Breslow (1970) generalized the Wilcoxon rank-sum test to allow for censored data. This test uses weights equal to the number of subjects at risk at each survival time, $w_i = n_i$, and is called the Wilcoxon or generalized Wilcoxon test by most software packages.

SAS's lifetest procedure provides two ways of obtaining the same test, but different variance estimators are used. In SAS, if we define the grouping variable to be a stratification variable, the variance estimator $\hat{v}_{1i}$ is used. If we use SAS's test option, then the variance estimator

$$\hat{v}_{1i}^* = \frac{n_{1i} n_{0i}}{n_i^2}$$

is used, which assumes that $d_i = 1$; there are no tied failure times. Thus, in any one example, we may obtain test statistics of similar magnitude but with slightly different values. Because survival time is often recorded in discrete units that may lead to ties, we recommend that the variance estimator $\hat{v}_{1i}$ be used.

The choice of weight influences the type of differences in the survivorship function the test is most apt to detect. The generalized Wilcoxon test, since it uses weights equal to the number at risk, will put relatively more weight on differences between the survivorship functions at smaller values of time. The log-rank test, since it uses weights equal to one, will place more emphasis than does the generalized Wilcoxon test on differences between the functions at larger values of time. Other tests have been proposed that use weight functions intermediate between these, for example, Tarone and Ware (1977) suggested using $w_i = \sqrt{n_i}$.

Peto and Peto (1972) and Prentice (1978) suggested using a weight function that depends more explicitly on the observed survival experience of the combined sample. The weight function is a modification of the Kaplan-Meier estimator and is defined in such a way that its value is known just prior to the observed failure. The value of any estimated survivorship function at a particular observed failure time is known only after the observation is made. The property of having the value known in advance of the actual observed failure is referred to as *predictable* in counting process terminology. This theory is needed to prove results concerning the distribution of the test statistics. The modified estimator of the survivorship function is

$$\tilde{S}(t) = \prod_{t_{(j)} \le t} \left(\frac{n_j + 1 - d_j}{n_j + 1} \right) \tag{2.21}$$

and the weight used is

$$w_i = \tilde{S}(t_{(i-1)}) \times \frac{n_i}{n_i + 1}. \tag{2.22}$$

Note that when $d_i = 1$ the weight is equal to the modified estimator, that is, $w_i = \tilde{S}(t_{(i)})$, which is an assumption made in the implementation of this test in BMDP. In the example demonstrating the calculations, we will use both the correct version of the weight given in (2.22) as well as BMDP's implementation. In subsequent examples, only the BMDP version of the Peto–Prentice test will be discussed, as it is the only software package providing this test.

Harrington and Fleming (1982) suggested a class of tests that incorporates features of both the log-rank and the Peto and Prentice tests. They suggest using the Kaplan–Meier estimator raised to a power, as the weight, namely

$$w_i = \left[\hat{S}(t_{(i-1)}) \right]^{\rho}.$$

If the power is $\rho = 0$ then $w_i = 1$ and the test is the log-rank test. However, if $\rho = 1$ then the weight is the Kaplan–Meier estimator at the previous survival time, a weight similar to that of the Peto and Prentice test. This test has been implemented in the S-PLUS software package.

The principle advantage of the Peto–Prentice and Harrington-Fleming tests over the generalized Wilcoxon test is that they weight relative to the overall survival experience. The generalized Wilcoxon test uses the size of the risk set and hence weights depend both on the censoring as well as the survival experience. If the pattern of censoring is markedly different in each of the groups, then this test may either reject or fail to reject, not on the basis of similarity or differences in the survivorship functions, but on the pattern of censoring. For this reason most software packages will provide information as to the pattern of censoring in each of the two groups. This information should be checked for comparability—especially when the results of several of these tests are provided and yield markedly different significance levels.

A problem can occur if the estimated survivorship functions cross one another. This means that in some time intervals one group will have a more favorable survival experience, while in other time intervals the other group will have the more favorable experience. This situation is analogous to having interaction present when applying Mantel–Haenszel methods to a stratified contingency table. Unfortunately, tests for the homogeneity across strata may not be used in most survival time applications, because data in tables like Table 2.7 will be too thin to satisfy the necessary large sample criteria. Fleming, Harrington and O'Sullivan (1987) proposed a test that addresses the problem by using, as a test statistic, the maximum observed difference between the two survivorship functions. This test has not been implemented in any software package. We consider methods based on regression modeling to address this issue in Chapter 7. For the time being, our only check is via a visual examination of the plot of the Kaplan–Meier estimator for the two groups being compared. If one or more of the various tests fails to reject a difference, and if we see that the curves cross, then this "interaction" may be present.

It is not possible to provide a categorical rank ordering of the values of the test statistics. The actual calculated values will depend on the observed survival and censoring times.

In order to illustrate the computation of each of the tests, we have chosen a small subset of subjects in each of the two drug use groups in the HMO-HIV+ study. These data are listed in Table 2.8. Column 1 of Table 2.9 lists the eight distinct survival times. Columns 2 through 5 present the quantities defined by the notation shown in Table 2.7, and columns 6 and 7 present quantities defined in equations (2.18) and (2.19). Columns 8 through 11 present values for the weight functions for the four tests, where "LR" stands for log-rank test weights, "WL" stands for generalized Wilcoxon test weights, "TW" stands for Tarone–Ware weights and "PP" stands for Peto–Prentice weights. The calculated values of the test statistics and their respective *p*-values are shown in Table 2.10. The difference between the values of the log-rank and generalized Wilcoxon tests in Table 2.10 reflects the fact that the two groups differed most at the later observed survival times. The significance levels in Table 2.10 are provided only for the purpose of illustrating the calculations since, with only 4 events in each group and an expected number of events in group 1 of 5.45, the assumption that the sample sizes are large is a bit tenuous.

Recall the Kaplan–Meier estimates of the survivorship functions for the two drug groups in the whole HMO-HIV+ study, shown in Figure 2.7. Note that the two curves do not cross at any point, indicating that the previously described problem of "interaction" may not be present. An inspection of the proportion of values that are censored and the pattern of censoring (not shown) indicates that the censoring experience of the two groups is similar. Thus it would appear that the assumptions necessary for using the tests for equality of the survivorship functions seem to hold. Table 2.11 presents the values of the test statistics.

In Table 2.11, all tests are highly significant and support the impression from Figure 2.7 that those with a prior history of drug use tended

Table 2.8 Listing of Data from the Two Drug Use Groups in the HMO-HIV+ Study Used to Illustrate the Tests for the Comparison of Two Survivorship Functions

Drug Use Group	Ordered Observed Survival Times
No	3, 4*, 5, 22, 34
Yes	2, 3, 4, 7*, 11

* Denotes a censored observation.

Table 2.9 Listing of Quantities Needed to Calculate the Tests for the Equality of Two Survivorship Functions

Time	d_{1i}	n_{1i}	d_i	n_i	$\hat{e}_{1i}$	$\hat{v}_{1i}$	LR	WL	TW	PP
							\multicolumn{4}{c}{Weights}
2	0	5	1	10	0.500	0.250	1	10	3.16	0.909
3	1	5	2	9	1.110	0.432	1	9	3.00	0.818
4	0	4	1	7	0.571	0.245	1	7	2.64	0.636
5	1	3	1	5	0.600	0.240	1	5	2.23	0.530
11	0	2	1	3	0.667	0.222	1	3	1.73	0.398
22	1	2	1	2	1.000	0	1	2	1.41	0.265
34	1	1	1	1	1.000	0	1	1	1.00	0.133

to die sooner than those who did not have a history of drug use. In practice, one could provide additional support for this conclusion by presenting the estimates of the within-group median survival times along with confidence interval estimates.

Each of the tests used to compare the survivorship experience in two groups may be extended to compare more than two groups. For example, the survivorship experience of three or four racial groups could be compared. In the HMO-HIV+ study, it was hypothesized that age might be related to survival. Since age is a continuous variable, one approach to assessing a potential relationship is to use regression modeling. This is discussed in detail in Chapter 3. An approach used in practice, for preliminary analyses that can yield easily understood summary measures, is to break a continuous variable into several groups of interest and use methods for grouped data on the categorized variable. We use this approach with groups based on the following intervals for age: $\{[20-29], [30-34], [35-39], [40-54]\}$. Table 2.12 presents the number of subjects, the number of deaths, the median survival time and

Table 2.10 Listing of the Test Statistics and p-Values for the Equality of Two Survivorship Functions Computed from Table 2.9

Statistic	Value	p-Value
Log-rank	1.512	0.219
Generalized Wilcoxon	1.250	0.264
Tarone–Ware	1.363	0.243
Peto-Prentice (Correct wt.)	1.327	0.249
Peto-Prentice (BMDP)	1.423	0.233

Table 2.11 Test Statistics and p-Values for the Equality of the Survivorship Functions for the Two Drug Use Groups in the HMO-HIV+ Study

Statistic	Value	p-Value
Log-rank	11.856	<0.001
Generalized Wilcoxon	10.910	<0.001
Tarone–Ware	12.336	<0.001
Peto–Prentice (BMDP)	11.497	<0.001

associated 95 percent confidence interval for each age group.

The estimated median survival time is 43 months for the youngest age group in Table 2.12, which is considerably larger than the estimated median in each of the other three groups. This suggests that these young subjects may have a more favorable survival experience than older subjects. However, the estimated standard error of the estimated median is 32.8 and the symmetric normal theory confidence interval covers the entire observed range of time. This problem arises because there are only 12 subjects in this age group, the minimum value of the estimated survivorship function is 0.24 at 58 months and the largest observations are two censored values at 60 months. The medians and confidence intervals for the other three groups suggest that survival experience worsens with age. The goal in the four-group comparison will be to evaluate whether trends seen in the medians persist when the entire survival experience of the groups is compared. Before presenting the graphs of the Kaplan–Meier estimates of the survivorship functions for the four age groups, we present the details of the extension of the two-group tests to the multiple-group situation.

If we assume that there are K groups, then the calculations of the test statistics are based on a two by K table for each observed survival time. The general form of this table is presented in Table 2.13. In a manner

Table 2.12 Number of Subjects, Events and Estimated Median Survival Time in Four Age Groups in the HMO-HIV+ Study

Age Group	Freq	Deaths	Median	95% CIE
20–29	12	8	43	*
30–34	34	29	9	6.3, 11.7
35–39	25	20	7	4.5, 9.5
40–54	29	23	4	2.5, 5.5

* Estimated standard error too large to compute a CIE.

similar to the two-group case, we estimate the expected number of events for each group under an assumption of equal survivorship functions as

$$\hat{e}_{ki} = \frac{d_i n_{ki}}{n_i}, \quad k = 1, 2, \ldots, K. \tag{2.23}$$

We compare the observed and expected numbers of events for $K-1$ of the K groups. The reason for this will be explained shortly. The easiest way to denote the $K-1$ comparisons is to use vector notation to represent both observed and estimated expected number of events as follows:

$$\mathbf{d}_i' = (d_{1i}, d_{2i}, \ldots, d_{K-1i}),$$

and

$$\hat{\mathbf{e}}_i' = (\hat{e}_{1i}, \hat{e}_{2i}, \ldots, \hat{e}_{K-1i}).$$

The difference between these two vectors is

$$(\mathbf{d}_i - \hat{\mathbf{e}}_i)' = (d_{1i} - \hat{e}_{1i}, d_{2i} - \hat{e}_{2i}, \ldots, d_{K-1i} - \hat{e}_{K-1i}). \tag{2.24}$$

For convenience, we have used the first $K-1$ of the K groups, but any collection of $K-1$ groups could equally well be used.

To obtain a test statistic, we need an estimator of the covariance matrix of $\mathbf{d}_i$. The elements of this matrix are obtained assuming that the observed number of events follows a multivariate central hypergeometric distribution [see Johnson and Kotz (1997)]. The diagonal elements of the $(K-1) \times (K-1)$ matrix, denoted $\hat{\mathbf{V}}_i$, are

$$\hat{v}_{kki} = \frac{n_{ki}(n_i - n_{ki})d_i(n_i - d_i)}{n_i^2(n_i - 1)}, \quad k = 1, 2, \ldots, K-1, \tag{2.25}$$

and the off-diagonal elements are

$$\hat{v}_{kli} = -\frac{n_{ki}n_{li}d_i(n_i - d_i)}{n_i^2(n_i - 1)}, \quad k, l = 1, 2, \ldots, K-1, k \neq l. \tag{2.26}$$

The various multiple-group versions of the two-group test statistics are obtained by computing a weighted difference between the observed and expected number of events. The weights used at each distinct survival time can be any of the weights used in the two-group test, denoted in general at time $t_{(i)}$ by w_i. To obtain a formula for the test statistic, we

Table 2.13   Table Used for the Test for the Equality of the Survivorship Function in K Groups at Observed Survival Time $t_{(i)}$

Event/Group	1	2	$\cdots$	k	$\cdots$	K	Total
Die	d_{1i}	d_{2i}	$\cdots$	d_{ki}	$\cdots$	d_{Ki}	d_i
Not Die	$n_{1i} - d_{1i}$	$n_{2i} - d_{2i}$	$\cdots$	$n_{ki} - d_{ki}$	$\cdots$	$n_{Ki} - d_{Ki}$	$n_i - d_i$
At Risk	n_{1i}	n_{2i}	$\cdots$	n_{ki}	$\cdots$	n_{Ki}	n_i

define a $K-1$ by $K-1$ diagonal matrix denoted $\mathbf{W}_i = \text{diag}(w_i)$. This matrix has the value of the weight, w_i, at time $t_{(i)}$ in all $K-1$ positions along the diagonal of the matrix. The test statistic to compare the survivorship experience of the K groups is

$$Q = \left[\sum_{i=1}^{m} \mathbf{W}_i (\mathbf{d}_i - \hat{\mathbf{e}}_i) \right]' \left[\sum_{i=1}^{m} \mathbf{W}_i \hat{\mathbf{V}}_i \mathbf{W}_i \right]^{-1} \left[\sum_{i=1}^{m} \mathbf{W}_i (\mathbf{d}_i - \hat{\mathbf{e}}_i) \right]. \qquad (2.27)$$

The reason we use only $K-1$ of the K possible observed to expected comparisons is to prevent the matrix in the center of the right-hand side of (2.27) from being singular. The value of the test statistic in (2.27) is the same, regardless of which collection of $K-1$ groups are used.

The expression on the right-hand side of (2.27) may look intimidating to those not familiar with matrix algebra calculations, but when $K = 2$ it simplifies to the more easily understood statistic defined in (2.20). Most software packages providing statistics for several definitions of the weight use (2.27). These packages typically provide only the test statistic and a p-value. One exception is SAS's lifetest procedure, which provides the individual elements in (2.24)–(2.25) for the log-rank and generalized Wilcoxon tests when the group variable is defined as a stratum variable. Under the hypothesis of equal survival functions, and if the summed estimated expected number of events is large, then Q will be approximately distributed as chi-square with $K-1$ degrees-of-freedom, and the p-value is $p = \Pr(\chi^2(K-1) \geq Q)$. The remarks made earlier about how the choice of weights in the two-group case can affect the ability of the test to detect differences apply to the multiple-group case as well.

The log-rank test, $w_i = 1$, has the following easily computed, conservative, approximation:

$$Q_c = \sum_{k=1}^{K} \frac{(d_{k+} - \hat{e}_{k+})^2}{\hat{e}_{k+}} < Q,$$

where

$$d_{k+} = \sum_{i=1}^{m} d_{ki},$$

and $\hat{e}_{k+}$ is defined similarly. If we calculate Q_c and reject the hypothesis of equal survival experience, then we would reject using Q.

The estimated survivorship functions for the four age groups are shown in Figure 2.8. The figure confirms our preliminary observations based on estimates of median survival times. We see that the survivorship function for the youngest group lies completely above those of the other three groups. It has a long right tail and does not go to zero since two observations are censored at 60 months. For the first 15 months, the estimated survivorship functions for the youngest three age groups follow the trend observed in the medians. In this interval, the three functions are, for the most part, inversely ordered by age. The functions for the middle two age groups cross four times between 15 and 45 months, suggesting that the survival experience for these two age groups may be similar in this range. The estimated survivorship function for the oldest age group lies completely below that of the other three groups for 34 months. This suggests that we should begin our analysis with a test for the overall equality of the survivorship experience. If we find that the experience of at least one group is different from the others, we should construct single degree-of-freedom contrasts to examine between-group differences, as is typically done in analysis of variance methods.

The values of the four test statistics using their respective weights in (2.27) are given in Table 2.14. Since each statistic is significant at beyond the 1 percent level, we reject the hypothesis that the survivorship functions for the four age groups are the same. We follow the test for overall group differences in survival experience with contrasts to try and describe more precisely the source(s) of the significance of the overall test. The BMDP package, program 1L, offers this option by allowing the user to specify a trend test and to input a set of coefficients to test for trend when the groups are not equally spaced. The SAS package lifetest procedure has a test option that provides a trend test for a numeric covariate. The test does not yield the same numeric value as the trend test in BMDP. We describe the test used in BMDP as it follows directly from the multiple group test in (2.27). The null hypothesis is that the survivorship functions are equal and the alternative is that they are rank-ordered and follow the trend specified by the coefficients denoted by the vector $\mathbf{c}' = (c_1, c_2, \ldots, c_{K-1})$. If the groups are equally

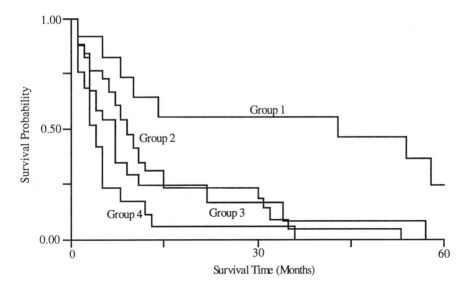

Figure 2.8 Estimated survivorship functions for the four age groups in the HMO-HIV+ study.

spaced, we may use $c_k = k$. The age groups we used in the HMO-HIV+ study are not equally spaced so we will use a vector of coefficients whose values are the midpoints of the four groups, i.e.,

$$\mathbf{c}' = (25, 32.5, 37.5, 47.5).$$

Any linear transformation of these coefficients would yield the same value of the test statistic. The statistic to test for trend, with one degree-of-freedom, is

$$Q_{\text{trend}} = \frac{\left[\mathbf{c}' \displaystyle\sum_{i=1}^{m} \mathbf{W}_i(\mathbf{d}_i - \hat{\mathbf{e}}_i)\right]^2}{\mathbf{c}'\left[\displaystyle\sum_{i=1}^{m} \mathbf{W}_i \hat{\mathbf{V}}_i \mathbf{W}_i\right]\mathbf{c}}. \tag{2.28}$$

The p-value is computed using the chi-square distribution with one degree-of-freedom, i.e., $p = \Pr(\chi^2(1) \geq Q_{\text{trend}})$. Table 2.15 presents the statistics and their p-values for the test of trend among the four age groups in the HMO-HIV+ study. These values are each just slightly

Table 2.14 Test Statistics, Degrees-of-Freedom and p-Values for the Equality of the Survivorship Functions for the Four Age Groups in the HMO-HIV+ Study

Statistic	Value	df	p-Value
Log-rank	19.906	3	<0.01
Generalized Wilcoxon	14.143	3	<0.01
Tarone–Ware	16.956	3	<0.01
Peto–Prentice (BMDP)	15.665	3	<0.01

smaller than the values in Table 2.14, providing strong evidence for a trend in survival experience that is inversely related to age. We explore this relationship in more detail when we consider regression modeling in the next chapter.

In the examples we have used from the HMO-HIV+ study to illustrate the comparison of the survivorship functions over groups, the magnitude of the test statistics has not varied too dramatically with the choice of weight, and the significance or non-significance of all test statistics has been consistent. However, this is not always the case and to illustrate this we use some data provided to us by our colleagues Drs. Carol Bigelow and Penny Pekow (at the University of Massachusetts) and Dr. Kathy Meyer (at Baystate Medical Center in Springfield, Massachusetts). These data were used as part of Ms. Shiaw-Shyuan Yuan's Masters degree project [Yuan (1993)]. The purpose of the study was to determine factors which predict the length of time low birth weight infants (<1500 grams) with bronchopulmonary dysplasia (BPD) were treated with oxygen. The data were collected retrospectively for the period December 1987 to March 1991. Beginning in August 1989, the treatment of BPD changed to include the use of surfactant replacement therapy. This was done with parental permission since, at the time, this therapy was considered experimental. A total of 78 infants met the study criteria, with 35 receiving surfactant replacement therapy and 43

Table 2.15 Trend Test Statistics, Degrees-of-Freedom and p-Values for the Equality of the Survivorship Functions among the Four Age Groups in the HMO-HIV+ Study

Statistic	Value	df	p-Value
Log-rank	19.066	1	<0.01
Generalized Wilcoxon	14.080	1	<0.01
Tarone–Ware	16.673	1	<0.01
Peto–Prentice (BMDP)	15.536	1	<0.01

not receiving this therapy. Five babies were still on oxygen at their last follow-up visit and represent censored observations. We refer to this study as the BPD study.

The outcome variable is the total number of days the baby required supplemental oxygen therapy. Figure 2.9 presents the Kaplan–Meier estimates of the survivorship functions for two groups defined by use of surfactant replacement therapy. The estimated median number of days of therapy for those babies who did not have surfactant replacement therapy (group 0) is 107 {95 percent CIE: (55.3, 158.7)}, and the estimated median number of days for those who had the therapy (group 1) is 71 {95 percent CIE: (33.3, 108.7)}. The median number of days of therapy for the babies not on surfactant is about 1.5 times longer than those using the therapy, but there is considerable overlap in the confidence intervals. The plots of the survivorship functions in Figure 2.9 indicate a progressively larger difference in the survivorship experience between the two groups over time. Table 2.16 presents test statistics and associated p-values for the equality of the survivorship functions. The Wilcoxon test is not significant at the 5 percent level, but the log-rank test is significant. The difference in the magnitude of the test statistics is due to the difference in the weights used. The Wilcoxon test uses a weight equal to the size of the risk set and thus is more likely to detect early differences. The log-rank test uses a weight equal to one and is more likely to detect later differences in the survivorship functions.

In any statistical analysis in which more than one test can be used, we need to make a decision about which results we will report. The log-rank test is the most frequently used and reported test for the comparison of survivorship functions. For most analyses, at least when each test has roughly the same level of significance, reporting only the results of the log-rank test is appropriate. When the tests give different results, then more than one result should be reported. This will provide the reader with a clearer picture as to where the survivorship functions are different. The current example demonstrates the importance of computing several of the tests. Most packages have both the log-rank and generalized Wilcoxon tests, and we recommend that both be computed. To our knowledge, only BMDP computes the Tarone–Ware and Peto–Prentice tests. The pattern of censoring can influence the magnitude of the tests, but the values of the Tarone–Ware and Peto–Prentice tests tend to be intermediate between the log-rank and Wilcoxon tests.

We conclude our presentation of the tests for comparison of survivorship functions with a brief discussion of the assumptions underlying the tests and the types of alternative hypotheses the tests have the power

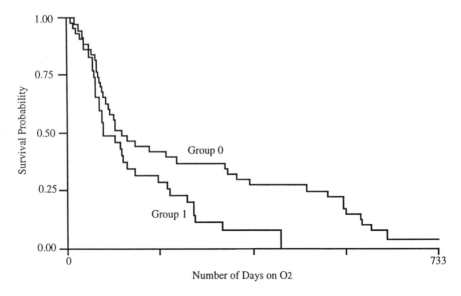

Figure 2.9 Estimated survivorship functions defined by surfactant use in the BPD study (0 = No surfactant, 1 = Surfactant).

to detect. Recall that the Kaplan–Meier estimator assumes that censoring is independent of survival time. In addition, the tests assume that the censoring is independent of the group. Problems in study design and data collection can lead to differential effects due to censoring, and the best protection is a carefully designed study. However, it is good practice to examine the censoring pattern in the data.

In general, we cannot over-emphasize the importance of a careful study of the plot of the Kaplan–Meier estimates of the survivorship functions. Any tests comparing these functions, and within-group point estimates of quantiles, should support what is seen in the plot. The plot is also the basic diagnostic tool to determine whether the tests described

Table 2.16 Test Statistics and p-Values for the Equality of the Survivorship Functions for Two Groups Defined by Surfactant Use in the BPD Study

Statistic	Value	df	p-Value
Log-rank	5.618	1	0.018
Generalized Wilcoxon	2.490	1	0.115
Tarone–Ware	3.698	1	0.055
Peto–Prentice (BMDP)	2.534	1	0.111

previously should be used or, if used, have any chance of detecting a difference. The alternative hypothesis that the tests are most likely to detect is a monotonic ordering of the survivorship functions (e.g., they lie one above another). The tests have little to no power to detect differences when the survivorship functions cross one another. An example of a worst-case scenario is when the survivorship functions for two groups have the same median and cross each other once at that value. For the early times one group has the more favorable survival experience, but for later times the other group does. None of the tests described in this section are able to detect this kind of difference. This is a situation analogous to the presence of interaction in a Mantel–Haenszel analysis of stratified contingency tables. Unfortunately, tests for interaction used with a Mantel–Haenszel analysis, such as the Breslow–Day test [Breslow and Day (1980)], can't be used, due to small cell frequencies in tables such as Table 2.13. In this case, one approach that can be used is to subdivide the sample on the basis of the stratification variable and then test for group differences within the strata. This approach is limited by the study size, as we can spread the data over only so many strata. Eventually there are too few subjects per stratum to reliably estimate the survivorship function. However, in practice, there may be one or two clinically plausible variables to use for stratification purposes. These types of differences, or interactions, between survivorship functions are much more clearly addressed using the regression modeling approach to be discussed in Chapter 3.

2.5 OTHER FUNCTIONS OF SURVIVAL TIME AND THEIR ESTIMATORS

The Kaplan–Meier estimator of the survivorship function has been, and continues to be, the most frequently used estimator, largely due to the fact that it is routinely calculated by most software packages. To motivate the discussion of another estimator, we begin by presenting a different representation of the survivorship function. If we assume that the underlying time random variable is absolutely continuous, then we may express the survivorship function as

$$S(t) = e^{-H(t)}, \tag{2.29}$$

where $H(t) = -\ln(S(t))$. The expression in (2.29) suggests that estimators of the survivorship function could be based on an estimator of $S(t)$

(e.g., the Kaplan–Meier estimator) or via an estimator of $H(t)$. Aalen (1975, 1978), Nelson (1969, 1972) and Altshuler (1970) have proposed an easily computed estimator of $H(t)$, which we refer to as the Nelson-Aalen estimator.

The work by Aalen is considered to be one of the landmark contributions to the field, as virtually all recent statistical developments for the analysis of survival time have been based on the counting process approach he used to derive his version of the estimator of $H(t)$. The statistical theory and use of this estimator in various applied settings are discussed in detail in Andersen, Borgan, Gill and Keiding (1993) and in Fleming and Harrington (1984, 1991). We will use results derived from the counting process theory to justify various techniques discussed in this text. We will not present the counting process approach in any detail since fully appreciating and understanding it requires having had calculus-based courses in mathematical statistics and probability theory.

Without providing any details as to its derivation (a heuristic argument is given later in this section), the Nelson–Aalen estimator of $H(t)$ is

$$\tilde{H}(t) = \sum_{t_{(i)} \le t} \frac{d_i}{n_i}. \tag{2.30}$$

An estimator of the survivorship function, based on (2.30), is

$$\tilde{S}(t) = e^{-\tilde{H}(t)}. \tag{2.31}$$

One theoretical problem is that the expression in (2.29) is valid for continuous time, but the estimator in (2.31) is discrete. However, the estimator in (2.31) provides the basis for the estimator of the survivorship function used with the proportional hazards regression model discussed in Chapter 3. For this reason, we consider it in some detail.

Even though packages may not provide the Nelson–Aalen estimator of the survivorship function, it is remarkably easy to compute. In the absence of ties, one merely sorts the data into ascending order on the time variable. The size of the risk set at $t_{(i)}$ is $n - i + 1$ and the estimator, $\tilde{H}(t)$ in (2.30), is obtained as the cumulative sum of the zero-one censoring indicator variable divided by the size of the risk set. The Nelson-Aalen estimator of the survivorship function is obtained by evaluating the expression in (2.31). When ties are present, one sorts the data into ascending order on time and into descending order on the censoring variable within values of time. Sorting in this way places the censored observations after the events when ties occur. One then calculates a variable equal to $n-i+1$, and uses a procedure such as STATA's collapse command, or the means procedure in SAS, to provide summary statistics at each value of time observed. One needs to obtain the maximum value of $n-i+1$ among the tied time values and the total number of events and/or censored observations. This reduced data set is used to calculate the Nelson-Aalen estimator using the cumulative sum described for the case where there are no ties.

Peterson (1977) proposed another estimator, which is based on the Kaplan–Meier estimator of the cumulative hazard function, as follows:

$$\hat{H}(t) = -\ln\left(\hat{S}(t)\right) = -\ln\left(\prod_{t_{(i)} \le t}\left(\frac{n_i - d_i}{n_i}\right)\right) = -\ln\left(\sum_{t_{(i)} \le t}\left(\frac{n_i - d_i}{n_i}\right)\right)$$

$$= \sum_{t_{(i)} \le t} -\ln\left(1 - \frac{d_i}{n_i}\right).$$

One may show, by using a Taylor series expansion (see Appendix 1), that $d_i/n_i \le -\ln(1 - d_i/n_i)$ for each survival time. Thus, the Nelson–Aalen estimator of the survivorship function will always be greater than or equal to the Kaplan–Meier estimator. If the size of the risk sets relative to the number of events is large, then $d_i/n_i \cong -\ln(1 - d_i/n_i)$ and there will be little practical difference between the Nelson-Aalen and the Kaplan-Meier estimators of the survivorship function.

The HMO-HIV+ study provides a good illustration of a situation in which there is little practical difference between the two estimators. Table 2.17 presents the results of collapsing the sample of 100 observations to obtain the necessary within-time summary statistics at each observed value of time: the frequency of occurrence (freq), the number of events (d), the size of the risk set (n), the Nelson–Aalen estimator, $\tilde{H}(t)$, the Nelson–Aalen estimator of the survivorship function, $\tilde{S}(t)$ and, for comparison, the Kaplan–Meier estimator, $\hat{S}(t)$. For example, at 3 months the values of the estimators are

$$\tilde{H}(3) = \frac{15}{100} + \frac{5}{83} + \frac{10}{73} = 0.347,$$

$$\tilde{S}(3) = e^{-0.347} = 0.707,$$

and

$$\hat{S}(3) = \left(1 - \frac{15}{100}\right) \times \left(1 - \frac{5}{83}\right) \times \left(1 - \frac{10}{73}\right) = 0.689.$$

Table 2.17 Summary Table Used to Calculate the Nelson-Aalen Estimator of the Survivorship Function for the HMO-HIV+ Study

Time	freq	d	n	$\tilde{H}(t)$	$\tilde{S}(t)$	$\hat{S}(t)$
1	17	15	100	0.150	0.861	0.850
2	10	5	83	0.210	0.810	0.799
3	12	10	73	0.347	0.707	0.689
4	5	4	61	0.413	0.662	0.644
5	7	7	56	0.538	0.584	0.564
6	3	2	49	0.579	0.561	0.541
7	7	6	46	0.709	0.492	0.470
8	4	4	39	0.812	0.444	0.422
9	3	3	35	0.897	0.408	0.386
10	4	3	32	0.991	0.371	0.350
11	3	3	28	1.098	0.333	0.312
12	4	2	25	1.178	0.308	0.287
13	1	1	21	1.226	0.294	0.273
14	1	1	20	1.276	0.279	0.260
15	2	2	19	1.381	0.251	0.232
19	1	0	17	1.381	0.251	0.232
22	1	1	16	1.444	0.236	0.218
24	1	0	15	1.444	0.236	0.218
30	1	1	14	1.515	0.220	0.202
31	1	1	13	1.592	0.204	0.187
32	1	1	12	1.675	0.187	0.171
34	1	1	11	1.766	0.171	0.156
35	1	1	10	1.866	0.155	0.140
36	1	1	9	1.977	0.138	0.125
43	1	1	8	2.102	0.122	0.109
53	1	1	7	2.245	0.106	0.093
54	1	1	6	2.412	0.090	0.078
56	1	0	5	2.412	0.090	0.078
57	1	1	4	2.662	0.070	0.058
58	1	1	3	2.995	0.050	0.039
60	2	0	2	2.995	0.050	0.039

The values at other times are obtained in a similar manner. Figure 2.10 presents graphs of the the Nelson–Aalen and Kaplan–Meier estimators. We see little practical difference between the two estimators, even though $\tilde{S}(t) \geq \hat{S}(t)$ at every observed value of time.

The function $H(t)$ is an important analytic tool for the analysis of survival time data. In much of the survival analysis literature it is called the *cumulative hazard function*, but in the counting process literature it is related to a function called the *cumulative* or *integrated intensity process*. The term "hazard" is used to describe the concept of the risk of "failure" in an interval after time t, conditional on the subject having survived to time t. The word "cumulative" is used to describe the fact that its value is the "sum total" of the hazard up to time t. At this point we focus on the hazard function itself, as it plays a central role in regression modeling of survival data.

Consider a subject in the HMO-HIV+ study who has a survival time of 7 months. For this subject to have died at 7 months, he/she had to be alive at 6 months. The hazard at 7 months is the failure rate "per month," conditional on the fact that the subject has lived 6 months. This is not the same as the unconditional failure rate "per month" at 7

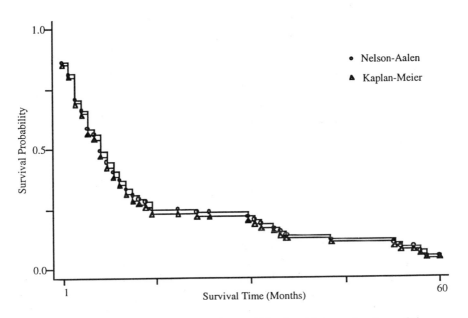

Figure 2.10 Graphs of the Nelson-Aalen and Kaplan-Meier estimators of the survivorship function from the HMO-HIV+ study.

months. The unconditional rate applies to subjects at time zero and, as such, does not use the information available as the study progresses about the survival experience in the sample. This accumulation of knowledge, over time, is generally referred to as *aging*. For example, of 100 subjects who enroll in a study, what fraction is expected to die at 7 months? The conditional failure rate applies only to that subset of the sample that has survived to a particular time, thus it accounts for the aging that has taken place in the sample.

The data from the HMO-HIV+ study can be used to demonstrate the difference between the conditional and the unconditional failure rate. If we assume that there were no censored observations in the study, the "freq" column in Table 2.17 gives the number of deaths. The first two columns of Table 2.17 are a typical presentation of grouped data. A histogram based on these data provides a graphical estimator of the un-conditional failure rate.

To construct the histogram, we divide the follow-up time into 10 intervals, each of width 6 months. Each interval is represented graphi-cally by a rectangle with height equaling the frequency drawn over the interval. To construct a relative histogram we divide each frequency by the total sample size. At this point we must decide what we wish to use as the appropriate unit of time. If we do nothing, we implicitly let 6 months denote "one unit" of time. If we wish to have "one unit" equal "one month" then we must further divide by 6. For other inter-vals of time, we would divide by the correct multiple of interval width and unit. If we divide by 6, the heights of the rectangles give us the relative proportions of the *total* number of subjects beginning at time "zero" who had a survival time in each interval, and the area of each rectangle is the observed unconditional failure rate per month in that interval.

For each time, t, the histogram estimator, $\hat{f}(t)$, is

$$\hat{f}(t) = \frac{(\text{freq})/(\text{width})}{n}, \qquad (2.32)$$

where "freq" denotes the number of survival times in the interval, "width" denotes the width of the interval relative to the definition of "one unit" and n is the total sample size. The fact that the numerator of the estimator is expressed relative to the total sample size makes it an unconditional estimator. This is further reflected by the fact that the total area of the histogram rectangles is one, meaning that each subject has been counted once and only once in the presentation of the data.

The interval grouped-data estimator of the hazard function is, for all values of time, t, in an interval,

$$\hat{h}(t) = \frac{(\text{freq})/(\text{width})}{n(t)}, \tag{2.33}$$

where the quantity $n(t)$ is used somewhat imprecisely to denote the number of subjects still alive (at risk) at the beginning of the current interval. The area of the rectangle formed by graphing $\hat{h}(t)$ versus t estimates the conditional, on $n(t)$, per-month failure rate in the interval. The sum of the areas of the rectangles up to and including an interval is an estimate of the cumulative hazard. Since subjects are at risk until they actually die or are censored, they may be counted more than once and the sum of the areas of the rectangles may be greater than one.

Figure 2.11 presents the graphs of the histogram and hazard function estimators of the unconditional and conditional failure rates, computed from the data in Table 2.17, using 6-month intervals (e.g., (0,6], (6,12],...,(54,60]). The shaded rectangles of the histogram, which estimate the overall, unconditional per-month failure rate, are initially high and then drop rapidly, staying consistently low to 60 months. This pattern reflects the many early deaths; relatively few subjects had survival times throughout the period of follow-up. This was described by the Kaplan–Meier estimator in Figure 2.2. On the other hand, the open rectangles of the hazard function estimate the failure rate in the current interval, given that a subject is alive at the beginning of the interval. This pattern is not as consistent as that seen in the shaded histogram due to the fact that each rectangle is based on fewer subjects than the previous one. In other words, the variability is greater in the estimator of the hazard than the histogram. The graph indicates a relatively high initial failure rate which drops and then rises again.

The histogram estimator in (2.32) is useful for providing an estimate of the unconditional rate only when there are no censored observations. It may be modified to handle censored observations by using the difference between the values of the Kaplan–Meier (or Nelson–Aalen) estimator of the survivorship function at the two endpoints of the interval. The hazard function estimator in (2.33) may be modified to accommodate censored observations by having censored values of time contribute to the count in the denominator but not in the numerator. To provide a better approximation of the number at risk over the whole interval in settings in which there are large numbers of subjects and/or the inherently continuous time variable has been recorded at a few dis-

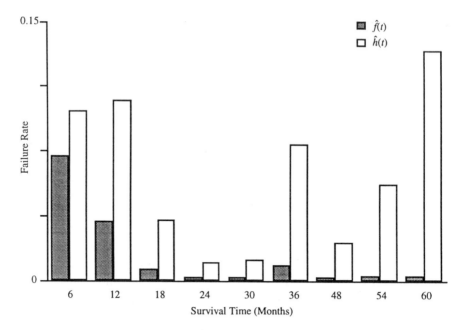

Figure 2.11 Graphs of the histogram estimator (shaded) of the unconditional failure rate and the hazard function estimator (open) of the conditional failure rate from the HMO-HIV+ study.

crete time points, the estimator of the hazard may use a denominator in which the number at risk at the beginning of the interval is reduced by one-half the number of subjects who failed, were censored or were lost for other reasons [see Lee (1992)].

Considering Figure 2.11, it is logical to postulate a function of time that describes, in a concise fashion, the form of either the unconditional or conditional failure rate, which may then be used to express the survivorship function as a function of time. If we can answer this question, then we have taken an important first step toward a more comprehensive analysis that will enable us to study which factors affect survival, namely parametrizing this function with a regression-like model.

As we think about the problem of trying to develop a function to describe survival time in the presence of censored data, we focus attention on the hazard function since it incorporates any aging that might take place. Figure 2.11 may be useful for general descriptive purposes but it is, in a sense, too discrete to be of use in developing a more precise function of time to describe the hazard function. What we would like is a more "continuous" time analysis. If we let the interval width shrink to the point where it is one measurement unit wide (i.e., one month in the HMO-HIV+ study), then the right-hand side of the estimator of the hazard function in (2.33) is d_i/n_i at observed survival times and is zero elsewhere.

Figure 2.12 presents a scatterplot of the pairs $\left(t_{(i)}, d_i/n_i\right)$, $i = 1, 2, \ldots, 31$ and a lowess smooth[2] of the plot [see StataCorp (1997), ksm command]. The smoothing done here is for illustrative purposes [see Andersen, Borgan, Gill and Keiding (1993) for a more complete discussion of smoothed estimators of the hazard function]. One difficulty with the plot in Figure 2.12 is that the hazard function should be estimated to be 0 at times when no deaths occurred. The smoothed curve in Figure 2.12 does not incorporate these 0 values. However, the goal in this section is to begin to make the transition from fully nonparametric to regression models discussed in subsequent chapters. Figure 2.12, while not totally correct, does serve to guide the reader in the direction of these regression models.

The smooth of the pointwise estimates of the hazard agrees with our original impression drawn from Figure 2.11 that the conditional risk is relatively high, drops and then rises. On the basis of this observation, we might postulate that the hazard function is a quadratic function of time,

$$h(t) = \theta_0 + \theta_1 t + \theta_2 t^2.$$

Suppose for the moment that we have a parametric form for the hazard function. We need to link the hazard function in a more direct way to the survivorship function. Since we assume the time variable is absolutely continuous, the cumulative hazard is, by methods of calculus,

$$H(t) = \int_0^t h(u) \, du, \qquad (2.34)$$

and by (2.29)

$$S(t) = e^{-\int_0^t h(u) \, du}. \qquad (2.35)$$

Those readers familiar with calculus will recognize the right-hand side of (2.34) as the integral of the hazard function over the time interval $[0,t]$. For readers not familiar with calculus, the estimator in (2.30) can

---

[2] For those unfamiliar with scatterplot smoothing methods, the purpose is to remove some of the "noise" in the plot by computing, for each y in the plot, a weighted average of the other y's near it.

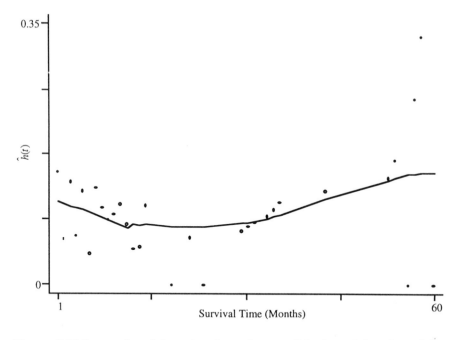

Figure 2.12 Scatterplot of the pointwise estimator of the hazard function, d_i/n, and its lowess smooth from the HMO-HIV+ study.

serve as a convenient mental model of what is being computed in (2.34). Another representation of the hazard function may be obtained by taking the log of (2.35) and then differentiating with respect to t yielding

$$h(t) = \frac{f(t)}{S(t)}, \qquad (2.36)$$

where $f(t)$ denotes the probability density function for the time random variable. Those not familiar with methods of calculus may think of the function $f(t)$ as what the histogram estimator in (2.32) becomes if we use larger and larger sample sizes and the width of each interval used in its construction becomes quite small. A similar intuitive argument may be applied to the hazard function estimator in (2.33) to motivate the expression in (2.36).

As noted above, one way to envision the hazard function is to think of it as a limiting, $n \rightarrow \infty$, version of the estimator in (2.33). In this argument, we let the width of each interval become quite small and, in the end, we have a function which describes the failure rate in the next instant following t. The expressions in (2.35)–(2.36) show that if we can specify the hazard function, then it is, in principle, relatively easy to obtain an expression for any of the other functions of survival time. The advantage of using the hazard function is that it characterizes the aging process as a function of time.

To obtain a better understanding of the hazard function and how it specifies the survivorship function, we consider various possible parametric models. A discussion of parametric survival time models is presented in Chapter 8. The goal here is see how this function describes the aging process.

The simplest possible model is for the hazard function to be constant, not depending on time [i.e., $h(t) = \theta$]. This hazard function states that at any particular time the chance that a subject "dies" in the next instant does not depend on how long the subject has survived. For example, in Figure 2.12 the average value of the plotted pointwise estimates of the hazard function is about 0.1. Thus, the constant hazard model is $\hat{h}(t) = 0.1$. The interpretation of this hazard function is that there is about a 10 percent chance that a subject will die in the next month, regardless of how long he/she has already survived. This model for the hazard may be clinically plausible in some studies of human populations when the follow-up time is relatively short. For example, the chance that a "healthy" 35-year-old person dies in the next year is about the same as that of a healthy 36- or 37- or 38- or 39-year-old subject.

The next simplest model is for the hazard to be a linear function of time, $h(t) = \theta_0 + \theta_1 t$. For example, an approximate straight-line fit to the plotted points in Figure 2.12 yields the model $\hat{h}(t) = 0.07 + 0.001t$. The interpretation is that at the beginning of the study subjects had about a 7 percent chance of dying in the next month, and this increases at about 0.1 percent per month. Since the hazard function must be greater than zero, the values of the parameters are constrained. For example, the model $h(t) = 0.12 - 0.004t$ describes the hazard in the first 30 months in Figure 2.12, but yields negative values after 30 months. This leads to the clinically implausible situation of positive probability of infinite physical life. Therefore, we have to use special methods when fitting hazard functions to observed data, since simple least squares regression methods will not be appropriate. We discuss these methods in detail in the next chapter.

On the basis of the lowess smooth in Figure 2.12, we postulated a quadratic function for the hazard function for the HMO-HIV+ study.

This is a more complicated function than the linear or constant model, but a life process of decreasing risk followed by increasing risk is clinically plausible. If one conceptualizes the risk of death in the next "instant" from birth to age 80, the function decreases for the first 5 or so years, remains fairly constant for 40 or so years and then begins to rise rapidly. This is more of a "bathtub" shape and requires a more complex function to describe it than a simple quadratic [see Lawless (1982)].

The major point is that the hazard function itself says a great deal about the fundamental underlying life-length process being studied. Specifying a fully parametric model leads to a specific life-length process. In some settings we may need this level of specificity, but in others it may not be necessary or flexible enough. This point will be dealt with directly in the next chapter.

The univariate descriptive methods discussed in this chapter, computed for the whole study or within a few subgroups, are an important first step in any analysis of survival time; however, these methods cannot be used to address the more sophisticated questions that can typically be addressed through regression modeling techniques. In Chapter 1 we discussed the general similarities and differences between regressions using dependent variables such as weight or disease status and regressions using survival time (with and without censoring) as the dependent variable. At this point, we are in a position to consider the regression methods for survival data in more detail.

Other texts presenting descriptive as well as other methods for survival data include: Collett (1994), Cox and Oakes (1984), Klein and Moeschberger (1997), Kleinbaum (1996), Le (1997), Lee (1992), Miller (1981), Marubini and Valsecchi (1995) and Parmar and Machin (1995).

EXERCISES

1. Listed below are values of survival time (length of follow-up) for 6 males and 6 females from the WHAS. Right-censored times are denoted by a "+" as a superscript.

   Males: 1, 3, 4^+, 10, 12, 18
   Females: 1, 3^+, 6, 10, 11, 12^+

Using these data, compute by hand (and verify hand calculations when possible with a software package) the following:

(a) The Kaplan-Meier estimate of the survivorship function for each gender.

(b) Pointwise 95 percent confidence intervals for the survivorship functions estimated in problem 1(a).

(c) The Hall and Wellner 95 percent confidence bands for the survivorship functions estimated in problem 1(b).

(d) Point and 95 percent confidence interval estimates of the 25th, 50th and 75th percentiles of survival time distribution for each gender.

(e) The mean survival time for each gender using all available times.

(f) A graph of the estimated survivorship functions for each gender computed in problem 1(a) along with the pointwise and overall 95 percent limit computed in problems 1(b) and 1(c).

2. Repeat problem 1 using data from grouped cohort 1 (1975 and 1978) from the Worcester Heart Attack Study. All calculations for this problem should be done using a software package.

3. Repeat problem 1 using grouped cohort 1 (1975 and 1978) from the WHAS with four groups defined by the age intervals: [24, 60], [61, 65], [66, 75] and [76, 99]. In this subgroup of the data, 60, 65 and 75 are approximately the three quartiles of the age distribution.

4. Compute by hand, and verify hand calculations with a software package, the log-rank, generalized Wilcoxon, and Peto–Prentice tests for the equality of two survivorship functions estimated in problem 1(a).

5. Repeat problem 4 using data from grouped cohort 1 (1975 and 1978) of the WHAS. Do the results of the test support what is seen in the graphs of the estimated survivorship functions?

6. Repeat problem 4 using data from grouped cohort 1 (1975 and 1978) of the WHAS with four groups defined by the age intervals: [24, 60], [61, 65], [66, 75] and [76, 93]. Using the midpoints of the four age intervals, test for trend using the test statistic defined in (2.28). In addition, test whether the survivorship experience for the middle two age groups is the same or different from the youngest and oldest age groups.

7. For the purposes of this problem restrict analyses to WHAS data from grouped cohort 1 (1975 and 1978). Prepare a table of descriptive statistics for survival time (length of follow-up) for each of the patient characteristic variables in Table 1.4. For age use the four groups in problem 6 above, and for CPK use two groups defined by the median.

8. Expand the analyses in problem 7 to include estimates from all 3 cohort groups combined. Note that in this problem the final age interval should be [76, 99].

CHAPTER 3

Regression Models for Survival Data

3.1 INTRODUCTION

In considering regression modeling of survival data, the first question we have to answer is: What are we going to model? Specifically, what will play the role of the systematic component in a regression model? The inherent aging process that is present when subjects are followed over time is what distinguishes survival time from other dependent variables. The presence of censoring in the data makes the study of survival time more interesting from a statistical research perspective, but from a practical point of view, it is an annoying technical detail that must be dealt with when we fit models. Of the functions describing the distribution of survival time discussed in Chapter 2, the hazard function best and most directly captures the essence of the aging process. Thus, a natural place to begin is to explore how to incorporate the hazard function into the heuristic approach to regression modeling presented in Chapter 1.

In Chapter 1 we used a scatterplot of data to motivate a regression model in which the log of survival time had a linear systematic component and an extreme minimum value error component. Assuming that the value of the covariate, x, is fixed and does not change over time, the model, as shown in (1.3), is

$$y = \beta_0 + \beta_1 x + \sigma \times \varepsilon^*, \qquad (3.1)$$

where $y = \ln(t)$ and $\varepsilon^* = \ln(\varepsilon)$. Expressed on the time scale, the model is multiplicative and of the form

$$t = \left(e^{\beta_0 + \beta_1 x}\right) \times \varepsilon^{\sigma}. \tag{3.2}$$

As expressed in (3.1) and (3.2), survival time is determined by a systematic component (the $\beta_0 + \beta_1 x$ part) and by an error component (the ε part). When we choose a particular parametric distribution for the error component in (3.1) or (3.2), we have also chosen a specific parametric structure for the hazard function. For example, if we assume that the value of the shape parameter in (3.2) is $\sigma = 1$, then the distribution of the error component in (3.2) is exponential with parameter equal to one, and the hazard function for a subject with covariate equal to x is

$$h(t, x, \boldsymbol{\beta}) = e^{-(\beta_0 + \beta_1 x)}. \tag{3.3}$$

Two points should be noted: (1) the hazard function does not depend on time; its value is determined by the covariate x and the unknown parameters β_0 and β_1, and (2) the hazard function and systematic component in the regression model are inversely related.

The fact that the hazard does not depend on time means that the risk of "failure" is the same no matter how long the subject has been followed. In Chapter 1, we considered the age of the subject in the HMO-HIV+ study as the covariate. The hazard function in (3.3) states that the risk of dying is determined solely by the age of the subject at the time of HIV+ diagnosis, and not by the time that has elapsed since enrollment in the study, $t = 0$. This assumption of a constant hazard may be unrealistic in many applied settings and should be examined carefully. We discuss this and other methods for model checking in Chapter 6.

One simple way to provide for a nonconstant hazard function is to assume that the shape parameter, σ, in (3.2) is not equal to 1. In this case, the error component has a Weibull distribution with parameters 1 and σ. Survival time has a Weibull distribution with one parameter equal to the systematic component in (3.1) and the second parameter equal to σ. The equation for the hazard function for (3.2) is

$$h(t, x, \boldsymbol{\beta}, \lambda) = \frac{\lambda t^{\lambda - 1}}{\left(e^{\beta_0 + \beta_1 x}\right)^{\lambda}}, \tag{3.4}$$

where we have set $\lambda = 1/\sigma$ to obtain a more concise expression. Considered as a function of survival time, the hazard function in (3.4) increases over time if $\lambda > 1$ and decreases if $\lambda < 1$. Because it can increase or decrease, the hazard function in (3.4) is more flexible than the constant hazard in (3.3). However, the change in the hazard function must be monotonic. For example, it would not be a good model if the hazard function first decreases and then increases (as is the case for human life over a many year period). Therefore, it still may not be suitable in certain applied settings.

The inverse relationship between the parameterization of the hazard and the systematic component is a result of the assumption that the distribution of the error component is exponential or Weibull. For example, if the value of the hazard function is 0.10, then the mean survival time is 10. Most software packages fit exponential regression models using the parameterization in (3.1).

In essence, the models described by (3.1)–(3.4) indicate that we are trying to accomplish two goals simultaneously. The model must describe the basic underlying distribution of survival time (error component), but it must also characterize how that distribution changes as a function of the covariates (systematic component). In some applied settings it is important to use a model that accomplishes both goals, but in other settings a model that addresses only the latter one is sufficient.

If we want a model to predict the life-length of a particular brand of computer hard disk as a function of temperature and relative humidity, we need it to address both goals. The desired end product of the statistical modeling is an equation that may be used to predict survival time of the hard disk for specific operating conditions. Fully parametric models such as those in (3.1)–(3.4) may be required, and a comprehensive study of such models is provided in the texts by Lawless (1982) and Nelson (1982). We consider several of these in Chapter 8.

On the other hand, we are often in a setting where we may wish to see if a combination of drug therapies improves survival of HIV+ patients when compared to a single drug therapy. In this case, a complete description of survival time is of secondary importance to a description of how the new therapy modifies the survival experience relative to the old one. In this example, we need to estimate parameters that can be used to compare the survival experience of the two treatment groups, and this comparison may need to be adjusted for other patient characteristics such as age or IV drug use. The regression models in (3.1)–(3.4) could be used to accomplish this goal. However, the assumptions required for their error components may be unnecessarily stringent, given that the desired inferences will be based solely on the parameters in the systematic portion of the model. Models used to describe survival time in a comparative sense are often called *semiparametric regression models* and are the major focus of this text.

3.2 SEMIPARAMETRIC REGRESSION MODELS

We noted in the previous chapter that we can describe the distribution of survival time in one of two equivalent ways. We can specify the density function of a parametric distribution or we can specify the hazard function. The advantage of the latter approach is that we directly address the aging process; but, as shown previously, it does not easily lend itself to the use of scatterplots to motivate regression models. The latter approach may also be preferred in a setting where the end products of the statistical analysis are estimated parameters that compare the survival experience of selected subgroups. By specifying a model through the hazard function, we may address specific questions such as how survival is related to the treatments under study and other subject characteristics.

Suppose we wish to compare the survival experience of cancer patients on two different therapies adjusting for age and gender, patient characteristics known to be associated with survival time. A natural place to begin is to put a regression model type structure on the hazard function. In general we specify the hazard function as a function of time and the covariates. In the hypothetical example there are three covariates: treatment, age and gender. For ease of notation assume for the remainder of this section and the next that there is one covariate denoted x. A regression model for the hazard function that addresses the study goal is

$$h(t, x, \beta) = h_0(t) r(x, \beta). \tag{3.5}$$

The hazard function, as expressed in (3.5), is the product of two functions. The function, $h_0(t)$, characterizes how the hazard function changes as a function of survival time. The other function, $r(x, \beta)$, characterizes how the hazard function changes as a function of subject covariates. The functions must be chosen such that $h(t, x, \beta) > 0$. Note that $h_0(t)$ is the hazard function when $r(x, \beta) = 1$. When the function $r(x, \beta)$ is such that $r(x = 0, \beta) = 1$, $h_0(t)$ is frequently referred to as the *baseline hazard function*. Under the model in (3.5) the ratio of the hazard functions for two subjects with covariate values denoted x_1 and x_0 is

$$\text{HR}(t, x_1, x_0) = \frac{h(t, x_1, \beta)}{h(t, x_0, \beta)},$$

so

$$\text{HR}(t, x_1, x_0) = \frac{h_0(t)r(x_1, \beta)}{h_0(t)r(x_0, \beta)}$$

$$= \frac{r(x_1, \beta)}{r(x_0, \beta)} \quad . \tag{3.6}$$

The hazard ratio (HR) depends only on the function $r(x, \beta)$. If the ratio function $\text{HR}(t, x_1, x_0)$ is easily interpreted, then the actual form of the baseline hazard function is of little importance.

Cox (1972) was the first to propose the model in (3.5) when he suggested using $r(x, \beta) = \exp(x\beta)$. With this parameterization the hazard function is

$$h(t, x, \beta) = h_0(t)e^{x\beta} \tag{3.7}$$

and the hazard ratio is

$$\text{HR}(t, x_1, x_0) = e^{\beta(x_1 - x_0)}. \tag{3.8}$$

This model is referred to in the literature by a variety of terms, such as the *Cox model*, the *Cox proportional hazards model* or simply the *proportional hazards model*. Part of the appeal of the Cox model is the interpretation of (3.8) as a "relative risk"-type ratio. For example, when a covariate is dichotomous, such as gender, with a value of $x_1 = 1$ for males and $x_0 = 0$ for females, the hazard ratio in (3.8) becomes

$$\text{HR}(t, x_1, x_0) = e^{\beta}.$$

If the value of the coefficient is $\beta = \ln(2)$, then the interpretation is that males are "dying" at twice the rate of females. We defer further discussion of the interpretation of the ratio in (3.8) as a function of the coefficients to Chapter 4.

The Cox model in (3.7) is the most frequently used form of the hazard function in (3.5). The term *proportional hazards* refers to the fact that in (3.7) the hazard functions are multiplicatively related, that is, their ratio is constant over survival time. This is an important assumption and methods for assessing its validity are presented in Chapter 6. Other parametrizations have been considered, most notably additive models. One example of an additive model is the *additive relative hazard model* whose hazard function is

$$h(t, x, \beta) = h_0(t)(1 + x\beta). \qquad (3.9)$$

Software packages, such as BMDP and EGRET, offer the user the choice of using (3.7) or (3.9) or a mix of the two. We discuss these and other additive models in Chapter 9. Other more generally parametrized positive functions have been suggested [see Andersen, Borgan, Gill and Keiding (1993, Chapter VII)], but none are in wide practical use. We focus primarily on (3.7), the proportional hazards model, as it is the most frequently used model in applied settings.

The hazard functions in (3.5), (3.7) and (3.9) are called semi-parametric functions since they do not explicitly describe the baseline hazard function, $h_0(t)$. It was noted at the beginning of this chapter that one way to specify the distribution of survival time is through the hazard function. Thus, a natural question is: What is the survivorship function for a model with hazard function (3.5)? If we use the relationship shown in (2.29), then the survivorship function is

$$S(t, x, \beta) = e^{-H(t, x, \beta)} \qquad (3.10)$$

where $H(t, x, \beta)$ is the cumulative hazard function at time t for a subject with covariate x. We have assumed that survival time is absolutely continuous, in which case the value of the cumulative hazard function may be expressed, using methods of calculus, as

$$H(t, x, \beta) = \int_0^t h(u, x, \beta)\, du$$

$$= r(x, \beta)\int_0^t h_0(u)\, du$$

$$= r(x, \beta)H_0(t). \qquad (3.11)$$

For those not comfortable with the methods of calculus, the expression in (3.11) may be thought of as a measure of the cumulative baseline risk, $H_0(t)$, which is modified by the function, $r(x, \beta)$, for a subject with covariate x. Substituting the result (3.11) into (3.10), the survivorship function for the general semiparametric hazard function is

$$S(t, x, \beta) = e^{-r(x, \beta)H_0(t)}.$$

Thus it follows that

$$S(t,x,\beta) = \left[e^{-H_0(t)}\right]^{r(x,\beta)}$$
$$= \left[S_0(t)\right]^{r(x,\beta)}, \tag{3.12}$$

where $S_0(t) = e^{-H_0(t)}$ is the baseline survivorship function.

Under the Cox model, the survivorship function is

$$S(t,x,\beta) = \left[S_0(t)\right]^{\exp(x\beta)}. \tag{3.13}$$

The form of the expression for the survivorship function in (3.13) is a consequence of the multiplicative relationship between the baseline hazard function and the exponential function that describes the effect of the covariates. The value of the baseline survivorship function is always between zero and one (true of any survivorship function). Suppose the covariate is age, denoted a, which we model using $x = a - \bar{a}$. The baseline survivorship function corresponds to a subject whose age is equal to the mean age, $\bar{a}$, of the data. Assuming that the risk associated with age is positive (as is usually the case), then $\beta > 0$, and for $a > \bar{a}$ it follows that $x > 0$, $\exp(x\beta) > 1$ and $S(t,x,\beta) < S_0(t)$. The interpretation is that the survivorship experience is less favorable for age a than at the mean age. In other words, at any point in time, the proportion of subjects alive at age a is smaller than the proportion alive at age $\bar{a}$. Similarly, if age is $a < \bar{a}$, then $x < 0$, $\exp(x\beta) < 1$ and $S(t,x,\beta) > S_0(t)$, implying that the survivorship experience is more favorable at age a than at the mean age.

In the next section, we consider estimation of the parameters in the proportional hazards model.

3.3 FITTING THE PROPORTIONAL HAZARDS REGRESSION MODEL

A brief introduction to the use of maximum likelihood to fit regression models to survival time data was provided in Chapter 1. The models fit in Chapter 1 correspond to those given in (3.1) (3.4), where both the systematic and, more importantly, the error components are fully specified. This complete specification allowed for an explicit expression for the likelihood function. We noted in Chapter 1 that the maximum likelihood approach described was used by most software packages to fit these models. The natural place to begin is with an exploration of whether the likelihood equation given in (1.5) can be used with the proportional hazards model in (3.7).

Assume we have n independent observations each containing information on the length of time a subject was observed, a single covariate whose value is determined at the time observation begins and remains at that value throughout the follow-up of the subject, and whether the observation was a survival time or was right censored. The data are denoted by the triplet (t_i, x_i, c_i), $i = 1,2,...,n$. In order to apply the likelihood function given in (1.5) to the survivorship function in (3.13), we need to obtain an expression for the density function. An application of methods from calculus shows that the density function is the ratio of the hazard function to the survivorship function [see (2.36)], yielding the expression

$$f(t, x, \beta) = h(t, x, \beta) \times S(t, x, \beta). \tag{3.14}$$

Substituting (3.14) into the likelihood equation in (1.5) yields

$$l(\beta) = \prod_{i=1}^{n} \left\{ \left[h(t_i, x_i, \beta) \times S(t_i, x_i, \beta) \right]^{c_i} \times \left[S(t_i, x_i, \beta) \right]^{1-c_i} \right\},$$

and further algebraic simplification yields

$$l(\beta) = \prod_{i=1}^{n} \left\{ \left[h(t_i, x_i, \beta) \right]^{c_i} \times \left[S(t_i, x_i, \beta) \right] \right\}. \tag{3.15}$$

As noted in Chapter 1, the estimate of the parameter, β, is the value that maximizes the log-likelihood function. The log-likelihood function, obtained by taking the log of the likelihood (3.15) and substituting expressions for the hazard function in (3.7) and the survivorship function in (3.13), is

$$L(\beta) = \sum_{i=1}^{n} \left\{ c_i \ln\left[h_0(t_i) \right] + c_i x_i \beta + e^{x_i \beta} \ln\left[S_0(t_i) \right] \right\}. \tag{3.16}$$

Full maximum likelihood requires that we maximize (3.16) with respect to the unknown parameter of interest, β, and the unspecified baseline hazard and survivorship functions. The proportional hazards model in (3.7) is chosen in order to avoid having to explicitly specify the error component of the model; therefore, it is not possible to use the log-likelihood function in (3.16). This problem is discussed in some detail in Kalbfleisch and Prentice (1980).

Cox (1972) proposed using an expression he called a "partial likelihood function" that depends only on the parameter of interest. He speculated that the resulting parameter estimators from the partial likelihood function would have the same distributional properties as full maximum likelihood estimators. Rigorous mathematical proofs of this conjecture came later, and the counting process approach based on martingales, as detailed in Andersen, Borgan, Gill and Keiding (1993, Chapter VII) and Fleming and Harrington (1991, Chapter 4), simplified earlier work. At this point, it is not vital that one understand the mathematics of these details. An intermediate level of presentation of the construction of the partial likelihood is provided in Collett (1994). The essential idea is similar to the one used to generate the conditional logistic regression model for matched case-control studies or other stratified designs that introduce a large number of nuisance parameters into the model [see Hosmer and Lemeshow (1989, Chapter 7)]. In the present setting, the partial likelihood is given by the expression

$$l_p(\beta) = \prod_{i=1}^{n} \left[\frac{e^{x_i\beta}}{\sum_{j \in R(t_i)} e^{x_j\beta}} \right]^{c_i}, \tag{3.17}$$

where the summation in the denominator is over all subjects in the risk set at time t_i, denoted by $R(t_i)$. Recall that the risk set consists of all subjects with survival or censored times greater than or equal to the specified time.

The expression in (3.17) assumes that there are no tied times, and it is often modified to exclude terms when $c_i = 0$, yielding

$$l_p(\beta) = \prod_{i=1}^{m} \frac{e^{x_{(i)}\beta}}{\sum_{j \in R(t_{(i)})} e^{x_j\beta}}, \tag{3.18}$$

where the product is over the m distinct ordered survival times and $x_{(i)}$ denotes the value of the covariate for the subject with ordered survival time $t_{(i)}$. The log partial likelihood function is

$$L_p(\beta) = \sum_{i=1}^{m} \left\{ x_{(i)}\beta - \ln\left[\sum_{j \in R(t_{(i)})} e^{x_j\beta} \right] \right\}. \tag{3.19}$$

We obtain the maximum partial likelihood estimator by differentiating the right hand side of (3.19) with respect to β, setting the derivative equal to zero and solving for the unknown parameter. The derivative of (3.19) with respect to β is

$$\frac{\partial L_p(\beta)}{\partial \beta} = \sum_{i=1}^{m} \left\{ x_{(i)} - \frac{\sum\limits_{j \in R(t_{(i)})} x_j e^{x_j\beta}}{\sum\limits_{j \in R(t_{(i)})} e^{x_j\beta}} \right\}$$

$$= \sum_{i=1}^{m} \left\{ x_{(i)} - \sum_{j \in R(t_{(i)})} w_{ij}(\beta) x_j \right\}$$

$$= \sum_{i=1}^{m} \left\{ x_{(i)} - \bar{x}_{w_i} \right\}, \tag{3.20}$$

where

$$w_{ij}(\beta) = \frac{e^{x_j\beta}}{\sum\limits_{l \in R(t_{(i)})} e^{x_l\beta}}$$

and

$$\bar{x}_{w_i} = \sum_{j \in R(t_{(i)})} w_{ij}(\beta) x_j.$$

We note that equation (3.20) looks different from the corresponding equation for the exponential regression model (1.11). The main difference is that equation (3.20) does not incorporate the actual values of survival time. In fact, the estimator obtained when setting the derivative in (3.20) equal to zero and solving for β yields the value such that the sum of the risk-set-weighted means of the covariate is equal to the sum of the covariate over the non-censored subjects.

Another expression for the derivative in (3.20) is obtained by taking the log of (3.17) and differentiating with respect to β, yielding

$$\frac{\partial L_p(\beta)}{\partial \beta} = \sum_{i=1}^{n} c_i \left\{ x_i - \frac{\sum\limits_{j \in R(t_i)} x_j e^{x_j \beta}}{\sum\limits_{j \in R(t_i)} e^{x_j \beta}} \right\}. \tag{3.21}$$

Most software packages provide the maximum partial likelihood estimator. We denote the solution to (3.20) and (3.21) as $\hat{\beta}$.

The estimator of the variance of the estimator of the coefficient is obtained in the same manner as variance estimators are obtained in most maximum likelihood estimation applications. The estimator is the inverse of the negative of the second derivative of the log partial likelihood at the value of the estimator. In particular, taking the derivative of (3.20) we obtain the following expression:

$$\frac{\partial^2 L_p(\beta)}{\partial \beta^2} = -\sum_{i=1}^{m} \left\{ \frac{\left[\sum\limits_{j \in R(t_{(i)})} e^{x_j \beta} \right]\left[\sum\limits_{j \in R(t_{(i)})} x_j^2 e^{x_j \beta} \right] - \left[\sum\limits_{j \in R(t_{(i)})} x_j e^{x_j \beta} \right]^2}{\left[\sum\limits_{j \in R(t_{(i)})} e^{x_j \beta} \right]^2} \right\}. \tag{3.22}$$

The form of this expression may be simplified by using the definition of $w_{ij}(\beta)$ following (3.20). The simplified expression is

$$\frac{\partial^2 L_p(\beta)}{\partial \beta^2} = -\sum_{i=1}^{m} \sum_{j \in R(t_{(i)})} w_{ij} \left(x_j - \bar{x}_{w_i} \right)^2. \tag{3.23}$$

The negative of the second derivative of the log partial likelihood in (3.22) or (3.23) is called the *observed information*, and we will denote it as

$$I(\beta) = -\frac{\partial^2 L_p(\beta)}{\partial \beta^2}. \tag{3.24}$$

Later in this chapter we will consider models containing more than one covariate and the result in (3.24) will be called the *observed information*

Table 3.1   Estimated Coefficient, Standard Error, z-Score, Two-Tailed p-Value and 95% Confidence Interval for the Proportional Hazards Model Containing Age

Variable	Coeff.	Std. Err.	z	$P>|z|$	95% CIE
AGE	0.0814	0.0174	4.67	<0.001	0.047, 0.116

matrix. The estimator of the variance of the estimated coefficient is the inverse of (3.24) evaluated at $\hat{\beta}$ and is

$$\hat{\text{Var}}(\hat{\beta}) = \mathbf{I}(\hat{\beta})^{-1}. \tag{3.25}$$

The estimator of the standard error, denoted $\hat{\text{SE}}(\hat{\beta})$, is the positive square root of the variance estimator in (3.25).

As an example, we can use the data from the HMO-HIV+ study to fit a model containing age of the subject as the covariate. The results are shown in Table 3.1. The value of the estimated coefficient is $\hat{\beta} = 0.0814$, and the estimated standard error of the estimated coefficient is $\hat{\text{SE}}(\hat{\beta}) = 0.0174$.

Typically, the first steps following the fit of a regression model are the assessment of the significance of the coefficient and the formation of a confidence interval. We discuss methods that can be used for each of these tasks.

We begin by presenting three different tests to assess the significance of the coefficient: the partial likelihood ratio test, the Wald test and the score test.

The partial likelihood ratio test, denoted G, is calculated as twice the difference between the log partial likelihood of the model containing the covariate and the log partial likelihood for the model not containing the covariate. Specifically,

$$G = 2\{L_p(\hat{\beta}) - L_p(0)\}, \tag{3.26}$$

where

$$L_p(0) = -\sum_{i=1}^{m} \ln(n_i), \tag{3.27}$$

and n_i denotes the number of subjects in the risk set at observed survival time $t_{(i)}$.

Under the null hypothesis that the coefficient is equal to zero (along with other mathematical conditions), this statistic will follow a chi-square distribution with 1 degree-of-freedom. This distribution can be used to obtain p-values to test the significance of the coefficient. The mathematical details using a counting process approach to the partial likelihood may be found in Andersen, Borgan, Gill and Keiding (1993) and Fleming and Harrington (1991). In practice, the "sufficiently" large sample size cited for likelihood ratio tests translates in this case to having the number of observed noncensored survival times be large.

Software packages fitting the proportional hazards model typically provide the value of the log partial likelihood for the fitted model and the value of G. For the example in Table 3.1, these values are $L_p(\hat{\beta}) = -288.518$ and $G = 21.350$. We can use (3.26) to obtain the log partial likelihood of model zero[1] as

$$L_p(0) = L_p(\hat{\beta}) - G/2 = (-288.518) - (21.35/2) = -299.195.$$

The significance level for the test is $\Pr(\chi^2(1) \geq 21.35) < 0.001$, so we reject the null hypothesis and conclude that age is significantly related to survival time. We defer discussion of the interpretation of the coefficient until the next chapter.

Another test for significance of the coefficient can be computed from the ratio of the estimated coefficient to its estimated standard error. This ratio is commonly referred to as a Wald statistic. Under the same mathematical assumptions required for the log partial likelihood ratio test, the Wald statistic will follow a standard normal distribution. The Wald statistic and its p-value are typically reported by software packages. Some statistical packages report the square of the Wald statistic, which follows a chi-square distribution with one degree-of-freedom. Unlike normal errors linear regression where the square of the t-statistic for the coefficient in a univariable model is equal to the F-test for significance, the Wald and log partial likelihood ratio test are not numerically related. The equation for the Wald statistic is

$$z = \frac{\hat{\beta}}{\widehat{SE}(\hat{\beta})} \tag{3.28}$$

---

[1] This will be useful later when we extend the partial likelihood ratio test to the mutivariable regression setting.

and the value shown in Table 3.1 is

$$z = (0.0814/0.0174) = 4.67.$$

The two-tailed p-value is $\Pr(|z| > 4.67) < 0.001$.

The third test one is likely to encounter is the score test. The test statistic is the ratio of the derivative of the log partial likelihood, equation (3.20), to the square root of the observed information, equation (3.24), all evaluated at $\beta = 0$. The equation for the score test is

$$z^* = \frac{\partial L_p / \partial \beta}{\sqrt{\mathbf{I}(\beta)}}\bigg|_{\beta=0}. \tag{3.29}$$

Under the hypothesis that the coefficient is equal to zero and the same mathematical conditions required for the Wald and partial likelihood ratio tests, this statistic follows a standard normal distribution. The value of the score test for the example in Table 3.1 is $z^* = 4.69$ and the two-tailed p-value is $\Pr(|z^*| > 4.69) < 0.001$. The score test, when computed by a software package such as SAS, may be reported as the square of the value of (3.29), which will follow a chi-square distribution with one degree-of-freedom under the null hypothesis.

In practice, the numeric values of the three tests $\left(\sqrt{G}, z \text{ and } z^*\right)$ should be quite similar and thus lead one to draw the same conclusion about the significance of the coefficient. In situations where there is disagreement, making it necessary to choose one test, the partial likelihood ratio test is the preferred choice.

A clear advantage of the score test is that it may be computed without evaluating the maximum partial likelihood estimator of the coefficient. For this reason, the score test has gained some favor as a test to use in model building applications in which evaluation of the estimator is computationally intensive. We return to consider this point further when we discuss variable selection in Chapter 5.

The confidence interval for the coefficient shown in Table 3.1 is called the Wald-statistic-based interval. Its endpoints are based on the same assumptions as the Wald test for significance, i.e., that the estimator is distributed normally with standard error estimated by the square root of (3.25). The endpoints of a $100(1 - \alpha)$ percent confidence interval for the coefficient are

$$\hat{\beta} \pm z_{1-\alpha/2} \hat{SE}(\hat{\beta}).$$

The endpoints of the 95 percent confidence interval shown in Table 3.1 are computed as

$$0.0814 \pm 1.96 \times 0.0174,$$

yielding the interval $0.047 \le \beta \le 0.116$. The interval does not include zero and is consistent with the results of all three tests of significance. We conclude that age is associated with survival time.

Up to this point we have considered models in which only one covariate is of interest. One advantage of using regression in any statistical analysis is the ability to include multiple covariates in the model simultaneously. The proportional hazards model may be formulated to include a variety of covariates. We now focus on the extension of the model to include a collection of p covariates whose values are measured on each individual at the time follow-up begins and remain fixed over time. Covariates whose values change over time, often referred to as *time-dependent* or *time-varying* covariates, as well as other covariate scenarios are discussed in Chapter 7.

Let the p covariates for subject i be denoted by the vector $\mathbf{x}_i' = (x_{i1}, x_{i2}, \ldots, x_{ip})$. This vector may be any collection of covariates: continuous covariates, design variables for nominal scale covariates, products of covariates (interactions) and other higher order terms. Denote the triplet of observed time, covariates and censoring variable as $(t_i, \mathbf{x}_i, c_i)$, $i = 1, 2, \ldots, n$. The partial likelihood for the multivariable model is obtained by replacing the single covariate, x, in (3.18) with the vector of covariates, $\mathbf{x}$. Its expression is so similar to (3.18) that it will not be repeated.

There are p equations, one for each covariate, similar to (3.20) which, when set equal to zero and solved, yield the maximum partial likelihood estimators. We denote the vector of coefficients as $\boldsymbol{\beta}' = (\beta_1, \beta_2, \ldots, \beta_p)$. The equation for the kth covariate is

$$\frac{\partial L_p(\boldsymbol{\beta})}{\partial \beta_k} = \sum_{i=1}^{m} \left\{ x_{(ik)} - \frac{\sum_{j \in R(t_{(i)})} x_{jk} e^{\mathbf{x}_j' \boldsymbol{\beta}}}{\sum_{j \in R(t_{(i)})} e^{\mathbf{x}_j' \boldsymbol{\beta}}} \right\}$$

$$= \sum_{i=1}^{m} \left\{ x_{(ik)} - \bar{x}_{w_i k} \right\}, \tag{3.30}$$

where

$$\bar{x}_{w,k} = \sum_{j \in R(t_{(i)})} w_{ij}(\boldsymbol{\beta}) x_{jk}$$

and

$$w_{ij}(\boldsymbol{\beta}) = \frac{e^{x'_j \boldsymbol{\beta}}}{\sum_{l \in R(t_{(i)})} e^{x'_l \boldsymbol{\beta}}}.$$

We use $x_{(ik)}$ to denote the value of covariate x_k for the subject with observed ordered survival time $t_{(i)}$. We denote the maximum partial likelihood estimator as $\hat{\boldsymbol{\beta}}' = (\hat{\beta}_1, \hat{\beta}_2, ..., \hat{\beta}_p)$.

The elements of the p by p information matrix are obtained by extending the definition in (3.24) to include all second-order partial derivatives, namely

$$\mathbf{I}(\boldsymbol{\beta}) = -\frac{\partial^2 L(\boldsymbol{\beta})}{\partial \boldsymbol{\beta}^2}.$$

The general form of the elements in this matrix is obtained from (3.23). The diagonal elements are

$$\frac{\partial^2 L_p(\beta)}{\partial \beta_k^2} = -\sum_{i=1}^{m} \sum_{j \in R(t_{(i)})} w_{ij}\left(x_{jk} - \bar{x}_{w,k}\right)^2 \tag{3.31}$$

and the off-diagonal elements are

$$\frac{\partial^2 L_p(\beta)}{\partial \beta_k \partial \beta_l} = -\sum_{i=1}^{m} \sum_{j \in R(t_{(i)})} w_{ij}\left(x_{jk} - \bar{x}_{w,k}\right)\left(x_{jl} - \bar{x}_{w,l}\right). \tag{3.32}$$

The estimator of the covariance matrix of the maximum partial likelihood estimator is obtained by extending (3.25) and is the inverse of the observed information matrix evaluated at the maximum partial likelihood estimator,

$$\hat{\text{Var}}(\hat{\boldsymbol{\beta}}) = \mathbf{I}(\hat{\boldsymbol{\beta}})^{-1}. \tag{3.33}$$

Software packages typically provide the value of the estimated standard error for all estimated coefficients in the model. Most packages provide the user with the option of obtaining the full estimated covariance matrix for the estimated parameters.

Consider a model for the HMO-HIV+ study that contains age, IV drug use and their product (interaction). This model may be used to determine whether the association of age with survival time is different for subjects with and without a history of IV drug use. The model is used here to present the results of fitting a multivariable model and to demonstrate how the partial likelihood ratio test may be used to assess the significance of subsets of parameters. We present the results of fitting the model in Table 3.2.

The log partial likelihood ratio test is not only the easiest test to compute, but is also the best of the three tests for assessing the significance of the fitted model. Its value is obtained from (3.26). The log partial likelihood for model 0 is the same for this example as in the univariable model in Table 3.1, $L_p(0) = -299.193$. The log partial likelihood for the fitted model is $L_p(\hat{\beta}) = -281.684$ and the value of the log partial likelihood ratio test is

$$G = 2[(-281.684)-(-299.193)] = 35.02.$$

Under the null hypothesis that all three coefficients are simultaneously equal to zero and, under the mathematical regularity and large sample conditions referred to above, G will follow a chi-square distribution with three degrees-of-freedom (one for each coefficient). The significance level for the test in this example is $\Pr(\chi^2(3) \geq 35.02) < 0.001$, providing evidence that at least one of the coefficients in the model is significantly associated with survival time.

The computation of both the score and Wald tests for the multiple

Table 3.2   Estimated Coefficients, Standard Errors, z-Scores, Two-Tailed p-Values and 95% Confidence Intervals for the Proportional Hazards Model Containing Age, History of IV Drug Use and Their Interaction

Variable	Coeff.	Std. Err.	z	$P>\vert z\vert$	95% CIE
AGE	0.094	0.0229	4.11	<0.001	0.049, 0.139
DRUG	1.186	1.2565	0.94	0.345	−1.277, 3.649
AGE×DRUG	−0.007	0.0337	−0.20	0.841	−0.073, 0.059

proportional hazards regression model requires matrix calculations. Specifically, we denote the vector of first partial derivatives whose elements are given in (3.29) as $\mathbf{u}(\boldsymbol{\beta})$. Under the hypothesis that all coefficients are equal to zero, and under the mathematical conditions needed for the partial likelihood ratio test, the vector of scores $\mathbf{u}(\mathbf{0}) = \mathbf{u}(\boldsymbol{\beta})\big|_{\boldsymbol{\beta}=0}$ will be distributed as multivariate normal with mean vector equal to zero and covariance matrix given by the information matrix evaluated at the coefficient vector equal to zero, $\mathbf{I}(\mathbf{0}) = \mathbf{I}(\boldsymbol{\beta})\big|_{\boldsymbol{\beta}=0}$. The elements in this matrix are obtained by evaluating the expressions in (3.31) and (3.32) with the coefficient vector equal to zero. The score test statistic is

$$\mathbf{u}'(\mathbf{0})[\mathbf{I}(\mathbf{0})]^{-1}\mathbf{u}(\mathbf{0}),$$

which is distributed asymptotically as chi-square with p degrees-of-freedom. The Wald test is obtained from equivalent theory which states that, under the null hypothesis, the estimator of the coefficient, $\hat{\boldsymbol{\beta}}$, will be asymptotically normally distributed with mean vector equal to zero and a covariance matrix that is estimated by the expression in (3.33). The multiple variable Wald test statistic is

$$\hat{\boldsymbol{\beta}}'\mathbf{I}(\hat{\boldsymbol{\beta}})\hat{\boldsymbol{\beta}},$$

which is also distributed asymptotically as chi-square with p degrees-of-freedom. Both the score and Wald test require matrix calculations that, while not difficult from a purely technical perspective, are inconvenient to perform in most packages. This is in contrast to the partial likelihood ratio test which is easily performed from readily available output. For this reason we will not make extensive use of the multiple variable score and Wald tests in this text. The values for the multiple variable score and Wald tests for the model in Table 3.2 are 35.146 and 32.167, respectively, each with p-value < 0.001.

In contrast to the multiple variable Wald test, the univariate Wald tests based on individual estimated coefficients can provide guidance, during the model building process, as to possible variables that might be eliminated from the model without compromising model performance. The individual significance levels in Table 3.2 suggest that age may be significant, but the picture is not as clear with respect to IV drug use and its interaction with age. To explore this further we fit a reduced model that excludes the interaction term. The results are shown in Table 3.3.

Table 3.3   Estimated Coefficients, Standard Errors, z-Scores, Two-Tailed p-Values and 95% Confidence Intervals for the Proportional Hazards Model Containing Age and History of IV Drug Use

Variable	Coeff.	Std. Err.	z	$P>|z|$	95% CIE
AGE	0.092	0.0185	4.97	0.001	0.056, 0.128
DRUG	0.941	0.2555	3.68	0.001	0.440, 1.442

The Wald tests for both remaining coefficients are significant. The partial likelihood ratio test for the excluded interaction term, keeping age and IV drug use in the model, is obtained by comparing the values of the log partial likelihood function for the models in Tables 3.2 and 3.3. This test is analogous to the partial F-test in linear regression, in that two models that have a common set of covariates are being compared. As in any multivariable analysis, we must make sure that both models have been fit to the same set of data. Since the HMO-HIV+ study does not have any missing data, this is not an issue in this example. The value of the log partial likelihood function for the reduced model is –281.704 which, when compared to that of the larger model, yields a test statistic whose value is

$$G = 2[(-281.684) - (-281.704)] = 0.04.$$

Under the null hypothesis that the interaction variable has a coefficient equal to zero, given that age and history of IV drug use are in the model, this statistic will follow a chi-square distribution with one degree-of-freedom. The significance level for the test in this case is $p = 0.841$, indicating that the interaction term does not contribute to the model.

In summary, the basic techniques for fitting the proportional hazards model are identical to those used in other modeling scenarios, such as the linear, logistic and Poisson regression models. Maximum likelihood methods are used to obtain estimators of the coefficients and their standard errors. We use log-likelihood functions in a standard manner to obtain test statistics that are used with the chi-square distribution to assess the overall significance of the model and to compare nested models. The only difference between the analysis of the proportional hazards model and other models is that the likelihood function is a partial, rather than a full, likelihood function.

3.4 FITTING THE PROPORTIONAL HAZARDS MODEL WITH TIED SURVIVAL TIMES

The partial likelihood function methods described in the previous section are based on the assumption that there were no tied values among the observed survival times. Since most, if not all, applied settings are likely to have some tied observations, modifications are needed. A number of approaches to handle tied data have been suggested and, of these, three are used by software packages: an exact expression that is derived in Kalbfleisch and Prentice (1980) and approximations due to Breslow (1974) and Efron (1977). The analyses presented in the previous section were all based on the Breslow approximation described below. An alternative to an approximate partial likelihood is to use one of the discrete time models discussed in Chapter 7.

We will not present the expression for the exact partial likelihood. The basis for its construction is to assume that the d ties at a particular survival time are due to lack of precision in measuring survival time. Thus the tied values could actually have been observed in any one of the $d!$ possible arrangements of their values. The exact partial likelihood is obtained by modifying the denominator of (3.18) to include each of these arrangements. The SAS software package includes the option of using the exact partial likelihood.

The approximations derived by Breslow (1974) and Efron (1977) are designed to provide expressions that are more easily computed than the exact partial likelihood, yet that still account for the fact that ties are present among the observed values of survival time. For ease of notation, we present the approximations to the exact partial likelihood for the case when the model contains a single covariate. The Breslow approximation uses as the partial likelihood

$$l_{p1}(\beta) = \prod_{i=1}^{m} \frac{e^{x_{(i)+}\beta}}{\left[\sum_{j \in R(t_{(i)})} e^{x_j \beta}\right]^{d_i}}, \tag{3.34}$$

where d_i denotes the number of subjects with survival time $t_{(i)}$ and $x_{(i)+}$ is equal to the sum of the covariate over the d_i subjects, that is, $x_{(i)+} = \sum_{j \in D(t_{(i)})} x_j$, where $D(t_{(i)})$ represents the subjects with survival times equal to $t_{(i)}$. The Efron approximation is a bit more complicated and yields a slightly better approximation to the exact partial likelihood than the Breslow approximation. It uses as the partial likelihood

$$l_{p2}(\beta) = \prod_{i=1}^{m} \frac{e^{x_{(i)+}\beta}}{\prod_{k=1}^{d_i}\left[\sum_{j\in R(t_{(i)})} e^{x_j\beta} - \frac{k-1}{d_i}\sum_{j\in D(t_{(i)})} e^{x_j\beta}\right]}. \tag{3.35}$$

Note that when $d_i = 1$, the terms in the numerators and denominators of (3.18), (3.34) and (3.35) are identical.

The maximum partial likelihood estimator for β in the presence of ties is obtained in the same manner as in the non-tied data case, with the exception that derivatives are taken with respect to the unknown parameter in the log of either the Breslow (1974) or Efron (1977) approximation to the partial likelihood. These equations are similar in form to (3.20)–(3.21). The estimator of the variance of the estimated coefficient is obtained from the second partial derivative evaluated at the value of the estimator, and results are similar to (3.23)–(3.25).

The HMO-HIV+ study provides a good setting for a comparison of the estimators obtained from the three forms of the partial likelihood in the presence of tied survival times. In this study, there are 31 distinct survival times among the 100 subjects, with the number of deaths at a particular time ranging from 1 to 17. If there are major differences in the estimators obtained from the three versions of the partial likelihood with ties, it should be apparent in this example because there are many tied survival times. The values of the estimator using each of the three methods are shown in Table 3.4 for the model containing age and IV drug use.

The results shown in Table 3.4 support the fact that the Efron (1977) method of correcting for tied survival times yields estimates closer to those obtained from the exact partial likelihood than estimates obtained from the Breslow (1974) approximation. While this is true in a strict numeric sense, all three point estimates are close to one another. The Breslow estimates differ from the exact estimates by 6–8 percent and the Efron estimates differ by 0.5 percent. The estimated standard errors are nearly identical. Hence, we would reach the same scientific conclusion using the estimates from the Breslow partial likelihood as we would using the estimates from the other two partial likelihoods. Thus, given a choice, one would prefer to use the Efron approximation, but in this example, the Breslow approximation yields acceptably close estimates.

Table 3.4   Estimated Coefficients and Standard Errors for Age and IV Drug Use Obtained from the Exact Partial Likelihood, Breslow and Efron Approximations

Method	AGE		DRUG	
	Coeff.	Std. Err.	Coeff.	Std. Err.
Exact	0.0977	0.0187	1.0226	0.2572
Breslow	0.0915	0.0185	0.9414	0.2555
Efron	0.0971	0.0186	1.0167	0.2562

The Breslow (1974) approximation is available in many software packages. The Efron (1977) approximation is available in the SAS and S-Plus packages. In many applied settings there will be little or no practical difference between the estimators obtained from the two approximations. Because of this, and since the Breslow approximation is more commonly available, unless stated otherwise, analyses presented in this text will be based on it.

3.5 ESTIMATING THE SURVIVORSHIP FUNCTION OF THE PROPORTIONAL HAZARDS REGRESSION MODEL

An estimator of the survivorship function of the proportional hazards model is available as an option in most software packages. This estimator may be used to describe the survival experience of subgroups of subjects of particular interest, adjusted for other covariates. This particular application is discussed in detail in Chapter 4. In this section we present how the estimator itself is obtained.

The expression for the survivorship function can be found in (3.13) and is repeated here for convenience:

$$S(t, \mathbf{x}, \boldsymbol{\beta}) = \left[S_0(t) \right]^{\exp(\mathbf{x}'\boldsymbol{\beta})}. \tag{3.36}$$

This indicates that once we have an estimator of the regression coefficients, all we need is an estimator of the baseline survivorship function. A likelihood-based approach, which assumes that the hazard is constant between observed survival times, is the foundation of the method. The details may be found in Lawless (1982) and are sketched here. A derivation of the estimator from the counting process approach is discussed by both Fleming and Harrington (1991) and Andersen, Borgan, Gill and Keiding (1993).

The essential idea of the likelihood approach is to mimic the arguments that lead to the Kaplan–Meier estimator of the survivorship function described in Chapter 2, equation (2.1). The key point in that development is the use of the quantity $\hat{\alpha}_i = 1 - d_i/n_i$ as an estimator of the conditional survival probability at observed ordered survival time $t_{(i)}$. The Kaplan–Meier estimator of the survivorship function is the product of estimators of the individual conditional survival probabilities. The expression for the conditional survival probability that leads to this estimator is $\alpha_i = S(t_{(i)})/S(t_{(i-1)})$. To extend this argument to the proportional hazards model, we define the conditional baseline survival probability as $\alpha_i = S_0(t_{(i)})/S_0(t_{(i-1)})$, and the conditional survival probability is

$$\frac{S(t_{(i)}, \mathbf{x}, \boldsymbol{\beta})}{S(t_{(i-1)}, \mathbf{x}, \boldsymbol{\beta})} = \left\{ \frac{\left[S_0(t_{(i)})\right]^{\exp(\mathbf{x}'\boldsymbol{\beta})}}{\left[S_0(t_{(i-1)})\right]^{\exp(\mathbf{x}'\boldsymbol{\beta})}} \right\} = \left\{ \frac{S_0(t_{(i)})}{S_0(t_{(i-1)})} \right\}^{\exp(\mathbf{x}'\boldsymbol{\beta})} = \alpha_i^{\exp(\mathbf{x}'\boldsymbol{\beta})}.$$

Maximum likelihood methods are employed conditional on the partial likelihood estimator of the regression coefficients in the model, $\hat{\boldsymbol{\beta}}$. In order to simplify the notation, we let $\hat{\theta}_l = \exp\left(\mathbf{x}'\hat{\boldsymbol{\beta}}\right)$, and the estimator of the conditional baseline survival probability is obtained by solving the equation

$$\sum_{l \in D_i} \frac{\hat{\theta}_l}{1 - \hat{\alpha}_i^{\hat{\theta}_l}} = \sum_{l \in R_i} \hat{\theta}_l, \tag{3.37}$$

where R_i denotes the subjects in the risk set at ordered observed survival time $t_{(i)}$ and D_i denotes the subjects in the risk set with survival times equal to $t_{(i)}$.

If there are no tied survival times, D_i contains one subject and the solution to (3.37) is

$$\hat{\alpha}_i = \left[1 - \frac{\hat{\theta}_i}{\sum_{l \in R_i} \hat{\theta}_l} \right]^{\hat{\theta}_i^{-1}}. \tag{3.38}$$

If there are tied survival times, the solution to (3.37) is obtained using iterative methods. The estimator of the baseline survivorship function is the product of the individual estimators of the conditional baseline survival probabilities

$$\hat{S}_0(t) = \prod_{t_{(i)} \le t} \hat{\alpha}_i,\tag{3.39}$$

where $\hat{\alpha}_i$ is the solution to (3.37). This estimator is used in some software packages, for example, SAS and STATA. Other packages may use an approximation to the solution for (3.36) due to Breslow (1974). To obtain this solution, one replaces $\alpha_i^{\hat{\theta}_l}$ on the left-hand side of (3.37) with the approximation $\alpha_i^{\hat{\theta}_l} \approx 1 + \hat{\theta}_l \ln(\alpha_i)$. The solution to (3.36) is then

$$\tilde{\alpha}_i = \exp\left[-d_i \Big/ \sum_{l \in R_i} \hat{\theta}_l \right],\tag{3.40}$$

and the estimator of the baseline survivorship function is again the product of the individual conditional survival probabilities. One uses (3.39) with the estimator in (3.40).

The estimator of the survivorship function in (3.36) is obtained by substituting the estimators of the baseline survivorship function and the estimator of the coefficients using covariate values of interest. Software packages typically provide the value of the estimator of the survivorship function using the observed time and covariates for all (noncensored as well as censored) subjects.

Some software packages provide an estimator of the baseline hazard function, which is a simple function of the estimator of the conditional survival probabilities, namely

$$\hat{h}_0(t_{(i)}) = 1 - \hat{\alpha}_i.$$

The individual pointwise estimators of the baseline hazard function will typically be too "noisy" or unstable (see Figure 2.12) to use themselves. However, by using smoothing methods referred to in Chapter 1, one may get a sense of the shape of the underlying baseline hazard function.

The estimator of the cumulative baseline hazard function is more practical to use since it is less noisy than the estimator of the baseline hazard function. Its estimator is obtained using the expression for the survivorship function shown in (2.29), namely

$$\hat{S}_0(t) = e^{-\hat{H}_0(t)},$$

thus the estimator of the cumulative baseline hazard function is

$$\hat{H}_0(t) = -\ln\left[\hat{S}_0(t)\right].$$

The estimator of the cumulative hazard function for a specific value of the covariates is

$$\hat{H}(t, \mathbf{x}, \hat{\boldsymbol{\beta}}) = -\ln\left[\hat{S}(t, \mathbf{x}, \hat{\boldsymbol{\beta}})\right]$$

$$= -e^{\mathbf{x}'\hat{\boldsymbol{\beta}}} \ln\left[\hat{S}_0(t)\right], \tag{3.41}$$

which, when graphed as a function of time, may provide a useful graphical descriptor of the "risk" experience.

We do not present an application of the estimators of the cumulative hazard function or survivorship function in this chapter. We defer it to Chapter 4, where we discuss the interpretation of the coefficients from a fitted proportional hazards model, the assumption of proportional hazards and graphical presentation of fitted models.

EXERCISES

1. Using the data from the WHAS for grouped cohort 1 (1975 and 1978), with length of follow-up as the survival time variable and status at last follow-up as the censoring variable, do the following:

(a) Fit the proportional hazards model containing age, sex, peak cardiac enzymes, left heart failure complications and MI order.

(b) Assess the significance of the model using the partial log likelihood ratio test. If it is possible in the software package, assess for the significance of the model using the score and Wald tests. Is the statistical decision the same for the three tests?

(c) Using the univariate Wald tests, which variables appear not to contribute to the model? Fit a reduced model and test for the significance of the variables removed using the partial log likelihood ratio test.

(d) Fit the reduced model in problem 1(c) using the Breslow, Efron and exact methods for tied survival times. Compare the estimates of the coefficients and standard errors obtained from the three methods for handling tied survival times. Are the results similar or different?

(e) Estimate the baseline survivorship function for the model fit in problem 1(c). Graph the estimated baseline survivorship function versus survival time. What covariate pattern is the "baseline" subject for the fitted model?

(f) Repeat problem 1(e) using age centered at the median age of 65 years. Explain why the range of the estimated survivorship functions in problems 1(e) and 1(f) are different.

(g) Using the model fit in problem 1(f) estimate the value of the survivorship function for each subject at his or her respective observed value of time. Graph the values of the estimated survivorship function versus survival time. Why is there scatter in this plot that was not present in the graphs in problems 1(e) and 1(f)?

2. Repeat problem 1 for each of the other grouped cohorts.

3. Repeat problem 1 using all the data from the WHAS (i.e., ignore cohort).

CHAPTER 4

Interpretation of a Fitted Proportional Hazards Regression Model

4.1 INTRODUCTION

The interpretation of a fitted proportional hazards model requires that we be able to draw practical inferences from the estimated coefficients in the model. We begin by discussing the interpretation of the coefficients for nominal (Section 4.2) and continuous (Section 4.3) scale covariates. In Section 4.4 we discuss the issues of statistical adjustment and the interpretation of estimated coefficients in the presence of statistical interaction. The chapter concludes with a discussion of the interpretation of fitted values from the model and covariate adjusted survivorship functions.

In any regression model, the estimated coefficient for a covariate represents the rate of change of a function of the dependent variable per-unit change in the covariate. Thus, to provide a correct interpretation of the coefficients, we must determine the functional relationship between the independent and dependent variables, and we must define the unit change in the covariate that is likely to be of interest.

In Chapter 3 we recommended that the hazard function be used in regression analysis to study the effect of one or more covariates on survival time. The first step in the process of interpreting the coefficients is to determine what transformation of the hazard function is linear in the coefficients. In the family of generalized linear models (i.e., linear, logistic, Poisson and other regression models) this linearizing transformation is known as the *link function* [see McCullagh and Nelder (1989)]. This same terminology can be applied to proportional hazards regression models.

The proportional hazards model can be used when the primary goal of the analysis is to estimate the effect of study variables on survival time. Suppose, for the moment, that we have a regression model containing a single covariate. Since the hazard function for the proportional hazards regression model is

$$h(t, x, \beta) = h_0(t) e^{x\beta},$$

it follows that the link function is the natural log transformation. We denote the log of a hazard function as $g(t, x, \beta) = \ln[h(t, x, \beta)]$. Thus, in the case of the proportional hazards regression model, the log-hazard function is

$$g(t, x, \beta) = \ln[h_0(t)] + x\beta. \tag{4.1}$$

The difference in the log-hazard function for a change from $x = a$ to $x = b$ is

$$[g(t, x = a, \beta) - g(t, x = b, \beta)] = \{\ln[h_0(t)] + a\beta\} - \{\ln[h_0(t)] + b\beta\}$$

$$= a\beta - b\beta$$

$$= (a - b)\beta. \tag{4.2}$$

Note that, since the baseline hazard function, $h_0(t)$, appears in both log hazards, it subtracts itself out. Thus, the difference in the log hazards does not depend on time. This critical *proportional hazards* assumption is examined in detail in Chapter 6, when we discuss methods for assessing model adequacy and assumptions.

The log hazard is the correct function to use to assess the effect of change in a covariate. However, it is not as easily interpreted as the expression we obtain when we exponentiate (4.2), namely

$$\mathrm{HR}(t, a, b, \beta) = \exp[g(t, x = a, \beta) - g(t, x = b, \beta)]$$

$$= \frac{h(t, a, \beta)}{h(t, b, \beta)}$$

$$= e^{(a-b)\beta}. \tag{4.3}$$

The quantity defined in (4.3) is the hazard ratio, and it plays the same role in interpreting and explaining the results of a survival analysis that the odds ratio plays in a logistic regression.[1] We return to this point in the next section.

The results in (4.2) and (4.3) are important as they provide the method that must be followed to interpret the coefficients in any proportional hazards regression model correctly. The presence of censored observations of survival time in the data does not alter the interpretation of the coefficients. Censoring is an estimation issue that was dealt with when we constructed the partial likelihood function, see (3.17). Once we have accounted for the censoring, we can ignore it.

4.2 NOMINAL SCALE COVARIATE

We begin by considering the interpretation of the coefficient for a dichotomous covariate. Dichotomous or binary covariates occur regularly in applied settings. They may be truly dichotomous (e.g., gender) or they may be derived from continuous covariates (e.g., age greater than 40 years).

Assume for the moment that we have a model containing a single dichotomous covariate, denoted X, coded 0 or 1. Following the procedure described in (4.2), the first step in interpreting the coefficient for X is to calculate the difference in the log hazard for a one unit change in the covariate. This yields

$$g(t,1,\beta) - g(t,0,\beta) = (1-0)\beta = \beta.$$

Thus, in the special case when the dichotomous covariate is coded zero and one, the coefficient is equal to the change of interest in the log hazard. We can exponentiate, following (4.3), the value of the difference in log hazards to obtain the hazard ratio

$$\mathrm{HR}(t,1,0,\beta) = e^\beta. \tag{4.4}$$

The form of the hazard ratio in (4.4) is identical to the form of the odds ratio from a logistic regression model for a dichotomous covariate. The difference is that, in the current context, it is a ratio of rates rather than of odds. In order to expand on this difference, suppose that we followed a large cohort of males and females for 5 years and noted

---

[1] See Hosmer and Lemeshow (1989) Chapter 3 for a detailed discussion of the interpretation of the coefficients in a logistic regression model.

whether a subject "died" during this period of time. In this hypothetical setting one might be tempted to analyze the end-of-study binary variable, death (yes = 1), using a logistic regression model. One should note that this binary variable is what we have defined as the censoring variable for the observation of time to death.[2] Suppose the value of the odds ratio for X, denoting gender (1 = male), is 2.0. This is interpreted to mean, under conditions where the odds-ratio approximates the relative risk, that the probability of death by the end of the study is 2 times higher for a male than for a female. A hazard ratio of 2 obtained from (4.4) means that, at any time during the study, the per-unit time rate of death among males is twice that of females. Thus, the hazard ratio is a comparative measure of survival experience over the entire time period, whereas the odds-ratio is a comparative measure of event occurrence only at the study endpoint. They are two different measures, and the fact that they may be of similar magnitude in an applied setting is, in a sense, irrelevant. Note that if one is able to observe the survival time for all subjects, logistic regression cannot be used at all.

In order to illustrate further the interpretation of the hazard ratio for a dichotomous covariate, survival times were created for a hypothetical cohort of 10,000 subjects, with 5,000 in each of two groups and a theoretical hazard ratio of 2.0. Subjects whose survival time exceeded 60 months were considered censored at 60 months. Each month the number at risk, the number of deaths, the estimated hazard rates, $h_k(t) = d_k(t)/n_k(t), k = 0,1$ and ratio, $HR(t) = h_1(t)/h_0(t)$ were computed. These quantities are listed in Table 4.1 for the first 12 months and the last 13 months of this study. The hazard ratio is graphed for the entire study period in Figure 4.1. The average estimated hazard ratio, $\overline{HR} = (1/58) \times \sum HR(t)$, has been added to Figure 4.1. The hazard rates and their ratios indicate that, during each month of the 60 months of follow-up, the death rate for group 1 is approximately twice that seen in group 0. The scatter about 2.0 is due to the randomness in the number of deaths observed at each time.

---

[2] Even though the two regression models can, under certain conditions, yield similar coefficients, see Hosmer and Lemeshow (1989, Chapter 8), we are not suggesting that a logistic regression of the censoring variable be used in place of a proportional hazards regression of survival time. We assume the reader has a clear understanding of the interpretation of coefficients from a logistic regression model and use it only to explain the difference between the interpretation of a coefficient under the two regression models.

The increase in the scatter over time in Figure 4.1 is due to the fact that the number in the risk sets decreases over time. There were no deaths at 58 months in group 1, so the hazard rate for group 1 and the rate ratio cannot be estimated, at least using the same estimator used for the other months. By design of the example, all values of time greater than or equal to 60 months are censored, so the point estimate of each hazard rate is zero and the estimate of the hazard ratio is undefined.

Table 4.1 Partial Listing of the Number of Deaths, the Number at Risk and the Estimated Hazard Rate in Two Hypothetical Groups and the Estimated Hazard Ratio at Time t

t	$d_0(t)$	$n_0(t)$	$h_0(t)$	$d_1(t)$	$n_1(t)$	$h_1(t)$	HR(t)
1	109	5000	0.022	207	5000	0.041	1.9
2	216	4891	0.044	378	4793	0.079	1.79
3	190	4675	0.041	370	4415	0.084	2.06
4	162	4485	0.036	367	4045	0.091	2.51
5	165	4323	0.038	262	3678	0.071	1.87
6	178	4158	0.043	250	3416	0.073	1.71
7	153	3980	0.038	245	3166	0.077	2.01
8	160	3827	0.042	227	2921	0.078	1.86
9	153	3667	0.042	226	2694	0.084	2.01
10	142	3514	0.04	189	2468	0.077	1.9
11	120	3372	0.036	185	2279	0.081	2.28
12	149	3252	0.046	199	2094	0.095	2.07
⋮	⋮	⋮	⋮	⋮	⋮	⋮	⋮
48	28	708	0.04	10	92	0.109	2.75
49	30	680	0.044	5	82	0.061	1.38
50	27	650	0.042	7	77	0.091	2.19
51	26	623	0.042	8	70	0.114	2.74
52	21	597	0.035	3	62	0.048	1.38
53	23	576	0.04	5	59	0.085	2.12
54	25	553	0.045	5	54	0.093	2.05
55	22	528	0.042	2	49	0.041	0.98
56	23	506	0.045	5	47	0.106	2.34
57	22	483	0.046	3	42	0.071	1.57
58	25	461	0.054	0	39	0	*
59	20	436	0.046	2	39	0.051	1.12
60	0	416	0	0	37	0	*

* Estimator undefined.

In most applied settings, there will be too much variability in the pointwise estimators of the hazard rates, $d(t)/n(t)$, for a figure like Figure 4.1 to be particularly informative about the value of the hazard rate or to determine whether it is constant over time. More sophisticated methods are considered in Chapter 6.

Table 4.2 presents the results of fitting the proportional hazards model containing the dichotomous variable for IV drug use in the HMO-HIV+ study. The point estimate of the coefficient is $\hat{\beta} = 0.779$. Since IV drug use was coded as 1 = yes and 0 = no, we know from (4.3) and (4.4) that we can obtain the point estimator of the hazard ratio by exponentiating the estimator of the coefficient. In this example the estimate is

$$\widehat{HR} = e^{0.779} = 2.18.$$

In the case of a dichotomous covariate coded zero and one, the hazard ratio depends only on the coefficient. Like the odds-ratio estimator

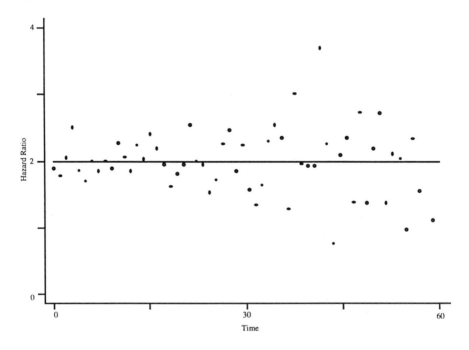

Figure 4.1 Graph of the estimated hazard ratios and the mean hazard ratio ($\overline{HR} = 2.0$) from Table 4.1.

Table 4.2   Estimated Coefficient, Standard Error, z-Score, Two-Tailed p-Value and 95% Confidence Interval for IV Drug Use from the HMO-HIV+ Study

Variable	Coeff.	Std. Err.	z	$P>|z|$	95% Conf. Int.
DRUG	0.779	0.2422	3.22	0.001	0.304, 1.254

in logistic regression, the sampling distribution of the estimator of the hazard ratio is skewed to the right, so confidence interval estimators based on the Wald statistic and its assumption of normality may not have good coverage properties unless the sample size is quite large. Comparatively speaking, the sampling distribution of the estimator of the coefficient is better approximated by the normal distribution than the sampling distribution of the estimated hazard ratio. As a result, its Wald statistic-based confidence interval will have better coverage properties. In this case, we obtain the endpoints of a 95 percent confidence interval for the hazard ratio by exponentiating the endpoints of the confidence interval for the coefficient. In the current example these are

$$\exp\left[\hat{\beta}\pm1.96\widehat{SE}\left(\hat{\beta}\right)\right]=\exp[0.779\pm1.96\times0.2422]=1.355, 3.504.$$

Alternative confidence interval estimators have been studied, one of which is based on the partial likelihood. To date, this method has not been implemented in most software packages.

The interpretation of the estimated hazard rate of 2.18 is that subjects with a history of IV drug use die at about twice the rate of those without a history of IV drug use, throughout the study period. The confidence interval suggests that ratios as low as 1.4 or as high as 3.5 are consistent with the observed data, at the $\alpha=0.05$ level.

As discussed in Chapter 3, the partial likelihood ratio test, the Wald test and the score test can be used to assess the significance of a coefficient. In the current example, the value of the partial likelihood ratio test is $G=10.20$, with a p-value equal to 0.001. The Wald test statistic is $z=3.22$, with a p-value also equal to 0.001. Both of these tests indicate that the coefficient for IV drug use is significant. One should note that the confidence interval for the hazard ratio does not include 1.0, another indicator of its significance. Most software packages provide output for the coefficients, but some, such as STATA, provide hazard ratios and/or coefficients.

Note that Table 4.2 contains no intercept term. This is the price one pays for choosing the semiparametric proportional hazards model. The intercept, were one present, would correspond to the log baseline hazard function, in this case drug group 0. The implication of this in practice is that we cannot, from the regression output of a proportional hazards model, reconstruct group-specific hazard rates. Only ratios can be estimated. If it is critical to have individual estimates of group-specific hazard rates, then one should use one of the fully parametric models discussed in Chapter 8.

Occasionally, the coded values for a dichotomous variable differ by more than 1 (e.g., +1, −1). In this case, it is not possible to obtain the estimator of the hazard ratio by simply exponentiating the estimator of the coefficient. One can always obtain the correct estimator by explicitly evaluating (4.2) and (4.3). If, as shown in (4.2) and (4.3), the two values are denoted as a and b, then the estimator of the hazard ratio is

$$\hat{\mathrm{HR}}(t,a,b,\beta) = e^{(a-b)\hat{\beta}} . \tag{4.5}$$

The endpoints of a $100(1-\alpha)$ percent confidence interval estimator for the hazard ratio can be obtained by exponentiating the endpoints of the confidence interval estimator for $(a-b)\beta$,

$$\exp\left[(a-b)\hat{\beta} \pm |a-b| z_{1-\alpha/2} \hat{\mathrm{SE}}\left(\hat{\beta}\right)\right], \tag{4.6}$$

where $|a-b|$ denotes the absolute value of $(a-b)$.

If a nominal scale covariate has more than two levels, denoted in general by K, we must model the variable using a collection of $K-1$ "design" (also known as "dummy" or "indicator") variables. The most frequent method of coding these design variables is to use *reference cell coding*. With this method, we choose one level of the variable to be the reference level, against which all other levels are compared. The resulting hazard ratios compare the hazard rate of each group to that of the referent group.

In Chapter 2, we considered an example in which age of subjects in the HMO-HIV+ study was categorized into four groups $[20-29]$, $[30-34]$, $[35-39]$ and $[40-54]$. Our goal was to describe, qualitatively, how survival experience in the cohort changes with age, through plots of estimated survivorship functions and a log-rank test. We can continue along these same lines by fitting a proportional hazards model

Table 4.3 Coding of the Three Design Variables for the Age Groups in the HMO-HIV+ Study

Age Group	AGE_2	AGE_3	AGE_4
1:[20 – 29]	0	0	0
2:[30 – 34]	1	0	0
3:[35 – 39]	0	1	0
4:[40 – 54]	0	0	1

to these data. The estimated hazard ratios provide a convenient and easily interpreted summary measure of the comparative survival experience of the four groups.

The methods discussed in this example may be applied to any covariate with multiple groups. The coding for the three design variables based on the four age groups, using the youngest age group as the referent group, are presented in Table 4.3. The results of fitting a proportional hazards model using these three design variables are presented in Table 4.4.

The value of the partial likelihood ratio test for the overall significance of the coefficients is $G = 19.56$ and the p-value, computed using a chi-square distribution with three degrees-of-freedom, is less than 0.001. This suggests that at least one of the three older age groups has a hazard rate that is significantly different from the youngest age group. The p-values of the individual Wald statistics indicate that the hazard rate in each of the three older groups is significantly different from that in the youngest (or reference) age group.

Before we can use (4.2) and (4.3) to obtain estimators of the hazard ratios, we need the equation for the log-hazard function. The log-hazard function, ignoring the log baseline hazard function, for the model fit in Table 4.4 is

$$g(t, \text{AGE\_GRP}, \beta) = \beta_1 \text{AGE\_2} + \beta_2 \text{AGE\_3} + \beta_3 \text{AGE\_4}.$$

The estimator of the hazard ratio comparing age group 2 to age group 1 is obtained by first calculating the difference in the estimators of the log-hazard functions, (4.2),

$$\left[g(t, \text{AGE\_GRP} = 2, \hat{\beta}) - g(t, \text{AGE\_GRP} = 1, \hat{\beta}) \right]$$
$$= \left(\hat{\beta}_1 1 + \hat{\beta}_2 0 + \hat{\beta}_3 0 \right) - \left(\hat{\beta}_1 0 + \hat{\beta}_2 0 + \hat{\beta}_3 0 \right) = \hat{\beta}_1 \, .$$

Exponentiating the result, we obtain

$$\hat{HR}(2,1) = e^{\hat{\beta}_1} \, .$$

We obtain the estimators of the other two hazard ratios by proceeding in a similar manner, and these are

$$\hat{HR}(3,1) = e^{\hat{\beta}_2}$$

and

$$\hat{HR}(4,1) = e^{\hat{\beta}_3} \, .$$

We calculate the value of the estimates in the example, shown in the second column of Table 4.5, by exponentiating the values of the coefficients, from Table 4.4.

When reference cell coding is used to create the design variables, the estimators of the hazard ratio comparing each group to the referent group are obtained by exponentiating the respective estimators of the coefficients.

We construct confidence interval estimators of the hazard ratios by exponentiating the endpoints of the confidence intervals for the individual coefficients. For example, the endpoints of the 95 percent confidence interval estimate for HR(2,1) shown in Table 4.5 are

$$\exp\left[\hat{\beta}_1 \pm 1.96\hat{SE}\left(\hat{\beta}_1\right)\right] = \exp[1.197 \pm 1.96 \times 0.4520] = 1.37, 8.01.$$

Similar calculations yield the endpoints for the other two confidence interval estimates.

The hazard ratios in Table 4.5 suggest: (1) subjects in their early thirties are dying at a rate which is about 3.3 times greater than subjects in their twenties, (2) subjects in their late thirties are dying at a rate

Table 4.4   Estimated Coefficients using Referent Cell Coding, Standard Errors, z-Scores, Two-Tailed p-Values and 95% Confidence Intervals for Age Categorized into Four Groups from the HMO-HIV+ Study

Variable	Coeff.	Std. Err.	z	P>|z|	95% Conf. Int.
AGE_2	1.197	0.451	2.65	0.008	0.313, 2.081
AGE_3	1.313	0.459	2.86	0.004	0.414, 2.213
AGE_4	1.860	0.469	3.96	<0.001	0.941, 2.780

which is about 3.7 times greater than subjects in their twenties and (3) subjects 40 or older have a mortality rate that is approximately 6 times greater than subjects in their twenties.

Given the similarity of the hazard ratios comparing each of the two age groups $[30-34]$ and $[35-39]$ to the referent group, it would make sense to test whether the survival experience in these two groups differs. We can estimate their hazard ratio and determine whether it is different from 1.0. We do this by using the general approach in (4.2) and (4.3). The specific difference in the hazard functions for the two groups is

$$\left[g\left(t, \text{AGE\_GRP}=3, \hat{\boldsymbol{\beta}}\right) - g\left(t, \text{AGE\_GRP}=2, \hat{\boldsymbol{\beta}}\right)\right]$$
$$= \left(\hat{\beta}_1 0 + \hat{\beta}_2 1 + \hat{\beta}_3 0\right) - \left(\hat{\beta}_1 1 + \hat{\beta}_2 0 + \hat{\beta}_3 0\right)$$
$$= \hat{\beta}_2 - \hat{\beta}_1 \ .$$

The estimator of the hazard ratio is

$$\hat{\text{HR}}(3,2) = e^{\left(\hat{\beta}_2 - \hat{\beta}_1\right)},$$

and its estimate is $\exp(1.313 - 1.197) = 1.123$. In order to obtain a confidence interval, we need an estimator for the variance of the difference between the two coefficients. The variance of the difference between two variables is

$$\hat{\text{Var}}(\hat{\beta}_2 - \hat{\beta}_1) = \hat{\text{Var}}(\hat{\beta}_2) + \hat{\text{Var}}\left(\hat{\beta}_1\right) - 2\hat{\text{Cov}}\left(\hat{\beta}_1, \hat{\beta}_2\right),$$

where $\hat{\text{Var}}$ denotes the estimator of the variance of the estimator in the parentheses and $\hat{\text{Cov}}$ denotes the estimator of the covariance of the two

Table 4.5 Estimated Hazard Ratios (HR) and 95% Confidence Intervals for Age Categorized into Four Groups from the HMO-HIV+ Study

Age Group	HR	95% Conf. Int.
1: $[20-29]$	1.00	
2: $[30-34]$	3.31	1.37, 8.01
3: $[35-39]$	3.72	1.51, 9.14
4: $[40-54]$	6.43	2.56, 16.12

estimators in the parentheses. These estimates may be obtained from most software packages by requesting the estimated covariance matrix of the estimated coefficients. Table 4.6 presents the covariance matrix for the estimated coefficients for the three age groups.

The estimated variances and covariance needed are $\widehat{\text{Var}}\left(\hat{\beta}_1\right) = 0.2034$, $\widehat{\text{Var}}\left(\hat{\beta}_2\right) = 0.2106$ and $\widehat{\text{Cov}}\left(\hat{\beta}_1,\hat{\beta}_2\right) = 0.1637$. The estimate of the variance of the difference in the two coefficients is

$$\widehat{\text{Var}}\left(\hat{\beta}_2 - \hat{\beta}_1\right) = 0.2034 + 0.2106 - 2 \times 0.1637 = 0.0867 \,,$$

and the estimated standard error is

$$\widehat{\text{SE}}\left(\hat{\beta}_2 - \hat{\beta}_1\right) = 0.2945 \,.$$

The endpoints of the 95 percent confidence interval estimate are

$$\exp\left[\left(\hat{\beta}_2 - \hat{\beta}_1\right) \pm 1.96\widehat{\text{SE}}\left(\hat{\beta}_2 - \hat{\beta}_1\right)\right]$$
$$= \exp\left[(1.313 - 1.197) \pm 1.96 \times 0.2945\right]$$
$$= 0.63,\ 2.00.$$

The confidence interval includes 1.0, indicating that the hazard rates for the two age groups may in fact be the same.

Instead of using the confidence interval, we could test the hypothesis of the equality of two coefficients via a Wald test. Many software packages allow the user to test whether specified contrasts of model coefficients are equal to zero. This is a convenient feature, especially when contrasts of interest are more complicated than simple differences. The Wald test for the contrast $\hat{\beta}_2 - \hat{\beta}_1$ is

$$z = \frac{\hat{\beta}_2 - \hat{\beta}_1}{\widehat{\text{SE}}\left(\hat{\beta}_2 - \hat{\beta}_1\right)} = \frac{1.313 - 1.197}{0.2945} = 0.395,$$

and the two-tailed p-value computed from the standard normal distribution is 0.69. Since the p-value is large, greater than 0.05, we fail to reject the hypothesis that the two coefficients are equal and conclude that the death rates in the two age groups may not be different.

Table 4.6   Estimated Variances and Covariances for the
Three Estimated Coefficients in Table 4.4

Variable (Coeff.)	AGE_2	AGE_3	AGE_4
AGE_2	0.2034	0.1637	0.1705
AGE_3	0.1637	0.2106	0.1666
AGE_4	0.1705	0.1666	0.2203

The test for a general contrast among the $K-1$ coefficients for a nominal scaled covariate with K levels is described as follows. Let the vector of estimators of the coefficients be denoted

$$\hat{\boldsymbol{\beta}}' = \left(\hat{\beta}_1, \hat{\beta}_2, ..., \hat{\beta}_{K-1}\right)$$

and the estimator of the covariance matrix be denoted $\hat{V}\left(\hat{\boldsymbol{\beta}}\right)$. Let the vector of constants specifying the contrast be denoted

$$\mathbf{c}' = \left(c_1, c_2, ..., c_{K-1}\right),$$

where the sum of the constants is zero. The single degree-of-freedom Wald test for the contrast is

$$Q = \frac{\mathbf{c}'\hat{\boldsymbol{\beta}}}{\sqrt{\mathbf{c}'\hat{V}\left(\hat{\boldsymbol{\beta}}\right)\mathbf{c}}} \, , \tag{4.7}$$

and the two-tailed p-value is obtained using the standard normal distribution. Most software packages will report the square of the Wald test and use the chi-square distribution to calculate the p-value. The equivalence of these two approaches follows from the fact that the distribution of the square of a $N(0,1)$ random variable follows a $\chi^2(1)$ distribution.

In the HMO-HIV+ study, it may be of interest to determine whether the average hazard ratio of the middle two age groups is equal to the hazard ratio for the oldest age group. The vector of constants for this contrast is $\mathbf{c}' = (0.5, 0.5, -1)$, the vector of estimated coefficients is given in Table 4.4, and the covariance matrix is shown in Table 4.6. We used STATA to perform the calculations, but other software packages, for example, SAS, could have been used. The value of the test statistic is $Q = 2.31$ with a p-value equal to 0.021. We conclude that the oldest age group has a hazard rate that is significantly greater than the average rate of the middle two age groups.

The method of using a contrast to compare coefficients can be especially useful when trying to pool categories of a nominal scale covariate recorded with more levels than can be practically used. Practical considerations are of primary importance in deciding which categories to combine, but contrasts may be used to judge whether the hazard rates of clinically similar groups are statistically similar.

Referent cell coding is the most frequently used scheme for coding design variables; however, it is just one of many possible methods. An alternative is deviation from means coding. This type of design variable coding may be used when one simply needs an overall assessment of differences in hazard rates. To illustrate the method, we apply it to the four age groups in the HMO-HIV+ study. This coding is obtained by replacing the first row of zeros in Table 4.3 with a row in which each value is equal to -1. The resulting estimated coefficient for an age group estimates the difference between the log hazard of the group and the arithmetic mean of the log hazards. The exponentiated estimated coefficient provides the ratio of the hazard rate of the particular group to the geometric mean of the hazard rates of all K groups.

The results of fitting a proportional hazards model using the deviation from means coding are shown in Table 4.7. The value of the partial likelihood ratio test for the overall significance of the coefficients is identical to that obtained using reference cell coding and is $G = 19.56$ with a p-value, computed using a chi-square distribution with three degrees-of-freedom, less than 0.001. The value of 0.104 for the estimated coefficient of design variable AGE_2 is equal to the estimate of the difference between the log-hazard rate for age group 2, [30, 34], and the estimate of the mean log-hazard rate. The Wald statistic has a p-value of 0.589, indicating that the log-hazard rate for this age group may not differ significantly from the average log-hazard rate. The coefficient for group 4 is 0.768 and its Wald statistic has a p-value less than 0.001. Thus, the log-hazard rate for this age group is significantly larger than the average log-hazard rate.

The coefficients in Table 4.7 are all positive, indicating that the average log-hazard rate falls between the log-hazard for age groups one and two. The estimated difference between the log hazard for the first age group and the average log-hazard rate is the negative of the sum of the coefficients in Table 4.7 and is -1.093. The easiest way to obtain an estimate of its standard error, Wald statistic, etc., is to make a small

Table 4.7   Estimated Coefficients Using the Deviation from Means Coding, Standard Errors, z-Scores, Two-Tailed p-Values and 95% Confidence Intervals for Age Categorized into Four Groups from the HMO-HIV+ Study

Variable	Coeff.	Std. Err.	z	$P>|z|$	95% Conf. Int.
AGE_2	0.104	0.192	0.54	0.589	–0.272, 0.481
AGE_3	0.221	0.206	1.07	0.285	–0.183, 0.624
AGE_4	0.768	0.209	3.67	<0.001	0.357, 1.178

change in the coding of the design variables and refit the model. We merely switch the row coded –1 with any other row. We do not recommend that hazard ratios be reported when using deviation from means coding, because the ratio cannot be interpreted in the same manner as the ratio from referent cell coding. The comparison is not a comparison of two distinct groups, but rather of one group to the geometric mean hazard rate of all groups combined.

Many other methods for coding design variables are possible. For example, coding that compares each group to the next largest group or each group to the average of the higher groups. These methods tend to be appropriate in special circumstances and will not be discussed further in this text. In general, the method of referent cell coding, perhaps followed by contrasts, should provide a useful and informative analysis in most circumstances.

4.3 CONTINUOUS SCALE COVARIATE

The interpretation of the coefficient for a continuous covariate is easier than that of a nominal scale variable in one sense, since indicator variables need not be introduced, but more difficult in another sense. Before we can use (4.2) and (4.3) to obtain an estimator of a hazard ratio we must do two things. First and foremost, we must verify that we have included the variable in its correct scale in the model. In this section we will assume that the log hazard is linear in the covariate of interest. Methods to assess the scale are discussed in Chapter 5. Second, we must decide what a clinically meaningful unit of change in the covariate is. Once these two steps are accomplished we may apply (4.2) and (4.3).

We illustrate the method using the HMO-HIV+ study and age as the covariate. The results of fitting a proportional hazards model containing age are shown in Table 4.8. The estimated coefficient in Table 4.8

gives the change in the log hazard for a 1-year change in age. Often a 1-year change in age is not of clinical interest. The HMO physicians conducting the study may be more interested in a 5-year change in age.

We obtain the correct change in the log-hazard function for a change of c units in a continuous covariate by using (4.2) and (4.3) with $a = x + c$ and $b = x$. This yields the following change in the log hazard:

$$\left[g(t, x+c, \beta) - g(t, x, \beta) \right] = \left\{ \ln[h_0(t)] + (x+c)\beta \right\} - \left\{ \ln[h_0(t)] + x\beta \right\}$$
$$= (x+c)\beta - x\beta$$
$$= c\beta \tag{4.8}$$

The change is simply equal to the value of the change of interest times the coefficient for a one-unit change. The estimator of the hazard ratio is

$$\widehat{HR}(c) = e^{c\hat{\beta}} \tag{4.9}$$

and the endpoints of a $100(1-\alpha)$ percent confidence interval estimator of the hazard ratio are

$$\exp\left[c\beta \pm z_{1-\alpha/2} |c| \widehat{SE}(\hat{\beta}) \right] . \tag{4.10}$$

Applying (4.9) and (4.10) for a 5-year change in age in the HMO-HIV+ study, we obtain an estimated hazard ratio of

$$\widehat{HR}(5) = e^{5 \times 0.081} = 1.50$$

and the endpoints of a 95 percent confidence interval are

$$\exp[5 \times 0.081 \pm 1.96 \times 5 \times 0.0174] = 1.264, \ 1.778.$$

Alternatively, we could have calculated the endpoints of the 95 percent confidence interval by multiplying the endpoints in Table 4.8 by 5 and then exponentiating. We suggest, for continuous covariates, that the hazard ratio for the clinically interesting unit of change, along with its confidence interval, be reported in any table of results. The unit of change should be indicated in the table heading or in a footnote.

Table 4.8   Estimated Coefficient, Standard Error, z-Score, Two-Tailed p-Value and 95% Confidence Interval for Age in the HMO-HIV+ Study

Variable	Coeff.	Std. Err.	z	$P>|z|$	95% Conf. Int.
AGE	0.081	0.0174	4.67	<0.001	0.047, 0.116

The interpretation of an estimated hazard ratio of 1.5 is that the hazard rate increases by 50 percent for every 5-year increase in age and is independent of the age at which the increase is calculated. The independence of the increase in age is due to the fact that the log hazard was assumed to be linear in age and subtracts itself out of the calculation in (4.8). The confidence interval estimate suggests that an increase in the hazard rate of between 30 and 80 percent is consistent with the data.

In summary, we wish to emphasize that the interpretation of the estimated hazard ratio for a continuous covariate depends not only on the assumption of linearity in the log hazard but also on the basic premise of a proportional hazards model. Methods for checking these assumptions are considered in detail in Chapters 5 and 6, respectively.

4.4 MULTIPLE-COVARIATE MODELS

The primary asset of any regression model is its ability to include multiple covariates and thereby statistically adjust for possible imbalances in the observed data before making statistical inferences. This process of adjustment has been given different names in various fields of study. In traditional statistical applications it is called *analysis of covariance*, while in clinical and epidemiological investigations it is often called *control of confounding*. A statistically related issue is the inclusion of higher order terms in a model representing interactions between covariates. These are also called *effect modifiers*. The strengths and limitations of statistical adjustment and inclusion of interactions in generalized linear models apply when using the proportional hazards regression model to analyze survival time. In this section we discuss these issues and establish a set of basic guidelines that we employ when discussing model development in the next chapter.

The UMARU IMPACT Study (UIS) is introduced in Section 1.3 (Table 1.3) and provides some excellent examples for demonstrating the statistical issues involved in adjustment and interaction. The analyses presented in this section are in no way definitive. They are used simply to demonstrate the interpretation of fitted proportional hazards models. Two variables collected on subjects in the UIS were age and IV drug use history. For demonstration purposes, we have recoded IV drug use history into a dichotomous variable, d, coded $0 =$ never and $1 =$ ever. We assume the log-hazard function is linear in the covariates and that our primary analysis goal is to estimate the hazard ratio associated with IV drug use, d.

Suppose we generate a proportional hazards model that contains only IV drug use. The log-hazard function of the model is

$$g(t,d,\theta_1) = \ln[h_0(t)] + d\theta_1 .$$

The difference in the log-hazard functions is

$$\left[g(t,d=1,\theta_1) - g(t,d=0,\theta_1)\right] = \left\{\ln[h_0(t)] + 1\theta_1\right\} - \left\{\ln[h_0(t)] + 0\theta_1\right\}$$
$$= \theta_1 . \qquad (4.11)$$

Suppose we generate a second model that contains both age and IV drug use. The log-hazard function for the larger model is

$$g(t,d,a,\boldsymbol{\beta}) = \ln[h_0(t)] + d\beta_1 + a\beta_2 , \qquad (4.12)$$

where a denotes age. The adjusted log-hazard ratio is obtained from (4.2) and (4.12), comparing a subject of age a who has a history of IV drug use to one of the same age a who does not have a history of IV drug use, and is

$$\left[g(t,d=1,a,\boldsymbol{\beta}) - g(t,d=0,a,\boldsymbol{\beta})\right]$$
$$= \left\{\ln[h_0(t)] + 1\beta_1 + a\beta_2\right\} - \left\{\ln[h_0(t)] + 0\beta_1 + a\beta_2\right\}$$
$$= \beta_1 + (a-a)\beta_2$$
$$= \beta_1. \qquad (4.13)$$

The results shown in (4.11) and (4.13) indicate that we have two estimators of the desired log-hazard ratio: (1) The so-called crude or unadjusted estimator $\hat{\theta}_1$ from (4.11), obtained from fitting the model that does not include age, and (2) the adjusted estimator $\hat{\beta}_1$ from (4.13), the coefficient of d obtained from fitting a model containing d and age. If the two estimators are similar, then adjustment for age was unnecessary, in a statistical sense. If the estimators are different, then adjustment was needed and the variable age is a confounder of the hazard ratio for d. The extent of adjustment, or difference, between $\hat{\theta}_1$ and $\hat{\beta}_1$ is a function of the difference in the distribution of age within the two IV drug use groups and the strength, $\hat{\beta}_2$, of the association between age and survival time.

Suppose that the model containing age, (4.12), is the correct model, and denote the average age of subjects with and without a history of IV drug use as $\bar{a}_1$ and $\bar{a}_0$, respectively. An approximation of the average log-hazard functions [see Fleming and Harrington (1991) page 134 for an exact expression] for the two drug use groups is

$$g(t, d = 0, \boldsymbol{\beta}) = \ln[h_0(t)] + \bar{a}_0 \beta_2$$

and

$$g(t, d = 1, \boldsymbol{\beta}) = \ln[h_0(t)] + \beta_1 + \bar{a}_1 \beta_2.$$

Taking the difference between these two expressions, the crude or unadjusted log-hazard ratio is approximately

$$\hat{\theta}_1 \approx \hat{\beta}_1 + (\bar{a}_1 - \bar{a}_0)\hat{\beta}_2. \tag{4.14}$$

Thus, the crude estimator will be approximately equal to the adjusted estimator if the difference in the mean age of the two drug use groups is zero or if the coefficient for age is zero. The two estimators will differ if at least one of the two is large or both are moderate in size. We recommend that the percent change in the adjusted estimate be computed as a measure of the amount of adjustment. The percent change estimator, in general, is defined as

$$\Delta\hat{\beta}\% = 100\frac{\hat{\theta} - \hat{\beta}}{\hat{\beta}}, \tag{4.15}$$

where $\hat{\theta}$ denotes the crude estimator from the model that does not contain the potential confounder and $\hat{\beta}$ denotes the adjusted estimator from the model that does include the potential confounder. We discuss the use of (4.15) in model building in Chapter 5.

If we assume (4.14) is true, then the approximate percent change is

$$\Delta\hat{\beta}_1\% = 100\frac{\hat{\theta}_1 - \hat{\beta}_1}{\hat{\beta}_1} \approx \frac{100(\bar{a}_1 - \bar{a}_0)\hat{\beta}_2}{\hat{\beta}_1}, \qquad (4.16)$$

an expression that isolates the two contributors to adjustment. In practice, one would evaluate only (4.15). The expressions in (4.14) and (4.16) are provided as a tool to assist in explaining why the crude and adjusted estimators could be different.

The results of fitting the two models to the UIS data are shown in Table 4.9. We note that AGE is missing for 5 subjects and DRUG is missing on an additional 18 subjects, so analyses have been restricted to the 605 subjects with complete data. The adjusted estimate for DRUG is 0.44 and the crude estimate is 0.32, a change of

$$\Delta\hat{\beta}_1\% = 100 \times \frac{0.32 - 0.44}{0.44} = -27\%.$$

The reasons the estimate changed are: (1) age is strongly associated with survival time, $p = 0.001$ in Table 4.9, and (2) the mean ages in the two drug use groups are different, 26.64 (never) and 31.05 (ever). Thus, both contributors to confounding on the right-hand side of (4.14) are large. In this example the right-hand side of (4.14) is equal to

$$0.32 = 0.44 - 0.026(31.05 - 26.64),$$

which is nearly identical to the crude estimate in Table 4.9.

A practical question is how large must the percent change in the co-efficient be to indicate that we need to include the potential confounder in the model. There are no rules, only suggestions. In practice, we have found that a change greater than 15–20 percent indicates that adjustment is needed.

The ability of the proportional hazards regression model to provide correct adjusted estimates of log-hazard ratios depends on having fit the correct model. In practice, this means that the proportional hazards model is correct and that we have fit a model containing the correct covariates, all of which are scaled correctly. These issues are discussed in detail in Chapters 5 and 6.

The derivation of the adjusted estimator in (4.13) implicitly assumes that the log-hazard ratio is constant for all ages. If this is not the case, then the two variables are said to interact; in other words, age modifies the effect of IV drug use. We address this question by determining

Table 4.9   Estimated Coefficients, Standard Errors, z-Scores, Two-Tailed p-Values and 95% Confidence Intervals for Two Models Fit to the UIS Data

Model	Variable	Coeff.	Std. Err.	z	$P>|z|$	95% Conf. Int.
Crude	DRUG	0.321	0.0948	3.39	0.001	0.135, 0.507
Adjusted	DRUG	0.439	0.1007	4.36	<0.001	0.242, 0.637
	AGE	-0.026	0.0078	-3.37	0.001	−0.042, −0.011

whether an interaction between the two variables (their product) contributes significantly to the model. The log-hazard function for the interactions model is obtained by adding the product of IV drug use and age to the model in (4.12) and is

$$g(t,d,a,\beta) = \ln[h_0(t)] + d\beta_1 + a\beta_2 + (d \times a)\beta_3 . \qquad (4.17)$$

If we assume for the moment that (4.17) is the correct model, then the only way we can obtain the correct expression for the log-hazard ratio for IV drug use is to apply (4.2). The log-hazard ratio is

$$\left[g(t,d=1,a,\beta) - g(t,d=0,a,\beta) \right]$$
$$= \left\{ \ln[h_0(t)] + 1\beta_1 + a\beta_2 + (1 \times a)\beta_3 \right\} - \left\{ \ln[h_0(t)] + 0\beta_1 + a\beta_2 + (0 \times a)\beta_3 \right\}$$
$$= \beta_1 + a\beta_3 . \qquad (4.18)$$

The implication of the result in (4.18) is that the log-hazard ratio for IV drug use depends on the age of the subject. Conversely, when there is no interaction (i.e., $\beta_3 = 0$), the log-hazard ratios in (4.13) and (4.18) are the same. In general, the reason we include interactions in a model is to better estimate the effects of the covariates since point and interval estimators obtained from (4.13) and (4.18) are different. This will happen only when the interaction term in (4.18) is statistically significant, as assessed via the partial likelihood ratio or an equivalent test. We recommend inclusion of interaction terms in a model only when they are statistically significant. We address this point in greater detail in the next chapter.

The primary goal of the UIS was to compare the effectiveness of two treatment interventions, "TREAT" in Table 1.3. Table 4.10 presents the results of fitting a proportional hazards model containing treatment, a second model containing treatment and age and a third model containing these variables along with their interaction. For illustrative purposes, we use a larger level of statistical significance, $p \leq 0.15$, than we might choose to use in an actual model building application. The coefficient for treatment in the crude model is significant. When we add age to the model, the value of the partial likelihood ratio test comparing the new model to the model that contains treatment only is $G = 3.42$ with a p-value equal to 0.064, and the percent change in the coefficient for treatment is -1.5 percent. Thus, we conclude that age is associated with survival time but is not a confounder of the treatment effect. The partial likelihood ratio test comparing the age adjusted model to the interactions model is $G = 2.57$ with a p-value equal to 0.109. Thus, from a statistical significance point of view, the best model is the interactions model. However, it appears from that model that the significant treatment and age effects seen in the adjusted model have disappeared. The estimator of the log-hazard function for the interactions model, ignoring the log baseline hazard function, is

$$g\left(t, \text{TREAT}, \text{AGE}, \hat{\boldsymbol{\beta}}\right) = \hat{\beta}_1 \text{TREAT} + \hat{\beta}_2 \text{AGE} + \hat{\beta}_3 \text{TREAT} \times \text{AGE}.$$

The estimators of the log-hazard functions for the two treatment groups are

$$g\left(t, \text{TREAT} = 0, \text{AGE}, \hat{\boldsymbol{\beta}}\right) = \hat{\beta}_2 \text{AGE}$$

and

$$g\left(t, \text{TREAT} = 1, \text{AGE}, \hat{\boldsymbol{\beta}}\right) = \hat{\beta}_1 + \left(\hat{\beta}_2 + \hat{\beta}_3\right) \text{AGE}.$$

The coefficient for age, $\hat{\beta}_2 = -0.002$ in Table 4.10, is the slope in age of the log-hazard function for treatment group 0. The fact that it is not significant implies that age is unrelated to survival time in treatment group 0. The slope in age for treatment group 1 is the sum of the age and interaction coefficients, $\hat{\beta}_2 + \hat{\beta}_3 = -0.002 + (-0.023) = -0.025$ in Table 4.10, and its significance can only be tested using the method of contrasts discussed in Section 4.2. This results in a Wald statistic $z = 2.41$ with a p-value equal to 0.016, which is significant. We conclude, therefore, that there is evidence of a significant association between age and survival time in treatment group 1.

   To obtain an estimator of treatment effect we can apply (4.2). The estimator of the difference in log-hazard functions is

$$\left[g\left(t, \text{TREAT}=1, \text{AGE}, \hat{\boldsymbol{\beta}}\right) - g\left(t, \text{TREAT}=0, \text{AGE}, \hat{\boldsymbol{\beta}}\right)\right]$$

$$= \left\{ \ln[h_0(t)] + \hat{\beta}_1 + \hat{\beta}_2\text{AGE} + \hat{\beta}_3\text{AGE}\right\} - \left\{ \ln[h_0(t)] + \hat{\beta}_2\text{AGE}\right\}$$

$$= \hat{\beta}_1 + \hat{\beta}_3\text{AGE}. \tag{4.19}$$

The magnitude of this estimator depends on the age of the subject. The estimator of the coefficient for treatment, $\hat{\beta}_1$, would be the estimator of treatment effect for a subject with age zero years. If we had centered the age data by subtracting the mean, then the coefficient, $\hat{\beta}_1$, is the estimator of treatment effect for a subject of age equal to the mean. To display the results of fitting such a model, we recommend that a table be presented containing point and interval estimates of treatment effect for a few key values of age. The point estimator is

$$\hat{\text{HR}}(\text{TREAT}, \text{AGE}) = \exp\left(\hat{\beta}_1 + \hat{\beta}_3\text{AGE}\right),$$

and the confidence interval estimator is

$$\exp\left[\left(\hat{\beta}_1 + \hat{\beta}_3\text{AGE}\right) \pm z_{1-\alpha/2}\hat{\text{SE}}\left(\hat{\beta}_1 + \hat{\beta}_3\text{AGE}\right)\right], \tag{4.20}$$

where

$$\hat{\text{SE}}\left(\hat{\beta}_1 + \hat{\beta}_3\text{AGE}\right) = \left\{ \hat{\text{Var}}\left(\hat{\beta}_1\right) + \text{AGE}^2\hat{\text{Var}}\left(\hat{\beta}_3\right) + 2\text{AGE}\hat{\text{Cov}}\left(\hat{\beta}_1, \hat{\beta}_3\right)\right\}^{0.5},$$

and the required estimated variances and covariance are obtained from the covariance matrix included in computer output. Table 4.11 contains values of the hazard ratio and associated 95 percent confidence

Table 4.10   Estimated Coefficients, Standard Errors, z-Scores, Two-Tailed p-Values and 95% Confidence Intervals for Three Models Fit to the UIS Data

Model	Variable	Coeff.	Std. Err.	z	$P>|z|$	95% Conf. Int.
Crude	TREAT	−0.220	0.089	−2.46	0.014	−0.395, −0.045
Adjusted	TREAT	−0.223	0.089	−2.50	0.013	−0.398, −0.048
	AGE	−0.013	0.007	−1.84	0.066	−0.027, 0.001
Interaction	TREAT	0.523	0.474	1.10	0.271	−0.407, 1.453
	AGE	−0.002	0.010	−0.18	0.861	−0.022, 0.018
	TREAT×AGE	−0.023	0.015	−1.60	0.110	−0.052, 0.005

Table 4.11 **Estimated Hazard Ratios (HR) and
95% Confidence Intervals for Treatment Effect
in the UIS at Ages 25, 30, 35 and 40**

Age	HR	95% Conf. Int.
25	0.944	0.723, 1.234
30	0.841	0.700, 1.011
35	0.749	0.617, 0.909
40	0.667	0.502, 0.886

interval for subjects of age 25, 30, 35 and 40 years.

The estimated hazard ratios in Table 4.11 are all less than one and decrease with age, indicating that the longer treatment period, TREAT = 1, is beneficial or protective for return to drug use and becomes increasingly beneficial the older the subject. The confidence intervals support a significant treatment effect for subjects 35 years and older.

Another form of an interactions model is one that contains continuous covariates that have been transformed, and one would like to estimate hazard ratios in the original measurement scale.

For example, suppose a log-hazard function contains both age and the square of age and we would like an estimate of the hazard ratio for a c year change in age. To obtain the correct expression for the difference in log-hazard functions for a c year change in age we must use (4.2) which yields

$$\left[g(t, \text{AGE} + c, \boldsymbol{\beta}) - g(t, \text{AGE}, \boldsymbol{\beta})\right] = \left\{\ln\left[h_0(t)\right] + \beta_1(\text{AGE} + c) + \beta_2(\text{AGE} + c)^2\right\}$$
$$- \left\{\ln\left[h_0(t)\right] + \beta_1\text{AGE} + \beta_2\text{AGE}^2\right\}$$
$$= \beta_1 c + \beta_2\left[2\text{AGE} \times c + c^2\right],$$

an expression which depends on both the change and the age at which the change is calculated. We obtain point and interval estimators of the hazard ratio by extending the result shown in (4.20). This yields the point estimator

$$\hat{\text{HR}}(\text{AGE} + c, \text{AGE}) = \exp\left[\hat{\beta}_1 c + \hat{\beta}_2\left(2\text{AGE} \times c + c^2\right)\right], \quad (4.21)$$

and the endpoints of the $100(1 - \alpha)$ percent confidence interval are

$$\exp\left\{\hat{\beta}_1 c + \hat{\beta}_2\left(2\text{AGE}\times c + c^2\right) \pm z_{1-\alpha/2}\widehat{\text{SE}}\left[\hat{\beta}_1 c + \hat{\beta}_2\left(2\text{AGE}\times c + c^2\right)\right]\right\},$$
(4.22)

where

$$\widehat{\text{SE}}\left[\hat{\beta}_1 c + \hat{\beta}_2\left(2\text{AGE}\times c + c^2\right)\right]$$

$$= \left[\begin{array}{c} c^2\widehat{\text{Var}}\left(\hat{\beta}_1\right) + \left(2\text{AGE}\times c + c^2\right)^2\widehat{\text{Var}}\left(\hat{\beta}_2\right) \\ + 2c\times\left(2\text{AGE}\times c + c^2\right)\widehat{\text{Cov}}\left(\hat{\beta}_1,\hat{\beta}_2\right)\end{array}\right]^{0.5}.$$

Expressions similar to (4.21) and (4.22) result when (4.2) is applied to other nonlinear transformations of a covariate.

Multiple variable proportional hazards regression models can be effective tools for sharpening estimates of hazard ratios for covariates. In the absence of interactions, one must be aware at each step in the model evaluation process of the amount of adjustment or confounding that is being controlled. If interactions are present in a model, then confounding is no longer an issue, as the estimate of effect depends on the value of other covariates and thus cannot be removed. In all cases, the interpretation depends on the assumption that the proposed model fits the data, the subject of Chapter 6.

4.5 INTERPRETATION AND USE OF THE COVARIATE-ADJUSTED SURVIVORSHIP FUNCTION

Methods for estimating the survivorship function following the fitting of a proportional hazards model were presented in Section 3.5. The key step presented in that section was the estimation of the baseline survivorship function, $\hat{S}_0(t)$, shown in (3.39). This estimator may be combined with the estimators of the coefficients in the model using (3.36) to obtain the estimator of the survivorship function, adjusting for the covariates, as follows:

$$\hat{S}(t,\mathbf{x},\hat{\boldsymbol{\beta}}) = \left[\hat{S}_0(t)\right]^{\exp\left(\mathbf{x}'\hat{\boldsymbol{\beta}}\right)}.$$
(4.23)

All software packages allow the user to request calculation of the estimator of the baseline survivorship function. The estimator may be used to derive other functions of survival time, for example, the estimator in (4.23), which is essential for graphical description of the results of the analysis and for other analyses, such as model assessment. We discuss graphical methods and estimation of quantiles and their interpretation in this section. Model assessment is discussed in Chapter 6.

We begin with the model containing IV drug use in the HMO-HIV+ study discussed in Section 4.2. In this section we use a dichotomous grouping variable, but the methods may be used with any nominal scale covariate. Table 4.2 presents the results of fitting the model. The estimator of the baseline survivorship function for this model is an estimator of the survivorship function for DRUG = 0. If we request that the baseline survivorship function be computed as part of the analysis, then the software evaluates (3.39), denoted

$$\hat{S}_0(t_i), \quad i = 1, 2, \ldots, n, \tag{4.24}$$

for each subject in the study, regardless of their survival status or value of IV drug use. It follows from (3.43) that the estimator $\hat{S}_0(t)$ is constant between observed survival times. Thus, the estimated value for subjects who were censored is equal to the value at the largest observed survival time for which they were still at risk.

We can compute an estimate of the survivorship function for DRUG = 1 by using the previously calculated value of the baseline survivorship function and evaluating

$$\hat{S}\left(t_i, \mathrm{DRUG} = 1, \hat{\beta}_1 = 0.779\right) = \left[\hat{S}_0(t_i)\right]^{\exp(0.779)}, \quad i = 1, 2, \ldots, 100, \tag{4.25}$$

where the value of the coefficient for DRUG is obtained from Table 4.2. The graphs of the two estimated survivorship functions, (4.24) and (4.25), are shown in Figure 4.2. The plot has been drawn with steps connecting the points rather than straight lines to emphasize the fact that the estimator is constant between observed survival times. It follows from (4.24) and (4.25) that each function has been plotted at exactly the same $n = 100$ values of time. The shape of the two curves is a consequence of the proportional hazards assumption. The ratio of the hazards at each point in Figure 4.2 is forced to be equal to $2.18 = \exp(0.779)$.

We presented a plot in Chapter 2, Figure 2.7, that is similar in appearance to Figure 4.2. There is an important distinction. The two curves in Figure 2.7 are based on separate, nonparametric Kaplan–Meier estimators. The Kaplan–Meier estimator uses only the data in each group and does not assume the hazards are proportional. The distinction between the curves in Figure 2.7 and Figure 4.2 is analogous to the distinction between a plot of the observed cumulative percent distribution as compared to a plot of the cumulative distribution function based on an assumption of normality (i.e., using the observed sample mean and variance). If the data are nearly normally distributed, the two curves will look alike, but the latter curve will be "smoother" (due to the normality assumption) than the former curve.

The difference between the curves in Figures 2.7 and 4.2 can be most clearly seen between 15 and 56 months. The lower curve in Figure 2.7 is constant between 15 and 56 months. No survival times in the IV drug use present group were observed in this interval and the largest observed time was a censored observation at 56 months. The lower curve in Figure 4.2, however, has jumps at each observed survival time,

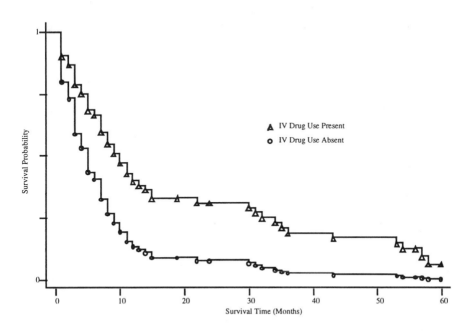

Figure 4.2   Graph of the estimated proportional hazards model survivorship function for IV drug use present (Δ) and absent (o) for the HMO-HIV+ study. Points are plotted at each observed time for both curves.

in the DRUG = 0 group between 15 and 56 months. Furthermore, the lower curve in Figure 4.2 does not end at 56 months with the value at 15 months, as was the case in Figure 2.7. The curve takes a downward jump as the hazard ratio is 2.18 at each point.

Another way one can think of the curves plotted in Figure 4.2 is to consider them as being like "fitted" or "predicted" regression lines. Here the "prediction" is on the survivorship probability scale. In this example, there is an implicit model-based extrapolation present in Figure 4.2. The lower curve, in the interval from 15 to 56 months, predicts or estimates the survivorship experience if: (1) the estimate of the baseline survivorship function correctly describes the survivorship experience in the IV drug use absent group, and (2) the proportional hazards model is correct.

The situation in Figure 4.2 is analogous to using linear regression to model weight as a function of height in males and females. It is likely that the shortest subjects are female and the tallest subjects are male. Once a model has been fit, the software may be used to graph the fitted model over the entire observed range of heights. A point on the line for females in the range of heights only observed for males is a prediction that depends on the unverifiable assumption that the fitted model is correct for females as tall as the tallest males. We have extrapolated the model beyond the observed range of data. The same type of extrapolation can occur in plots of survivorship functions.

As noted, the extrapolation in Figure 4.2 is in an interval between observed values. A more serious extrapolation problem would have occurred if the largest observed time in the IV drug use present group had been 12 months. These extrapolation issues suggest that one must give careful consideration to what points are used when plotting a covariate-adjusted survivorship function. A more conservative plot than Figure 4.2 is shown in Figure 4.3 where points are plotted only for observed values of time. The plot in Figure 4.2 has 200 plotted values while there are 100 plotted values in Figure 4.3.

The plot in Figure 4.3 is constant for the IV drug use present group between 15 and 56 months and thus better reflects the observed data. Figure 4.3, in conjunction with the analysis in Table 4.2, illustrates the significantly poorer survival experience of the IV drug use present group.

If the observed range of survival times is comparable for each group, we recommend using a plot like Figure 4.2 as it uses all the data and reflects best the fitted model and its assumptions. However, if there are clinically important differences in the observed range of survival

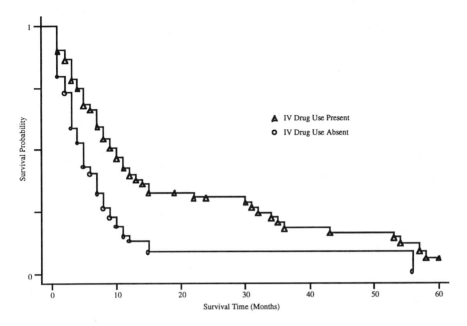

Figure 4.3 Graph of the estimated proportional hazards survivorship function for IV drug use present (Δ) and absent (o) for the HMO-HIV+ study. Points are plotted at observed times for each group.

times, then we recommend using a plot like Figure 4.3. One must use caution when reading the literature, as it may be difficult to determine whether plotted data involve inappropriate extrapolation of the fitted model. The best approach in practice is to provide results from both a thorough univariate analysis of survival experience in subgroups of special interest, as well as results from a regression analysis.

As noted in the previous section, the strength of regression modeling is the ability to adjust statistically for possible imbalances in the observed data. As an example of a more complicated model, suppose we fit a proportional hazards model containing age and IV drug use. The goal is to present survivorship functions for the two drug use groups, controlling for age. We must give some thought to what we mean by *controlling for age*.

We would like the estimated survivorship function to use the covariates in the same way that covariates are controlled for in a linear regression. In linear regression, a point on the regression line (or plane for a multiple variable model) is the model-based estimate of the conditional mean of the dependent variable among subjects with values of the covariates defined by the point. The analogy to linear regression can help our thinking, but the situation is a bit more complicated in a proportional hazards regression analysis. Since the model does not contain an intercept, we do not have a fully parametric hazard function and thus the model cannot predict an individual point estimate of the conditional "mean" survival time. The estimated survivorship function in (4.23) is the proportional hazards model-based estimator of the conditional statistical distribution of survival time. The word "conditional" here means restricting observation to a cohort with covariate values equal to values specified. Pursuing this notion further, suppose we were able to follow an extremely large cohort of subjects for 5 years. Suppose also that the cohort is large enough that we can perform a fully stratified analysis and compute the Kaplan–Meier estimator of the survivorship function for each possible set of values of the covariates, such as, IV drug users who are 40 years old. If the proportional hazards model is correct, then the estimator in (4.23) and the Kaplan–Meier estimator should be similar, within statistical variation. One may use this estimator, (4.23), to describe survival time graphically and to compute estimates of quantiles, such as the median, in the same way the Kaplan–Meier estimator was used in Chapter 2.

The most frequent use of estimated survivorship functions in applied settings is to provide curves, like those in Figure 4.2 or 4.3, which may be used to compare groups visually and control for other model covariates. If the model does not contain grouping variable by covariate interactions, then the resulting survivorship functions are in a sense "parallel" in a way similar to lines with the same slope in a linear regression. In practice, one would choose one set of "typical" values of the other covariates. For a continuous covariate like age, we usually choose the common value to be the mean, median or another central value. In the HMO-HIV+ study, the mean age is 36.02 and the median age is 35. Thus, 35 seems like a good value to use for age in this example. If we merely fit the proportional hazards model containing age and IV drug use and request computation of the baseline survivorship function, then the program would estimate survivorship experience for IV drug use absent and, although biologically impossible, age equal to zero years. To obtain the estimate of the two age-adjusted survivorship functions, we would have to evaluate the expression in (4.23) using the coefficients from the fitted model with $(DRUG = 0, AGE = 35)$ and $(DRUG = 1, AGE = 35)$. This approach, while algebraically correct, can cause unwanted round-off and computational error in some situations. We would like to avoid computations that involve exponentiating large

Table 4.12   Estimated Coefficients, Standard Errors, z-Scores, Two-Tailed p-Values and 95% Confidence Intervals for the Model Containing IV Drug Use and Age Centered at 35 Years in the HMO-HIV+ Study

Variable	Coeff.	Std. Err.	z	$P>|z|$	95% Conf. Int.
DRUG	0.941	0.2555	3.68	<0.001	0.441, 1.442
AGE_C	0.092	0.0185	4.95	<0.001	0.055, 0.128

positive or negative numbers. One way to do this is to center continuous covariates. In the current example, we fit the model using AGE_C $=$ AGE -35, and the results are shown in Table 4.12. These results are identical to ones which would have been obtained if we had used age uncentered, with the only difference being in the baseline survivorship function. When we center age, not only are our results computationally more accurate, but the estimate of the baseline survivorship function corresponds to IV drug use absent and age equal to 35 years, the zero value of the two covariates in the model. To obtain the second estimated survivorship function we compute

$$\hat{S}\left(t_j, \text{DRUG}=1, \text{AGE\_C}=0, \hat{\boldsymbol{\beta}}\right) = \left[\hat{S}_0\left(t_j\right)\right]^{\exp(1 \times 0.941)}, \; j=1,2,\ldots,100 \; .$$

Graphs of the two estimated survivorship functions, plotted at observed values of time in each group, are shown in Figure 4.4. In this example, we chose these points to plot because of the absence of data in the interval from 15 to 56 months in the IV drug present group. The curves in this graph provide proportional hazards estimates of the survivorship experience of two cohorts of 35-year-old subjects differing in their IV drug use. Each point on the two curves depends on the actual observed survival times and the proportional hazards assumption.

The shapes of the curves in Figure 4.5 are determined by the choice of age equal to 35 as the center or "zero" value and the proportional hazards assumption. In order to illustrate the parallelism in the plots for any value of age and the effect of the choice of the center, four sets of curves have been plotted in Figures 4.5a–4.5d. The plots use age equal to 30, 35, 40 and 45 years, respectively. Since these plots have been prepared to demonstrate the effect of centering and the proportional hazards assumption, the two curves in each plot use all observed survival times. The basic parallelism is present, as the hazard ratio at each point in each of the four plots is 2.18. The progressive steepness in the plots

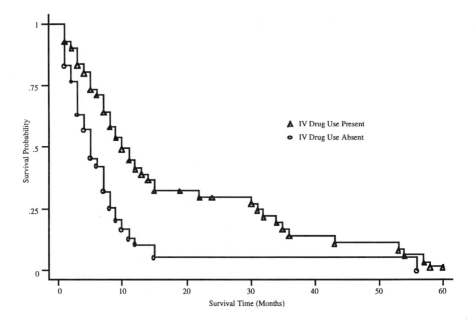

Figure 4.4 Graph of the age-adjusted estimated proportional hazards survivorship function for IV drug use present (Δ) and absent (o) for the HMO-HIV+ study. Points are plotted at observed survival time for each group.

is due to the fact that age is inversely related to survival, has a positive coefficient in the log-hazard function and is increasing from Figure 4.5a to 4.5d. For simply demonstrating the effect of IV drug use on survival controlling for age, any one of the four plots could have been used. Since the median age was 35, the plot in Figure 4.4 corresponds to Figure 4.5b, but with the noted difference in plotted values.

When the fitted model is even moderately complex, it may be difficult to decide what combination of covariate values best represents the middle of the data. We recommend that continuous covariates be centered to avoid potential numerical problems. In such complex situations, plots based on values of a quantity called the *risk score* are frequently used in practice. The risk score is the value of the linear portion of the proportional hazards model. In a model containing p covariates, the estimator is

$$\hat{r}\left(\mathbf{x}, \hat{\boldsymbol{\beta}}\right) = \sum_{k=1}^{p} \hat{\beta}_k x_k \, .$$

For ease of notation, we denote the value of the estimator of the risk score for the jth subject as

$$\hat{r}_j = \hat{r}\left(\mathbf{x}_j, \hat{\boldsymbol{\beta}}\right) = \sum_{k=1}^{p} \hat{\beta}_k x_{jk} , \qquad (4.26)$$

and, to be in agreement with notation used in Chapter 3, its exponentiated value as

$$\hat{\theta}_j = e^{\hat{r}_j} . \qquad (4.27)$$

Most software packages will provide calculated values of either or both (4.26) and (4.27). The baseline survivorship function corresponds to a risk score of zero which may or may not be of clinical interest. Typically, one can obtain the values of the quartiles of the risk score from a descriptive statistics routine and obtain the estimated survivorship function from (4.23) by evaluating

$$\hat{S}\left(t_j, \hat{r}_q, \hat{\boldsymbol{\beta}}\right) = \left[\hat{S}_0\left(t_j\right)\right]^{\exp\left(\hat{r}_q\right)}, \quad j = 1, 2, \ldots, n , \qquad (4.28)$$

where the $\hat{r}_q$, $q = 25, 50, 75$, correspond to the empirical quartiles of the risk score.

This procedure may be modified when we wish to graph the estimated survivorship functions for a grouping variable, controlling for a risk score based on the remaining covariates. In this setting, we subtract out the contribution of the grouping variable to the risk score, calculate the median value of what remains, and then add back in the contribution of the grouping variable when calculating the estimator of the survivorship function. Suppose the grouping variable is dichotomous and is the first of the p covariates in the model. The modified risk score, obtained by removing the effect of the grouping variable, is

$$\hat{rm}_j = \hat{r}_j - \hat{\beta}_1 x_{j1}, \quad j = 1, 2, \ldots, n .$$

If we denote the median of the modified risk scores as $\hat{rm}_{50}$, then the estimates of the survivorship functions for the two groups at this median are

$$\hat{S}\left(t_j, x_1 = 0, \hat{rm}_{50}, \hat{\boldsymbol{\beta}}\right) = \left[\hat{S}_0\left(t_j\right)\right]^{\exp\left(\hat{rm}_{50} + \hat{\beta}_1(0)\right)} \tag{4.29}$$

and

$$\hat{S}\left(t_j, x_1 = 1, \hat{rm}_{50}, \hat{\boldsymbol{\beta}}\right) = \left[\hat{S}_0\left(t_j\right)\right]^{\exp\left(\hat{rm}_{50} + \hat{\beta}_1(1)\right)} \tag{4.30}$$

for $j = 1, 2, \ldots, n$ subjects. Before plotting the graphs of (4.29) and (4.30), one should examine the range of observed survival times in the two groups for biologically important gaps.

The data from the UIS may be used to provide examples of plotting survivorship functions from a more complex model. This model has been chosen for demonstration purposes only. We have not attended to a number of important modeling details and, as a result, this model should not be construed as being the final model for assessing treatment effect. Table 4.13 presents the results of fitting a model containing treatment, age, history of IV drug use (0 = never, 1 = previous or recent) and the number of previous drug treatments. We centered age at 30 years and the number of previous drug treatments at 3. The results in Table 4.13 support a significant treatment effect after adjusting for the other variables in the model. The estimated hazard ratio for treat-

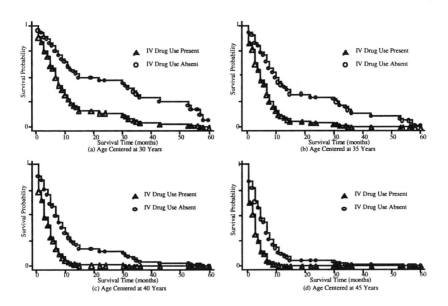

Figure 4.5 Graphs of the age-adjusted estimated proportional hazards survivorship function for IV drug use present (Δ) and absent (o) at four different ages in the HMO-HIV+ study.

ment and the 95 percent confidence interval are $\hat{HR} = \exp(-.227) = 0.80$ and $(0.666, 0.954)$, respectively. The estimate supports a protective effect for the longer treatment, with about a 20 percent reduction in the hazard rate for returning to drug use.

In this example, the equation for the estimated risk score for the jth subject is

$$\hat{r}_j = -0.227 \times TREAT_j - 0.031 \times \left(AGE\_C_j\right) + 0.343 \times DRUG_j$$
$$+ 0.031 \times \left(NDRGTX\_C_j\right)$$

and the modified estimated risk score is

$$\hat{rm}_j = \hat{r}_j - \left(-0.227 \times TREAT_j\right) .$$

The median value of the modified risk score is 0.1588 and the equations for the estimators of the modified risk score-adjusted survivorship functions obtained from (4.29) and (4.30) are

$$\hat{S}\left(t_j, TREAT = 0, \hat{rm}_{50}, \hat{\beta}\right) = \left[\hat{S}_0(t_j)\right]^{\exp(0.1588)} \qquad (4.31)$$

and

$$\hat{S}\left(t_j, TREAT = 1, \hat{rm}_{50}, \hat{\beta}\right) = \left[\hat{S}_0(t_j)\right]^{\exp(0.1588-0.227)} . \qquad (4.32)$$

Since the observed range of survival times in the two treatment groups is comparable, we chose to use all observed survival times to plot (4.31) and (4.32), which are shown in Figure 4.6. Since the analysis is based on a large number of subjects (593 of the 628 subjects had complete data on the four covariates), the use of separate plotting symbols has been suppressed to avoid an unnecessarily cluttered graph.

The two curves in Figure 4.6 reflect both the use of the median modified risk score and the assumption of proportional hazards. Use of other quantiles of the modified risk score would shift the curves to the left or right in a manner similar to Figure 4.5. The longer times until return to drug use for the longer duration treatment group are illustrated in the graph.

We can use adjusted estimated survivorship functions, such as those shown in Figures 4.5 and 4.6, to estimate the adjusted median survival

Table 4.13   Estimated Coefficients, Standard Errors, z-Scores, Two-Tailed p-Values and 95% Confidence Intervals for the Model Containing Treatment (TREAT), Age–30 (AGE_C), IV Drug Use (DRUG) and Number of Prior Drug Treatments–3 (NDRGTX_C) from the UIS, $n = 593$

Variable	Coeff.	Std. Err.	z	$P>|z|$	95% Conf. Int.
TREAT	−0.227	0.0916	−2.48	0.013	−0.407, −0.048
AGE_C	−0.031	0.0079	−3.87	<0.001	−0.046, −0.015
DRUG	0.343	0.1043	3.29	0.001	0.138, 0.547
NDRGTX_C	0.031	0.0080	3.87	<0.001	0.015, 0.047

times in the same manner as described in Section 2.3. If the graph is not too complicated, we can use the graphical approach illustrated in Figure 2.6. However, since this is not likely to be accurate enough in most applied settings, we determine the estimator as

$$\hat{t}_{50} = \min\left\{t : \hat{S}\left(t, \mathbf{x}, \hat{\boldsymbol{\beta}}\right) \le 0.50\right\}, \tag{4.33}$$

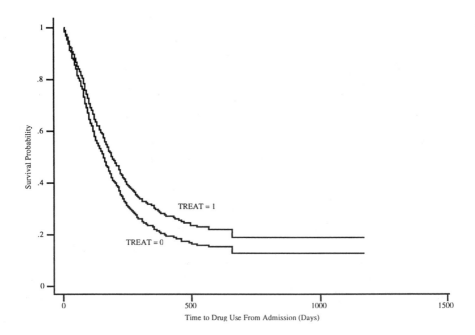

Figure 4.6 Graphs of the modified risk score-adjusted estimated survivorship function for treatment in the UIS.

where the estimator of the survivorship function in (4.33) is any one of the adjusted estimators. Application of (4.33) to the estimates graphed in Figure 4.6 yields adjusted estimated median time to return to drug use of 159 days and 190 days for the short and long treatments, respectively.

Confidence interval estimators for the covariate-adjusted estimator of the median survival time are discussed in the next section. In general, the methods are a bit more complex than those discussed to this point in the text. The methods have not been implemented in standard statistical software packages, but one could write a routine to calculate the confidence interval estimator.

The covariate-adjusted survivorship function in (4.23) is the one most frequently used in applied settings. An alternative estimator, first proposed by Makuch (1982) and further studied by Thomsen, Keiding and Altman (1991), is discussed and recommended by Marubini and Valsecchi (1995, Chapter 6). The estimator, called the "direct adjusted survival curve" by Makuch, is obtained by averaging the individual estimated survivorship functions over all subjects at each observed time. The average may be computed overall or within subgroups of subjects. As shown in Marubini and Valsecchi (1995), the direct adjusted estimator is easy to compute for a simple model containing one or two dichotomous covariates, but a more complicated model requires special programming. Evaluation of the direct estimator requires calculating an average over all subjects in specified groups at each observed value of time. In order to illustrate the method, we computed the directly adjusted estimator for the model shown in Table 4.12. It is presented in Figure 4.7, along with the estimator in (4.23) for the IV drug use groups present and absent shown in Figure 4.4. The two sets of curves in Figure 4.7 certainly convey the same message regarding the effect of treatment and, in this case, would yield similar estimates of the median time to drug relapse in each drug group. The difference between the two estimators is most easily explained if we focus on a single value of time, say 24 months. The value of the upper covariate-adjusted estimated survivorship function, (4.23), is just $\hat{S}_0(24)$, that is, the survivorship function for age equal to 35 years and IV drug use absent. The direct adjusted estimate for this group at 24 months is

$$\tilde{S}(24, \text{DRUG} = 0) = \frac{1}{51} \sum_{j=1}^{51} \hat{S}\left(24, \text{DRUG}_j = 0, \text{AGE}_j - 35\right), \quad (4.34)$$

where

$$\hat{S}\left(24, \text{DRUG}_j = 0, \text{AGE}_j - 35\right) = \left[\hat{S}_0(24)\right]^{\exp\left(0 \times \hat{\beta}_1 + \left(\text{AGE}_j - 35\right) \times \hat{\beta}_2\right)}.$$

The mean on the right-hand side of (4.34) is the average restricted to the 51 subjects in the HMO-HIV+ study who had no history of IV drug use. The point on the lower covariate-adjusted curve is

$$\hat{S}(24, \text{DRUG} = 1, \text{AGE\_C} = 0) = \left[\hat{S}_0(24)\right]^{\exp\left(1 \times \hat{\beta}_1\right)},$$

and the corresponding point on the directly-adjusted curve is

$$\tilde{S}(24, \text{DRUG} = 1) = \frac{1}{49} \sum_{j=1}^{49} \hat{S}\left(24, \text{DRUG}_j = 1, \text{AGE}_j - 35\right), \quad (4.35)$$

where

$$\hat{S}\left(24, \text{DRUG}_j = 1, \text{AGE}_j - 35\right) = \left[\hat{S}_0(24)\right]^{\exp\left(1 \times \hat{\beta}_1 + \left(\text{AGE}_j - 35\right) \times \hat{\beta}_2\right)}.$$

The mean on the right-hand side of (4.35) is the average restricted to the 49 subjects in the HMO-HIV+ study who had a history of IV drug use. In order to obtain the graph of the direct-adjusted survivorship functions in Figure 4.7, the expressions in (4.34) and (4.35) must be computed for each observed value of time.

Marubini and Valsecchi (1995) state that the directly adjusted estimator better takes into account the variability in the observed values of the covariates. The difference between the two adjusted curves is due to the fact that the survivorship function is a non-linear function of the covariate (i.e., age) in the example. Which of the two adjusted curves, covariate or direct, one should use in practice depends on the goal of the analysis. If the purpose is to provide a figure to be used to compare survivorship under different levels of a nominal scale covariate, the more easily calculated covariate-adjusted curve is likely to be adequate in most applied settings.

Model-based estimation of the survival probability at a fixed time point is another application of the estimator in (4.23). For example, we may use the results in Table 4.12 to obtain an estimator of the probability of survival to 18 months for a subject with specified age and history of IV drug use. This is analogous to using a fitted linear regression

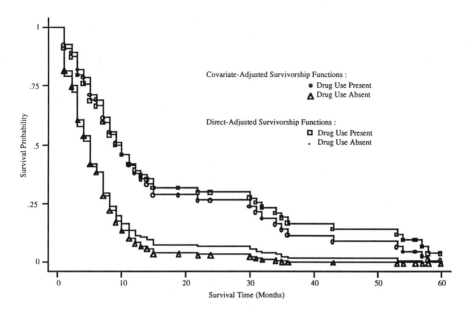

Figure 4.7 Graphs of the covariate age-adjusted and direct age-adjusted estimated survivorship function for IV drug use present and absent for the HMO-HIV+ study.

model to predict the outcome in a new subject. Suppose our new subject is 42 years old and has a previous history of IV drug use. The predicted 18-month survival probability for this subject is

$$\hat{S}(18, \text{DRUG} = 1, \text{AGE\_C} = 42 - 35) = \left[\hat{S}_0(18)\right]^{\exp(1 \times 0.941 + 7 \times 0.092)}$$

$$= \left[0.3239\right]^{\exp(1 \times 0.941 + 7 \times 0.092)} = \left[0.3239\right]^{4.879} = 0.0041,$$

where the value of $\hat{S}_0(18)$ is obtained from a tabulation of the estimated baseline survivorship function. Thus the fitted model[3] predicts less than a 1 percent chance of survival to 18 months. We discuss the confidence interval estimator for the predicted survival probability in the next section. The entire survivorship function may be obtained using the same

---

[3] We wish to remind the reader that the HMO-HIV+ study data are hypothetical and, as such, fitted models should not be used to draw substantive, real-world conclusions.

method discussed above for covariate-adjusted survival curves (i.e., considering (4.23) with the stated covariate values as a function of time).

The estimated survivorship function, (4.23), can be an effective tool to describe the results of a regression analysis of survival time. We wish to reemphasize the importance of giving careful thought to the plotted range of the curve and estimates of survival probabilities. It is all too easy, with current statistical software, to present graphs and predictions that may inappropriately extrapolate the fitted model.

4.6 CONFIDENCE INTERVAL ESTIMATION OF THE COVARIATE-ADJUSTED SURVIVORSHIP FUNCTION

In this section we present a confidence interval estimator for the covariate-adjusted survivorship function. The method and resulting formula are not difficult to understand as they follow from the methods discussed in Chapter 2 for the Kaplan–Meier estimator. However, application requires that one use matrices and matrix calculations, and these require greater computing expertise than is required in other sections.

Andersen, Borgan, Gill and Keiding (1993) present an estimator [equation (7.2.33), page 506] of the variance of the log of the covariate-adjusted survivorship function. Their estimator is identical to one presented by Marubini and Valsecchi (1995, in the Appendix to Chapter 6). As noted in Chapter 2, better coverage properties are obtained if a confidence interval for the survivorship function is based on the log-log transformation of the function. An expression for a variance estimator for this further transformation is given by Andersen, Borgan, Gill and Keiding (1993) following (7.2.33). The equation for the variance estimator for a fixed set of the p covariates, denoted $\mathbf{x}_0$, is

$$\hat{\mathrm{Var}}\left\{\ln\left[-\ln\left(\hat{S}\left(t,\mathbf{x}_0,\hat{\boldsymbol{\beta}}\right)\right)\right]\right\} = \hat{A}\left(t,\hat{\boldsymbol{\beta}}\right) + \hat{B}\left(t,\mathbf{x}_0,\hat{\boldsymbol{\beta}}\right), \qquad (4.36)$$

where

$$\hat{A}\left(t,\hat{\boldsymbol{\beta}}\right) = \frac{1}{\left\{\ln\left[\hat{S}_0(t)\right]\right\}^2} \sum_{t_{(i)} \leq t} \frac{d_i}{\left[\hat{C}\left(t_{(i)},\hat{\boldsymbol{\beta}}\right)\right]^2},$$

$$\hat{B}\left(t,\mathbf{x}_0,\hat{\boldsymbol{\beta}}\right) = \left(\mathbf{x}_0 - \overline{\mathbf{x}}(t)\right)' \hat{\mathrm{Var}}\left(\hat{\boldsymbol{\beta}}\right)\left(\mathbf{x}_0 - \overline{\mathbf{x}}(t)\right),$$

$$\hat{C}\left(t_{(i)}, \hat{\beta}\right) = \sum_{j \in R(t_{(i)})} e^{x'_j \beta},$$

and

$$\overline{x}(t) = \frac{1}{\ln\left[\hat{S}_0(t)\right]} \sum_{t_{(i)} \le t} d_i \left\{ \frac{\sum\limits_{j \in R(t_{(i)})} x_j e^{x'_j \beta}}{\left[\hat{C}\left(t_{(i)}, \hat{\beta}\right)\right]^2} \right\}.$$

We note that d_i represents the number of subjects with survival time equal to $t_{(i)}$ and $\hat{\text{Var}}\left(\hat{\beta}\right)$ denotes the estimator of the covariance matrix of the estimated coefficients. The endpoints of a $100(1-\alpha)$ percent confidence interval for the log-log function are

$$\ln\left[-\ln\left(\hat{S}(t, x_0, \hat{\beta})\right)\right] \pm z_{1-\alpha/2} \hat{\text{SE}}\left\{\ln\left[-\ln\left(\hat{S}(t, x_0, \hat{\beta})\right)\right]\right\}, \qquad (4.37)$$

where $\hat{\text{SE}}\{\ \}$ in (4.37) denotes the estimator of the standard error and, in this case, is the positive square root of the estimator in (4.36). If we denote the lower and upper endpoints obtained from (4.37) as $l\left(t, x_0, \hat{\beta}\right)$ and $u\left(t, x_0, \hat{\beta}\right)$, then the lower and upper endpoints of the confidence interval estimator of the survivorship function are obtained in a manner similar to (2.8) and are

$$\exp\left\{-\exp\left[u\left(t, x_0, \hat{\beta}\right)\right]\right\} \text{ and } \exp\left\{-\exp\left[l\left(t, x_0, \hat{\beta}\right)\right]\right\}. \qquad (4.38)$$

The expressions in (4.37) and (4.38) may be used to obtain a confidence interval for an individual predicted survival probability or to provide a pointwise confidence band for a covariate-adjusted survivorship function. In the previous section, we used the fitted model shown in Table 4.12 for the HMO-HIV+ study to predict the 18-month survival time for a 42-year-old subject with a history of prior drug use. The model-based prediction was 0.004. We use (4.37) and (4.38) to obtain the endpoints of a 95 percent confidence interval, giving us the interval (0.00005, 0.04894). The interpretation of this interval is that a predicted probability in this interval would be consistent with the observed data. Stated another way, the subject has at best a 4.9 percent chance of surviving 18 months.[4]

We obtain pointwise confidence bands for the survivorship function by evaluating (4.38) at each observed survival time. These can be plotted, along with the estimated survivorship function, similar to the plot in Figure 2.5. The Hall and Wellner joint confidence bands discussed in Chapter 2 have not been extended to the covariate-adjusted survivorship function.

We plotted, in Figure 4.4, age- (equal to 35) adjusted survivorship functions for both levels of the IV drug use variable. Confidence bands for each function are obtained through definition of the fixed covariate x_0. These have been calculated and are shown separately in Figures 4.8a and 4.8b. The confidence bands for the IV drug use groups absent and present in (4.37) and (4.38) are

$$x_0' = (DRUG = 0, AGE\_C = 0) \text{ and } x_0' = (DRUG = 1, AGE\_C = 0),$$

respectively. The graphs have been drawn in a manner similar to Figure 4.2, using each observed survival time. The general shape of the confidence bands is similar to those presented in Chapter 2, although they are wider and more skewed for longer survival times.

We can also use the graphs to obtain a confidence interval estimate of the median survival time (or other quantile) by following the same method described in Chapter 2. A more accurate determination may be obtained by applying (4.33) to the lower and upper confidence bands, respectively. Applying (4.33) to the actual values generating the curves in Figures 4.8a and 4.8b yields age-adjusted median survival times and 95 percent confidence limits for IV drug use present of 5 (4, 7) and for IV drug use absent of 10 (8, 14). These confidence intervals are essentially an extension of the Brookmeyer-Crowley method discussed in Chapter 2. We noted in Chapter 2 that the Brookmeyer–Crowley method assumes that there are no tied survival times. However, there are a number of ties in the HMO-HIV+ study and, as a result, the presented confidence interval estimate should be interpreted cautiously, as the ef-

---

[4] One should not infer from this statement or our presentation of the methods in this text that we advocate the use of fitted survival time models and possible subsequent predictions as tools for individual patient/subject decision-making. This is a difficult and sensitive subject, and a statistical model should be one small part of a much larger discussion.

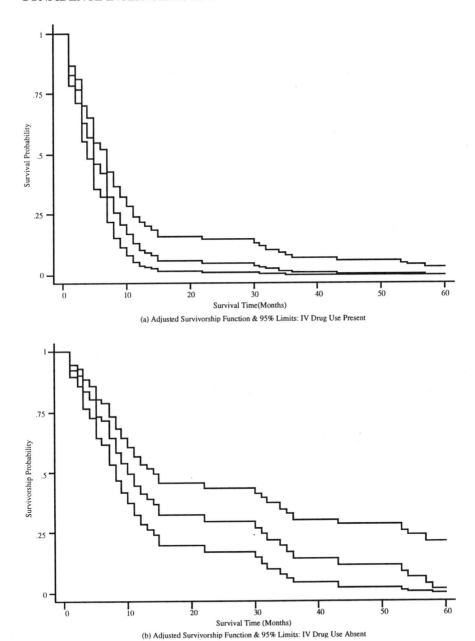

(a) Adjusted Survivorship Function & 95% Limits: IV Drug Use Present

(b) Adjusted Survivorship Function & 95% Limits: IV Drug Use Absent

Figure 4.8 Graphs of age-adjusted estimated survivorship function and 95% pointwise confidence limits for IV drug use present and absent for the HMO-HIV+ study.

fect of ties on the coverage properties of the interval has not been studied.

In applied settings, we recommend that adjusted point and interval estimates of median survival time be used for descriptive purposes only. One should avoid the temptation to use the confidence intervals to draw inferential conclusions about the equality of median survival times. This hypothesis should be tested via the partial likelihood ratio test for the significance of the coefficient for the grouping variable in the fitted proportional hazards model.

In Section 4.5 we presented a covariate-adjusted survivorship function that used the median risk score for a single curve and a modified risk score when adjusted survivorship functions for two groups were compared. Unfortunately, we cannot directly employ the variance estimator and confidence interval in (4.37) and (4.38), since the median value of the risk score may not correspond to a single fixed set of covariates. One solution is to rerun the analysis adjusting for the set of covariate values yielding a risk score that came closest to the median. However, it is possible, with a complex model, that several sets of covariate values will have risk scores equally close to the median value. In this case, one could choose the set that seems clinically closest to a middle set of values. The adjusted survival curves obtained from using the actual median risk score and the ones obtained from the set of covariates with risk score nearest the median should be quite similar and adequate for descriptive analyses. Again, this is an issue only if one wants to add confidence bands to the plot of a risk score-adjusted survivorship function.

EXERCISES

1. Using all the data from the WHAS (i.e., ignore cohort), with length of follow-up as the survival time variable and status at last follow-up as the censoring variable, do the following:

   (a) Fit the proportional hazards model containing sex and estimate the hazard ratio, pointwise and with a 90 percent confidence interval. Interpret the point and interval estimates.

   (b) Add age to the model fit in 1(a). Is age a confounder of the effect of sex? Explain the reasons for your answer.

   (c) Is there a significant interaction between age and sex? (Use $\alpha = 0.10$ for this problem).

(d) Using the model fit in 1(c) estimate the hazard ratio, pointwise and 90 percent confidence interval, for gender at age 50, 60, 65, 70 and 80.

(e) Using the model fit in 1(c), estimate (pointwise and with a 90 percent confidence interval) the hazard ratio for a 10-year increase in age for each gender.

(f) Using the model fit in 1(c), compute, and then graph, the estimated survivorship functions for 65-year-old males and females. Interpret the survivorship experience presented in this graph.

(g) Using the graph in 1(f), estimate the median survival time for 65-year-old males and females.

2. Repeat problem 1, parts (b)–(d), with age broken into four groups at its quartiles. In part (d) estimate hazard ratios for each age group.

3. Using the data from the WHAS (i.e., ignore cohort), with length of follow-up as the survival time variable and status at last follow-up as the censoring variable, do the following:

(a) Fit the proportional hazards model containing age centered at 65 years, sex, peak cardiac enzymes centered at 650, cardiogenic shock complications, left heart failure complications and MI order and obtain the estimated baseline survivorship function. (Note: In this problem, ignore the possible sex $\times$ age interaction investigated in problems 1 and 2.) Estimate hazard ratios (via point estimates and 95 percent confidence intervals) for each variable in the model.

(b) Using the methods for the modified risk score, compute and graph estimated survivorship functions for subjects with and without cardiogenic shock complications. Use the estimated survivorship functions to estimate the median survival time.

CHAPTER 5

Model Development

5.1 INTRODUCTION

In any applied setting, performing a proportional hazards regression analysis of survival data requires a number of critical decisions. It is likely that we will have data on more covariates than we can reasonably expect to include in the model, so we must decide on a method to select a subset of the total number of covariates. When selecting a subset of covariates, we must consider such issues as clinical importance and adjustment for confounding, as well as statistical significance. Once we have selected the subset, we must determine whether the model is "linear" in the continuous covariates and, if not, what transformations are suggested by the data and clinical considerations. Which interactions, if any, should be included in the model is another important decision. In this chapter we discuss these and other practical model development issues.

The end use of the estimated regression model will most often be a summary presentation and interpretation of the factors that have influenced survival. This summary may take the form of a table of estimated hazard ratios and confidence intervals and/or estimated covariate-adjusted survivorship functions. Before this step can be taken, we must critically examine the estimated model for adherence to key assumptions (e.g., proportional hazards) and determine whether any subjects have an undue influence on the fitted model. In addition, we may calculate summary measures of goodness-of-fit to support our efforts at model assessment. Methods for model assessment are discussed and illustrated in Chapter 6.

The methods available to select a subset of covariates to include in a proportional hazards regression model are essentially the same as those used in any other regression model. In this chapter we present three methods for selecting a subset of covariates. Purposeful selection is a method that is completely controlled by the data analyst, while stepwise and best subsets selection of covariates are statistical methods. These approaches to covariate selection have been chosen since use of one or more of them will yield, in the vast majority of model building applications, a subset of statistically and clinically significant covariates.

A word of caution: statistical software for fitting regression models to survival data is, for the most part, easy to use and provides a vast array of sophisticated statistical tools and techniques. One must be careful, therefore, not to lose sight of the problem and end up with the software prescribing the model to the analyst rather than the other way around.

Regardless of which method is used for covariate selection, any survival analysis should begin with a thorough bivariate analysis of the association between survival time and all important covariates. These methods are discussed in detail in Chapter 2. For categorical covariates, this should include Kaplan–Meier estimates of the group-specific survivorship functions, point and interval estimates of the median and/or other quantiles, survival time and use of one or more of the significance tests to compare survivorship experience across the groups defined by the variable. For descriptive purposes, continuous covariates could be broken into quartiles, or other clinically meaningful groups, and the methods for categorical covariates could then be applied. Alternatively, point and interval estimates of the hazard ratio for a clinically relevant change in the covariate could be used in conjunction with the significance level of the partial likelihood ratio test. These results should be displayed using the tabular conventions of the scientific field.

5.2 PURPOSEFUL SELECTION OF COVARIATES

Purposeful selection of covariates begins with a multivariable model that contains all variables significant in the bivariate analysis at the 20–25 percent level, as well as any other variables not selected with this criterion, but which are judged to be of clinical importance. If there are adequate data to fit a model containing all study covariates, this full model could be the beginning multivariable model. The rationale for choosing a relatively modest level of significance is based on recommendations for linear regression by Bendel and Afifi (1977), for discriminant analysis by Costanza and Afifi (1979), and for change in coefficient modeling in epidemiology by Mickey and Greenland (1989).

Use of this level of significance should lead to the inclusion, in the pre-liminary multivariable model, of any variable that has the potential to be either an important confounder, or is statistically significant. Following the fit of the initial multivariable model, we use the p-values from the Wald tests of the individual coefficients to identify covariates that might be deleted from the model. Some caution should be taken at this point not to reduce the size of the model by deleting too many seemingly nonsignificant variables at one time. The p-value of the partial likeli-hood ratio test should confirm that the deleted covariate is not signifi-cant. This is especially important when a nominal scale covariate with more than one design variable has been selected for deletion, since we typically make a rough guess about overall significance based on the significance levels of the individual coefficients of the design variables. Following the fitting of the reduced model, we assess whether or not re-moval of the covariate has produced an "important" change in the co-efficients of the variables remaining in the model. In general, we use a value of about 20 percent as an indicator of an important change in a coefficient. If the variable excluded is an important confounder, it should be added back into the model. This process continues until no covariates can be deleted from the model.

At this point, we recommend that any variable excluded from the initial multivariable model be added back into the model to confirm that it is neither statistically significant nor an important confounder. We have encountered situations in practice where a variable had a bivariate test p-value that exceeded 0.8 but it became highly significant when added to a multivariable model. At the conclusion of this step we have the "preliminary main effects model."

The next step is to examine the scale of the continuous covariates in the preliminary main effects model. A number of techniques are avail-able, all of which are designed to determine whether the data support the hypothesis that the effect of the covariate is linear in the log hazard and, if not, which transformation of the covariate is linear in the log hazard. The simplest method is to replace the covariate with design variables formed from the quartiles or other cutpoints that may have been used in the bivariate descriptive analysis. The estimated coefficients for the de-sign variables are plotted versus the midpoints of the groups and, at the midpoint of the first group, a point is plotted at zero. If the correct scale is linear in the log hazard, then the polygon connecting the points should be nearly a straight line. If the polygon departs substantially from a linear trend, its form may be used to suggest a transformation of the covariate. The advantage of the quartile method is that it does not require any special software. The disadvantage is that it is not powerful enough to detect subtle, but often important, deviations from a linear trend.

Another approach is to use the method of fractional polynomials, developed by Royston and Altman (1994), to suggest transformations. We wish to determine what value of x^p yields the best model for the covariate. In theory, we could incorporate the power, p, as an additional parameter in the estimation procedure. However, this would greatly increase the complexity of the estimation problem. Royston and Altman propose replacing full maximum likelihood estimation of the power by a search through a small but reasonable set of possible values. We will provide a brief description of the method and later demonstrate its use, along with the other methods, in an example.

The method of fractional polynomials may be used with a multivariable proportional hazards regression model, but, for sake of simplicity, we describe the procedure using a model with a single continuous covariate. The hazard function for the proportional hazards regression model shown in (3.7) is

$$h(t,x,\beta) = h_0(t)e^{x\beta},$$

and the log-hazard function, which is linear in the covariate, is

$$\ln[h(t,x,\beta)] = \ln[h_0(t)] + x\beta.$$

One way to generalize this log-hazard function is to specify it as

$$\ln[h(t,x,\beta)] = \ln[h_0(t)] + \sum_{j=1}^{J} F_j(x)\beta_j .$$

The functions $F_j(x)$ are a particular type of power function. The value of the first function is $F_1(x) = x^{p_1}$. In theory, the power, p_1, could be any number, but in most applied settings we would try to use something simple. Royston and Altman (1994) propose restricting the power to be among those in the set $\wp = \{-2, -1, -0.5, 0, 0.5, 1, 2, 3\}$, where $p_1 = 0$ denotes the log of the variable. The remaining functions are defined as

$$F_j(x) = \begin{cases} x^{p_j}, p_j \neq p_{j-1} \\ F_{j-1}(x)\ln(x), p_j = p_{j-1} \end{cases}$$

for $j = 2,\ldots,J$ and restricting powers to those in $\wp$. For example, if we chose $J = 2$ with $p_1 = 0$ and $p_2 = -0.5$, then the log-hazard function is

$$\ln[h(t, x, \boldsymbol{\beta})] = \ln[h_0(t)] + \ln(x)\beta_1 + \frac{1}{\sqrt{x}}\beta_2.$$

As another example, if we chose $J = 2$ with $p_1 = 2$ and $p_2 = 2$, then the log-hazard function is

$$\ln[h(t, x, \boldsymbol{\beta})] = \ln[h_0(t)] + x^2\beta_1 + x^2\ln(x)\beta_2.$$

The model is quadratic in x if $p_1 = 1$ and $p_2 = 2$. Again, we could allow the covariate to enter the model with any number of functions, J; but in most applied settings an adequate transformation may be found if we use $J = 1$ or 2. Implementation requires, for $J = 1$, fitting 8 models, that is, $p_1 \in \wp$. The best model is the one with the largest log partial likelihood. The process is repeated with $J = 2$ by fitting the 64 models obtained from all possible pairs of powers, that is, $(p_1, p_2) \in \wp \times \wp$, and the best model is again the one with the largest log partial likelihood. The relevant question is whether either of the two best models is significantly better than the linear model. Let $L(1)$ denote the log partial likelihood for the linear model, that is, $J = 1$ and $p_1 = 1$, and $L(p_1)$ denote the log partial likelihood for the best $J = 1$ model and $L(p_1, p_2)$ denote the log partial likelihood for the best $J = 2$ model. Royston and Altman (1994) suggest, and verify with simulations, that each term in the fractional polynomial model contributes approximately 2 degrees-of-freedom to the model, effectively one for the power and one for the coefficient. Thus, the partial likelihood ratio test comparing the linear model to the best $J = 1$ model,

$$G(1, p_1) = -2\{L(1) - L(p_1)\},$$

is approximately distributed as chi-square with 1 degree-of-freedom under the null hypothesis of linearity. The partial likelihood ratio test comparing the best $J = 1$ model to the best $J = 2$ model,

$$G[p_1,(p_1,p_2)] = -2\{L(p_1) - L(p_1,p_2)\},$$

is approximately distributed as chi-square with 2 degrees-of-freedom under the null hypothesis that the second function is equal to zero. Similarly, the partial likelihood ratio test comparing the linear model to the best $J = 2$ model is distributed approximately as chi-square with 3 degrees-of-freedom. Note that to keep the notation simple, we have used p_1 to denote the best power both when $J = 1$ and as the first of the two powers for $J = 2$. These are not likely to be the same numeric value in practice.

In an applied setting, the partial likelihood ratio tests are used to choose which of the three forms of the covariate is best. In general, we recommend that, if a more complicated model is selected for use, it should provide a statistically significant improvement over a simpler model, and the transformations should make clinical sense.

The only software package that has fully implemented the method of fractional polynomials is STATA. In addition to the method as described above, STATA's fractional polynomial routine offers the user considerable flexibility in expanding the set of powers searched; however, in most settings the default set of values should be adequate.

Graphical methods to check the scale of covariates may be performed in most software packages. The most easily used of these are similar to residual methods from linear regression; see Ryan (1997). A complete discussion of residuals is provided in Chapter 6. The reader wishing to know the details of residual construction is welcome to read Section 6.2 before proceeding, but it is not necessary for the purpose of using them in plots to assess the scale of a covariate. The components of the residual for the ith subject are the value of the censoring variable, c_i, and the estimated cumulative hazard $\hat{H}_i = \hat{H}(t_i, \mathbf{x}_i, \hat{\boldsymbol{\beta}})$; see (3.41), and these are used to calculate the martingale residuals, defined as

$$\hat{M}_i = c_i - \hat{H}_i.$$

Therneau, Grambsch and Fleming (1990) suggest fitting a model that excludes the covariate of interest. The results are used to calculate $\hat{M}_i$ and to generate smoothed values (e.g., the lowess smooth). These are then plotted versus the values of the excluded covariate, and the shape of the plot, and especially the smooth, provides an estimate of the functional form of the covariate in the model. Grambsch, Therneau and Fleming (1995) expand on their earlier work and suggest that one begin with a fit of the model containing all covariates. They demonstrate in simulations and examples that a plot of the log of the ratio of a smoothed c to a smoothed $\hat{H}$ versus the covariate has greater diagnostic power than their earlier proposed method. Both of these plots are illustrated in the example in this chapter. Descriptions of applications of these and other related methods may be found in Therneau (1995).

Gray (1992) suggests that spline functions may be used as a way of modeling a continuous covariate without meeting stringent assumptions of a linear scale. Ryan (1997) discusses the construction and use of spline functions in linear regression. Harrell et al. (1996) demonstrate the use of spline functions in a variety of modeling settings, including the proportional hazards model. Since spline functions are not readily available in most software packages, they will not be discussed further or used in the example.

The final step in the variable selection process is to determine whether interactions are needed in the model. In this setting, an interaction term is a new variable that is the product of two covariates in the model. There may be special considerations that dictate that a particular interaction term or terms be included in the model, regardless of the statistical significance of the coefficient(s). If this is the case, these interaction terms and their component terms should be added to the main effects model and the larger model fit before proceeding with a statistical evaluation of other possible interactions. However, in most settings, there will be insufficient clinical theory to justify automatic inclusion of interactions.

The selection process begins by forming a set of biologically plausible interaction terms from the main effects in the model. The significance of each separate interaction is assessed by adding it to the main effects model and using the partial likelihood ratio test. All interactions significant at the 5 percent level are then added jointly to the main effects model. Wald statistic p-values are used as a guide to selecting interactions that may be eliminated from the model, but significance should be checked by the partial likelihood ratio test.

Several important points should be kept in mind when selecting interaction terms. Since the reason for including interactions is to improve inferences and obtain a more realistic model, we feel that all interaction terms should be statistically significant at usual levels of significance, such as 5 or 10 percent, and perhaps as low as 1 percent in some settings. Inclusion of nonsignificant interactions in a model will needlessly increase standard error estimates, thus unnecessarily widening confidence interval estimates of hazard ratios.

When an interaction term is added to a model, large changes in the coefficients of the corresponding main effects are likely to occur. However, changes in the main effect coefficients induced by interaction terms are not relevant and, as a result, do not indicate confounding. When interaction terms are present, the corresponding main effect terms do not, in most cases, estimate hazard ratios of interest. In addition, when there is statistically significant interaction, we include the corresponding main effect terms in the model regardless of their statistical significance. We are interested in examining how the main effect and interaction terms combine to estimate hazard ratios.

At this point we have a preliminary model. Our next step would be to assess its fit and adherence to key assumptions. These methods are discussed in the next chapter.

We illustrate the method of purposeful selection of covariates using the data from the UIS. These data were introduced in Section 1.3 and the variables are defined in Table 1.3. Recall that the goal of the study was to compare the effectiveness of two treatments (of different lengths) for the prevention of return to drug use at two different sites. At this point we will not consider the covariate, length of stay, for inclusion in the model since it is related to the outcome variable, time to drug use as measured from admission date. We will use it when we consider extending the proportional hazards model to include time-dependent or varying covariates in Chapter 7.

A modification that is sometimes used in a clinical trial setting where there is a clear "treatment" variable is to exclude the treatment variable from the variable selection process. The treatment variable is then added to the preliminary main effects model containing all of the variables associated with outcome, irrespective of treatment. The rationale for this approach is that one obtains an estimate of the additional effect of treatment, adjusting for other covariates. This approach is in contrast to modeling in epidemiological studies where "treatment" would be the risk factor of interest. In these settings, selection of variables may be based on the change in the coefficient (estimate of effect) of the risk factor variable. Thus, rather than being the last variable to enter, the risk factor enters the model first. What this points out is that one must have clear goals for the analysis and proceed thoughtfully using a variety of statistical tools and methods. The variable selection methods discussed may be an integral part of this analysis. In the example, we include the treatment variable among those in the first multivariable model.

The results of the bivariate analysis of each covariate in relation to time to drug relapse are presented in Table 5.1 for discrete covariates

Table 5.1 Estimated Median Time to Drug Use with 95% Brookmeyer–Crowley Confidence Intervals, Log-Rank Test and Partial Likelihood Ratio Test p-Values for Categorical Covariates in the UIS ($n = 628$)

Variable	Category	Median Time to Drug Use (95% CIE)	Log-Rank Test p-Value	Partial Likelihood Ratio Test p-Value
HERCOC	Both	150 (106,196)	0.047	0.051
	Heroin	142 (110, 184)		
	Cocaine	183 (148, 226)		
	Neither	181 (154, 220)		
IVHX	Never	194 (171, 228)	<0.001	<0.001
	Previous	170 (130, 226)		
	Recent	147 (115, 168)		
RACE	White	152 (124, 174)	0.007	0.006
	Other	193 (164, 232)		
TREAT	Short	130 (113, 154)	0.009	0.010
	Long	190 (175, 226)		
SITE	A	156 (131, 174)	0.124	0.121
	B	198 (159, 231)		
AGE	20 - 27	154 (121, 198)	0.282	0.282
	28 - 32	148 (123, 180)		
	33 - 37	162 (121, 207)		
	38 - 56	189 (162, 242)		
BECKTOTA	0 - <10	211 (166, 245)	0.229	0.229
	10 - <15	169 (124, 208)		
	15 - <25	168 (136, 192)		
	25 - <55	147 (106, 187)		
NDRUGTX	0 - 1	170 (142, 227)	0.002	0.002
	2 - 3	177 (162, 207)		
	4 - 6	127 (106, 183)		
	7 - 40	123 (106, 184)		

and in Table 5.2 for continuous covariates. All variables, except age categorized in four groups, are significant at the 20 percent level and therefore are candidates for inclusion in the multivariable model. The discrete forms, in Table 5.1, of the continuous covariates in Table 5.2 are presented primarily for descriptive purposes. If a variable is significant with either coding scheme, the variable should be added to the list

Table 5.2 Estimated Hazard Ratio for Time to Drug Relapse with 95% Confidence Intervals, Wald Test and Partial Likelihood Ratio Test p-Values for Continuous Covariates in the UIS (n = 628)

Variable	Change	Hazard Ratio for Change (95% CIE)	Wald Test p-Value	Partial Likelihood Ratio Test p-Value
AGE	5 years	0.94 (0.87, 1.01)	0.074	0.072
BECKTOTA	10 points	1.12 (1.02, 1.22)	0.020	0.021
NDRUGTX	5 treatments	1.16 (1.08, 1.25)	<0.001	<0.001

for inclusion in the multivariable model and used in its continuous form. We assess the correct scale of the variable following the fitting of the preliminary main effects model.

Before we fit the multivariable model, we note the close agreement in Table 5.1 between the significance levels of the partial likelihood ratio test and the log-rank test. This is as expected since, for a discrete covariate, the score test is algebraically related to the log-rank test and the performance of the score test is quite similar to the partial likelihood ratio test. The implication is that the log-rank test is an acceptable choice for purposes of covariate selection for the initial multivariable model.

Table 5.3 presents the results of fitting the multivariable proportional hazards model containing all variables significant at the $p < 0.20$ level in the bivariate analysis. This analysis includes 575 subjects for whom complete information is available on all covariates. Examining the p-values for the Wald statistics with the goal of trying to simplify the model, we note that none of the design variables for previous heroin or cocaine use is significant. The coefficient for intervention site is also not significant but, due to its importance in the study design, we keep it in the model. We next fit a model excluding previous heroin or cocaine use. Table 5.4 presents the results of fitting this reduced model. The partial likelihood ratio test comparing the models in Tables 5.3 and 5.4 is $G = 1.39$ which, with 3 degrees-of-freedom, has a p-value of 0.71, supporting our decision to remove the variable. The maximum change in the coefficient for any variable remaining in the model is 18.5 percent for the design variable for recent IV drug use, IVHX_3. This is not judged to be an important enough change to warrant inclusion of the heroin and cocaine use design variables in the model, so we proceed with the simpler model.

Table 5.3 Estimated Coefficients, Standard Errors, z-Scores, Two-Tailed p-Values and 95% Confidence Intervals for the Proportional Hazards Model Containing Variables Significant at the 20% Level in the Bivariate Analysis for the UIS (n = 575)

Variable	Coeff.	Std. Err.	z	$P>\|z\|$	95% CIE
AGE	−0.029	0.008	−3.53	<0.001	−0.045, −0.013
BECKTOTA	0.008	0.005	1.68	0.094	−0.001, 0.018
NDRUGTX	0.028	0.008	3.42	0.001	0.012, 0.045
HERCO_2	0.065	0.150	0.44	0.663	−0.228, 0.359
HERCO_3	−0.094	0.166	−0.57	0.572	−0.418, 0.231
HERCO_4	0.028	0.160	0.18	0.861	−0.286, 0.342
IVHX_2	0.174	0.139	1.26	0.208	−0.097, 0.446
IVHX_3	0.281	0.147	1.91	0.056	−0.007, 0.569
RACE	−0.203	0.117	−1.74	0.082	−0.432, 0.026
TREAT	−0.240	0.094	−2.54	0.011	−0.425, −0.055
SITE	−0.102	0.109	−0.94	0.348	−0.317, 0.112

Log-likelihood = −2640.0305.

Examining the p-values for the Wald statistics in Table 5.4, we find that other than SITE (which, for practical reasons, will stay in the model) and BECKTOTA (which is marginally significant), the only non-significant variable is one of the pair of design variables IVHX_2 and IVHX_3, which together describe previous IV drug use. Two possible modeling strategies are: (1) Keep the design variables intact using all three codes, or (2) collapse the categories for "never" and "previous" to create a binary variable coded as "not recent" versus "recent." The

Table 5.4 Estimated Coefficients, Standard Errors, z-Scores, Two-Tailed p-Values and 95% Confidence Intervals for the Reduced Proportional Hazards Model for the UIS (n = 575)

Variable	Coeff.	Std. Err.	z	$P>\|z\|$	95% CIE
AGE	−0.028	0.008	−3.45	0.001	−0.044, −0.012
BECKTOTA	0.008	0.005	1.60	0.110	−0.002, 0.077
NDRUGTX	0.028	0.008	3.35	0.001	0.012, 0.044
IVHX_2	0.196	0.137	1.43	0.153	−0.073, 0.465
IVHX_3	0.333	0.120	2.78	0.006	0.098, 0.568
RACE	−0.209	0.116	−1.81	0.071	−0.436, 0.018
TREAT	−0.232	0.094	−2.47	0.013	−0.415, −0.048
SITE	−0.099	0.109	−0.92	0.359	−0.312, 0.113

Log-likelihood = −2640.7278.

Table 5.5 Estimated Coefficients, Standard Errors, z-Scores, Two-Tailed p-Values and 95% Confidence Intervals for the Reduced Proportional Hazards Model for the UIS (n = 575)

Variable	Coeff.	Std. Err.	z	$P>|z|$	95% CIE
AGE	-0.026	0.008	-3.25	0.001	-0.042, -0.010
BECKTOTA	0.008	0.005	1.70	0.090	-0.001, 0.018
NDRUGTX	0.029	0.008	3.54	<0.001	0.013, 0.045
IVHX_3	0.256	0.106	2.41	0.016	0.047, 0.464
RACE	-0.224	0.115	-1.95	0.051	-0.450, 0.001
TREAT	-0.232	0.093	-2.48	0.013	-0.416, -0.049
SITE	-0.087	0.108	-0.80	0.422	-0.298, 0.124

Log-likelihood = -2641.7294.

second choice yields a simpler model and may be preferred if the non-significant design variable does not confound the associations of the remaining variables in the model. Table 5.5 presents the results of fitting the model with IV drug use recoded as "not recent" versus "recent." The partial likelihood ratio test comparing the models in Tables 5.4 and 5.5 is $G = 2.00$ which, with one degree-of-freedom, yields a p-value of 0.157. The maximum percent change in a coefficient is -23.1 percent for the new binary variable, IVHX_3, but this change is uninterpretable since the reference group is different in the two models. Arguments could be given for the use of either the three-code version or the collapsed two-code version of the IV drug use variable. We will use the binary variable as it yields a simpler model and going from three to two codes has not changed the coefficients for any of the other variables, most notably treatment.

The next step in the modeling process is to examine the scale of the three continuous variables in the model: AGE, BECKTOTA and NDRUGTX. The first method we illustrate is the use of design variables. This approach to scale selection involves, for each of the three continuous variables, replacing the variable in the model with three design variables formed using the cutpoints shown in Table 5.1. Table 5.6 presents a summary of the resulting coefficients and group midpoints. The second step is to graph the coefficients against the group midpoints. These are shown in Figure 5.1.

The plots of the coefficients for age and especially Beck score support an assumption of linearity in the log hazard. The shape of the plot for number of previous drug treatments is more complicated. Analysis

Table 5.6   Estimated Coefficients for the Three Design Variables Formed from the Cutpoints Shown in Table 5.1 for the Variables AGE, BECKTOTA and NDRUGTX, in the UIS ($n = 575$)

AGE		BECKTOTA		NDRUGTX	
Midpoint	Coeff.	Midpoint	Coeff.	Midpoint	Coeff.
24.0	0.000	5.0	0.000	0.5	0.000
30.5	0.036	12.5	0.047	2.5	−0.070
35.5	−0.209	20.0	0.098	5.0	0.259
47.5	−0.391	40.0	0.216	23.5	0.399

of the Wald statistics shows that the coefficient for the second group is not significant, and the Wald test of the equality of the coefficients for the third and fourth groups is $z = 1.02$ with a p-value of 0.312, also not significant. This suggests that an alternative coding possibility is to form a binary covariate using the median (three previous treatments) as the cutpoint. The results of fitting this model are encouraging, in that the coefficient for the new binary variable was highly significant and the fitted model had a log partial likelihood that was only slightly smaller than that from the model with the three design variables. We defer discussing this model until we explore the use of the method of fractional polynomials for examining the scale of the three continuous covariates.

The results of the fractional polynomial analysis for age and Beck score confirm what was observed in the plot of the design variables in Figures 5.1a and 5.1b. An assumption of linearity in the log hazard seems quite reasonable for these two variables, so the computer output will not be presented.

The analysis of number of previous drug treatments (NDRUGTX) suggested that the log hazard is not linear. Table 5.7 presents the fractional polynomial results. The table contains four rows, and each row corresponds to a particular parametrization of the number of previous drug treatments. The first row represents a model containing all covariates in Table 5.5 except NDRUGTX, that is, the coefficient is set equal to zero. The model represented in the second row is the one shown in Table 5.5, as noted by the power of 1 in the last column. The significance level reported in the third column of the second row is for the partial likelihood ratio test of NDRUGTX entering the model as a linear term, that is,

$$G = [5294.497 - 5283.459] = 11.038$$

and $p = 0.00099$. The best power when NDRUGTX enters the model with a single, $J = 1$, term is $p_1 = 0.5$ (i.e., the square root of NDRUGTX). The approximate partial likelihood ratio test comparing the use of $p_1 = 1$ to $p_1 = 0.5$ is

$$G = [5283.459 - 5283.088] = 0.371,$$

and the reported p-value is $\Pr[\chi^2(1) \geq 0.371] = 0.543$. From this we conclude that a model using the square root of NDRUGTX is no better than a model using NDRUGTX as a linear term. The best powers when NDRUGTX enters the model with two terms, $J = 2$, is described by ($p_1 = -1$, $p_2 = -1$). The interpretation is that the two terms are x^{-1} and $(x^{-1})\ln(x)$. Since NDRUGTX is equal to zero for some subjects, the software fits the model using $x = (\text{NDRUGTX} + 1)/10$. The partial likelihood ratio test of this model versus the linear model is

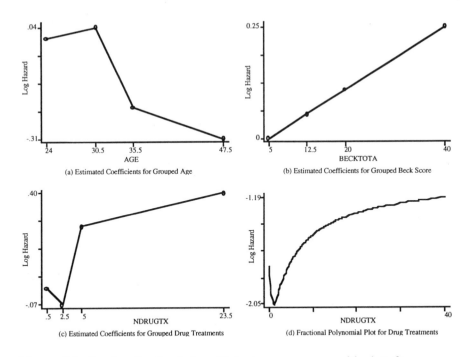

(a) Estimated Coefficients for Grouped Age

(b) Estimated Coefficients for Grouped Beck Score

(c) Estimated Coefficients for Grouped Drug Treatments

(d) Fractional Polynomial Plot for Drug Treatments

Figure 5.1 Graphs of estimated coefficients versus group midpoints for (a) AGE, (b) BECKTOTA, (c) NDRUGTX, and (d) the best two-term fractional polynomial model (−1,−1) for NDRUGTX.

Table 5.7   Summary of the Use of the Method of Fractional Polynomials for Number of Previous Drug Treatments for the UIS ($n = 575$)

	$-2\times$Log-like.	G for Model vs Linear	Approx. p-Value	Powers
Not in model	5294.497			
Linear	5283.459	0.000	0.001[*]	1
$J = 1$ (2 df)	5283.088	0.371	0.543[+]	0.5
$J = 2$ (4 df)	5276.543	6.916	0.038[#]	$-1, -1$

[*] Compares linear model to model without NDRUGTX.
[+] Compares the best $J = 1$ model to one with NDRUGTX linear.
[#] Compares the best $J = 2$ model to the best $J = 1$ model.

$$G = [5283.459 - 5276.543] = 6.916,$$

and its significance is $p = \Pr[\chi^2(3) \geq 6.916] = 0.075$. The partial likelihood ratio test of the best $J = 1$ model to the best $J = 2$ model is

$$G = [5283.088 - 5276.543] = 6.545$$

with $p = \Pr[\chi^2(2) \geq 6.545] = 0.038$. This test has 2 degrees-of-freedom since, when J is increased from 1 to 2, two additional terms (power and coefficient) are added to the model. To aid in the interpretation of the best two-term model, its graph is presented in Figure 5.1d. Even though the vertical scales are different in Figures 5.1c and 5.1d, there is a striking similarity in their shape, suggesting that the drop in the log-hazard function for a few previous drug treatments may be an important finding. This point is discussed in more detail in Chapter 6.

The two residual-based plots discussed earlier may be used as an alternative or adjunct to the method of fractional polynomials. The plots are shown in Figure 5.2 for age, in Figure 5.3 for the Beck score, and in Figure 5.4 for number of previous drug treatments. Each figure contains two plots. The top plot (a) is of the residuals and their smooth from a model that excludes the covariate of interest. The bottom plot (b) is of the log of the ratio of smoothed censor to smoothed cumulative hazard, called the expected in the figure headings. Since the scales of the two plots are different, plot (b) tends to overemphasize the shape, but the shapes in the two plots are consistent with each other.

For the sake of clarity we describe in some detail the steps used to produce the two plots in Figure 5.2. For Figure 5.2a, we fit a model containing the covariates in the model shown in Table 5.5, excluding

AGE. We requested that the values of $\hat{M}_i$, the martingale residuals, be calculated and saved. Figure 5.2a is a scatterplot of the $\hat{M}_i$ and their lowess smooth versus age.

To construct Figure 5.2b, we began by fitting the model in Table 5.5, including AGE, and requested that the martingale residuals be calculated and saved, also denoted as $\hat{M}_i$ for ease of notation. These residuals were used to calculate $\hat{H}_i = c_i - \hat{M}_i$, where c is the censoring variable. The values of c_i were plotted versus age and a lowess smooth was calculated and saved, denoted c_{ism}. The values of $\hat{H}_i$ were plotted versus age and a lowess smooth was calculated and saved, denoted $\hat{H}_{ism}$. The smoothed values were used to calculate

$$ f_i = \ln\left(\frac{c_{ism}}{\hat{H}_{ism}}\right) + \hat{\beta}_{AGE} \times AGE_i, $$

where $\hat{\beta}_{AGE} = -0.026$ from Table 5.5. Figure 5.2b is a plot of the pairs (f_i, AGE_i) connected by straight lines. The plots in Figure 5.3 and Figure 5.4 were obtained in an identical manner. The size of the plotting symbol for $\hat{M}_i$ in Figure 5.2a, 5.3a and 5.4a has been reduced to emphasize the smoothed values.

The smoothed values in Figures 5.2a and 5.2b are nearly straight lines, supporting our treatment of age as linear in the model. The plots in Figures 5.3a and 5.3b demonstrate the instability of smoothed values in areas where there are not many values. One subject had a Beck score of 54 and was censored at 621 days, and the next smallest Beck score was 43. Thus, one subject is causing the downturn seen in both parts of Figure 5.3. After eliminating the effect in the plot of this one subject, the smoothed values are nearly a straight line and support treating the Beck score as linear in the model.

The plots in Figure 5.4 show the same decline and then rise in the log hazard for number of previous drug treatments that was observed in Figure 5.1. The graphs clearly illustrate the nonlinear behavior of number of previous drug treatments in the model. However, even the most experienced analyst would be hard-pressed to come up with a parametric function describing this shape. An advantage of the method of fractional polynomials is that it suggests the functional form for non-linearly scaled continuous covariates.

Table 5.8   Estimated Coefficients, Standard Errors, z-Scores, Two-Tailed p-Values and 95% Confidence Intervals for the Proportional Hazards Model Using the Best Two-Term Fractional Polynomial Model for Number of Previous Drug Treatments for the UIS ($n = 575$)

Variable	Coeff.	Std. Err.	z	$P > \lvert z \rvert$	95% CIE
AGE	–0.028	0.008	–3.46	0.001	–0.044, –0.012
BECKTOTA	0.009	0.005	1.84	0.066	–0.001, 0.019
NDRUGFP1	–0.523	0.124	–4.20	<0.001	–0.767, –0.279
NDRUGFP2	–0.195	0.048	–4.04	<0.001	–0.289, –0.100
IVHX_3	0.259	0.108	2.39	0.017	0.047, 0.470
RACE	–0.242	0.116	–2.10	0.036	–0.468, –0.016
TREAT	–0.211	0.094	–2.25	0.024	–0.395, –0.027
SITE	–0.105	0.109	–0.97	0.335	–0.319, 0.109

Log-likelihood = –2638.272.

In summary, thoughtful model development should include the use of both the graphical methods described and the method of fractional polynomials to assess the scale of continuous covariates.

The final decision as to what scale to use comes down to a choice between a model with a single linear term, a binary variable using the median as the cutpoint, and the best two-term fractional polynomial model. The model with a single square root term is no better than the model with a single linear term. As noted in the discussion of the use of design variables, the model using a binary covariate with the median as the cutpoint was slightly better than the linear model, (log-likelihood = –2644.61 using a linear term versus a log-likelihood = –2643.58 using the binary coding). Given the simplicity and ease of interpretation of the binary coding, this model is the better of these two. The best fractional polynomial model is considerably more complicated than the binary model. However, consultation with the study team confirmed that the drop and rise in the log-hazard function seen in Figures 5.1c and 5.1d is not only plausible but of considerable interest. Thus, we will proceed using the scaling of NDRUGTX as selected by the method of fractional poly-nomials. Table 5.8 presents the results of fitting this preliminary main effects model. In this model

$$NDRUGFP1 = \left[(NDRUGTX + 1)/10 \right]^{-1}$$

and

$$NDRUGFP2 = \left[(NDRUGTX + 1)/10 \right]^{-1} \times \ln\left[(NDRUGTX + 1)/10 \right].$$

Figure 5.2a contains residuals from the model excluding AGE and Figure 5.2b the log of the ratio of smoothed values.

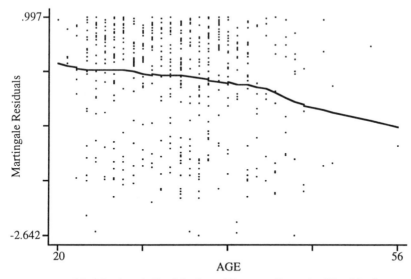

(a) Martingale Residuals and Lowess Smoothed Residuals.

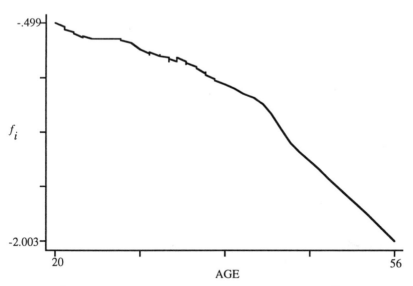

(b) Log(Smoothed Censor/Smoothed Expected) + $\beta_{AGE}{\times}$AGE.

Figure 5.2 Plots of two residual-based methods for selecting the scale of AGE in the UIS ($n = 575$).

Figure 5.3a contains residuals from the model excluding BECKTOTA and Figure 5.3b the log of the ratio of smoothed values.

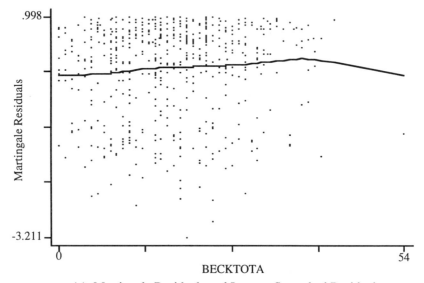

(a) Martingale Residuals and Lowess Smoothed Residuals.

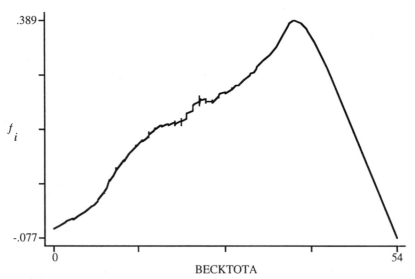

(b) Log(Smoothed Censor/Smoothed Expected) + $\beta_{\text{BECKTOTA}} \times$BECKTOTA.

Figure 5.3 Plots of two residual-based methods for selecting the scale of BECKTOTA in the UIS ($n = 575$).

Figure 5.4a contains residuals from the model excluding NDRUGTX and Figure 5.4b the log of the ratio of smoothed values.

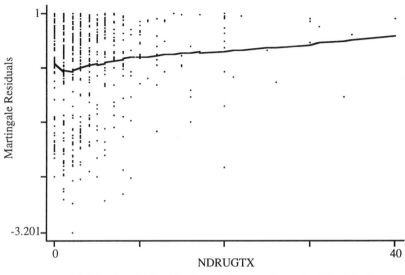

(a) Martingale Residuals and Lowess Smoothed Residuals.

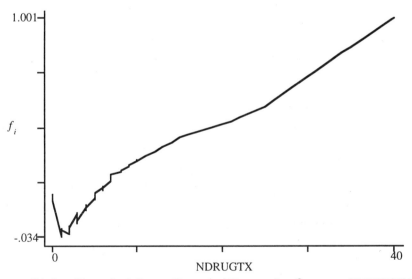

(b) Log(Smoothed Censor/Smoothed Expected) + $\beta_{NDRUGTX}\times$NDRUGTX.

Figure 5.4 Plots of two residual-based methods for selecting the scale of NDRUGTX in the UIS ($n = 575$).

Table 5.9 Interaction Variables, Degrees-of-Freedom (df) and *p*-Values for the Partial Likelihood Ratio Test for the Addition of the Interaction to the Model in Table 5.8

Interaction	Variables	df	*p*-Value
AGE	BECKTOTA	1	0.989
	NDRUGTX	2	0.028
	IVHX_3	1	0.460
	RACE	1	0.896
	TREAT	1	0.190
	SITE	1	0.028
BECKTOTA	NDRUGTX	2	0.316
	IVHX_3	1	0.241
	RACE	1	0.649
	TREAT	1	0.354
	SITE	1	0.912
NDRUGTX	IVHX_3	2	0.392
	RACE	2	0.746
	TREAT	2	0.214
	SITE	2	0.640
IVHX_3	RACE	1	0.568
	TREAT	1	0.534
	SITE	1	0.385
RACE	TREAT	1	0.310
	SITE	1	0.001
TREAT	SITE	1	0.247

The next step in the model building process is to add the design variables for heroin or cocaine use back into the model to be sure that they are neither significant in their own right nor confounders of the other main effects. The partial likelihood ratio test for the inclusion of heroin or cocaine use in the model is $G = 1.67$ which, with 3 degrees-of-freedom, yields a *p*-value of 0.644. The maximum percent change in a coefficient was less than 20 percent for all main effects in Table 5.8. We therefore conclude that heroin or cocaine use is not required in the model.

The final step in the model building process is the consideration of interaction terms. This step begins with the creation of a list of plausible interactions formed from the main effects in Table 5.8. Consultation with the study team determined that any pair of variables in the prelimi-nary main effects model could generate a clinically plausible inter-

Table 5.10   Estimated Coefficients, Standard Errors, z-Scores, Two-Tailed p-Values and 95% Confidence Intervals for the Preliminary Interactions Proportional Hazards Model for the UIS (n = 575)

Variable	Coeff.	Std. Err.	z	$P>\|z\|$	95% CIE
AGE	−0.054	0.028	−1.94	0.053	−0.109, 0.001
BECKTOTA	0.010	0.005	2.01	0.044	0.000, 0.020
NDRUGFP1	−0.674	0.644	−1.05	0.294	−1.938, 0.589
NDRUGFP2	−0.172	0.252	−0.68	0.496	−0.667, 0.322
IVHX_3	0.229	0.108	2.13	0.034	0.018, 0.441
RACE	−0.488	0.135	−3.62	<0.001	−0.752, −0.224
TREAT	−0.242	0.095	−2.56	0.010	−0.427, −0.057
SITE	−1.119	0.546	−2.05	0.040	−2.190, −0.049
AGEXSITE	0.026	0.017	1.60	0.111	−0.006, 0.059
RACEXSITE	0.863	0.248	3.48	0.001	0.376, 1.349
AGEXNDRUGFP1	0.002	0.019	0.08	0.933	−0.036, 0.040
AGEXNDRUGFP2	−0.002	0.008	−0.26	0.796	−0.017, 0.013

Log-likelihood = −2627.424

action. These are added, one at a time, to the preliminary main effects model. Table 5.9 presents the two variables forming the interaction, the degrees-of-freedom and the p-value for the partial likelihood ratio test comparing the models with and without the interaction. Thus Table 5.9

Table 5.11   Estimated Coefficients, Standard Errors, z-Scores, Two-Tailed p-Values and 95% Confidence Intervals for the Preliminary Final Proportional Hazards Model for the UIS (n = 575)

Variable	Coeff.	Std. Err.	z	$P>\|z\|$	95% CIE
AGE	−0.041	0.010	−4.18	<0.001	−0.061, −0.022
BECKTOTA	0.009	0.005	1.76	0.078	−0.001, 0.018
NDRUGFP1	−0.574	0.125	−4.59	<0.001	−0.820, −0.329
NDRUGFP2	−0.215	0.049	−4.42	<0.001	−0.310, −0.119
IVHX_3	0.228	0.109	2.10	0.036	0.015, 0.441
RACE	−0.467	0.135	−3.47	0.001	−0.731, −0.203
TREAT	−0.247	0.094	−2.62	0.009	−0.432, −0.062
SITE	−1.317	0.531	−2.48	0.013	−2.359, −0.275
AGEXSITE	0.032	0.016	2.02	0.044	0.001, 0.064
RACEXSITE	0.850	0.248	3.43	0.001	0.365, 1.336

Log-likelihood = −2630.418

contains all possible pairs of variables. The interaction terms are formed as the arithmetic product of the pair of variables. The interactions involving the number of previous drug treatments are formed using the two terms obtained from the method of fractional polynomials. Three interactions are identified as being significant, $p < 0.05$: age and number of previous drug treatments, age and site and race and site. These interactions were added to the preliminary main effects model in Table 5.8, and the resulting fitted model is shown in Table 5.10.

The p-values for the Wald tests suggest that the interaction between age and number of previous drug treatments may not be important in the larger interactions model. We note that the two coefficients for this interaction are of approximately the same magnitude, but with opposite signs. This suggests that these two variables may be highly colinear. To explore this, we fit the model containing only AGEXNDRUGFP1. The Wald statistic for its coefficient was significant ($p = 0.013$). However, the Wald statistic for AGEXSITE in this model was not significant ($p = 0.107$). To further explore the interactions with age, we fit a model containing only AGEXSITE. In this model, the Wald statistic for the coefficient was signifcant ($p = 0.044$). Thus it appears that we have a choice between two models, one containing only AGEXNDRUGFP1, the other containing only AGEXSITE. Since the latter model is simpler and easier to interpret, we define our preliminary final model as the one presented in Table 5.11. The model will not be identified as the final model until its fit and adherence to model assumptions has been checked. Before this important topic is considered in detail in Chapter 6, we consider stepwise and best subsets selection of covariates, two statistical methods for selection of main effect variables.

5.3 STEPWISE SELECTION OF COVARIATES

Covariates may be selected for inclusion in a proportional hazards regression model using stepwise selection methods that operate in an identical manner to those used in other regression models, such as linear or logistic regression. The statistical test used as a criterion is most often the partial likelihood ratio test. However, the score test and Wald test are also often used by software packages. From the conceptual point of view, it does not matter which test is actually used. However, the partial likelihood ratio test is the best of the three tests and should be used when there is a choice.

We assume familiarity with stepwise methods from either linear or logistic regression, thus the presentation here will not be detailed. Detailed descriptions of stepwise selection of covariates may be found in Hosmer and Lemeshow (1989), Chapter 4, for logistic regression and in Ryan (1997), Chapter 7, for linear regression.

We begin by describing the full stepwise selection process, which consists of forward selection followed by backward elimination. The forward selection process adds to the model the covariate that is most statistically significant among those not in the model. The backward elimination process checks each covariate in the model for continued significance. Two variations of the full stepwise procedure available in most software packages are to use forward selection only or backward elimination only.

Most software packages that have implemented stepwise selection of covariates for the proportional hazards model treat all the covariates available for selection as if they were continuous. This implies that to consider nominal scale covariates with more than two levels correctly, one must create and include in the list of covariates the individual design variables. An additional problem is that the individual design variables are not considered as a unit, and the program may select a subset of them. When this occurs, there has been an implicit recoding of the covariate and the user must make sure that the recoding makes clinical sense or must add the unselected design variables into the model when it is examined in more detail. We will return to this point in the example. The stepwise procedure will be described, as it is currently implemented by default, using single degree-of-freedom tests for entry and removal of covariates.

Step 0: Assume that there are p possible covariates, denoted x_j, $j = 1,2,...,p$. This list is assumed to include continuous covariates as well as all design variables for nominal scaled covariates. Thus, for example, a particular x_j might stand for age or for the design variable for IV drug use at level 2. At step 0 the partial likelihood ratio test and its p-value for the significance of each covariate is computed by comparing the log partial likelihood of the model containing x_j to the log partial likelihood of model zero (i.e., the model containing no covariates). This test statistic is

$$G^{(0)}(j) = -2[L^{(0)}(j) - L(0)], \quad j = 1,2,...,p, \qquad (5.1)$$

where $L(0)$ is equal to the log partial likelihood of model zero, the no covariate model, and $L^{(0)}(j)$ is equal to the log partial likelihood of the model containing covariate x_j. The test's significance level is

$$p^{(0)}(j) = \Pr\left[\chi^2(1) \geq G^{(0)}(j)\right] . \tag{5.2}$$

Evaluation of (5.1) and (5.2) requires fitting p separate proportional hazards models. The parenthesized superscript in (5.1) and (5.2) denotes the step, and j indexes the particular covariate. The candidate for entry into the model at step 1 is the most significant covariate and is denoted by x_{e_1}, where

$$p^{(0)}(e_1) = \min_{j}\left[p^{(0)}(j)\right] . \tag{5.3}$$

For the variable x_{e_1} to be entered into the model, its p-value must be smaller than some pre-chosen criterion for significance, denoted p_E. If the variable selected for entry is significant (i.e., $p^{(0)}(e_1) < p_E$), then the program goes to step 1; otherwise it stops.

Step 1: This step begins with variable x_{e_1} in the model. Then $p-1$ new proportional hazards models (each including one remaining variable along with x_{e_1}) are fit, and the results are used to compute the partial likelihood ratio test of the fitted two-variable model to the one-variable model containing only x_{e_1},

$$G^{(1)}(j) = -2\left[L^{(1)}(j) - L\left(x_{e_1}\right)\right], \quad j = 1, 2, \ldots, p \text{ and } j \neq e_1, \tag{5.4}$$

where the values of the deviance in (5.4) are -2 times the partial log likelihoods of the respective models. The p-value for the test of the significance of adding x_j to the model containing x_{e_1} is

$$p^{(1)}(j) = \Pr\left[\chi^2(1) \geq G^{(1)}(j)\right] . \tag{5.5}$$

The variable selected as the candidate for entry at step 2 is x_{e_2} where

$$p^{(1)}(e_2) = \min_{j \neq e_1}\left[p^{(1)}(j)\right] . \tag{5.6}$$

If the selected covariate x_{e_2} is significant, $p^{(1)}(e_2) < p_E$, then the program goes to step 2; otherwise it stops.

Step 2: This step begins with both x_{e_1} and x_{e_2} in the model. During this step, two different evaluations occur. The step begins with a backward elimination check for the continued contribution of x_{e_1}. That is, does x_{e_1} still contribute to the model after x_{e_2} has been added? This is essentially an evaluation of (5.4) and (5.5) with the roles of the two variables reversed. From an operational point of view, we choose a different significance criterion for this check, denoted p_R. We choose this value such that $p_R > p_E$ to eliminate the possibility of entering and removing the same variable in an endless number of successive steps. Assume the variable entered at step 1 is still significant.

The program fits $p - 2$ proportional hazards models (each including one remaining variable along with x_{e_1} and x_{e_2}) and computes the partial likelihood ratio test and its p-value for the addition of the new covariate to the model, namely

$$G^{(2)}(j) = -2\left[L^{(2)}(j) - L\left(x_{e_1}, x_{e_2}\right)\right], \quad j = 1, 2, \ldots, p \text{ and } j \neq e_1, e_2$$

and

$$p^{(2)}(j) = \Pr\left[\chi^2(1) \geq G^{(2)}(j)\right].$$

The covariate x_{e_3} selected for entry at step 3 is the one with the smallest p-value, that is,

$$p^{(2)}(e_3) = \min_{j \neq e_1, e_2}\left[p^{(2)}(j)\right].$$

The program proceeds to step 3 if $p^{(2)}(e_3) < p_E$; otherwise it stops.

Step 3: Step 3, if reached, is similar to step 2 in that the elimination process determines whether all variables entered into the model at earlier steps are still significant. The selection process then followed is identical to the selection part of earlier steps. This procedure is followed until the last step, step S.

Step S: At this step one of two things may happen: (1) all the covariates are in the model and none may be removed or (2) each covariate not in the model has $p^{(S)}(j) > p_E$. At this point, no covariates are selected for entry and none of the covariates in the model may be removed.

The number of variables selected in any application will depend on the strength of the associations between covariates and survival time and the choice of p_E and p_R. Due to the multiple testing that occurs, it is nearly impossible to calculate the actual statistical significance of the full stepwise process. Research in linear regression by Bendel and Afifi (1977) and in discriminant analysis by Costanza and Afifi (1979) indicates that use of $p_E = 0.05$ excludes too many important covariates and that one should choose a level of significance of 15 percent. In many applications it may make sense to use 25–50 percent to allow more variables to enter than will ultimately be used and then narrow the field of selected variables using $p < 0.15$ to obtain a multivariable model for further analysis. An unavoidable problem with any stepwise selection procedure is the potential for the inclusion of "noise" covariates and the exclusion of important covariates. One must always examine the variables selected and excluded for basic scientific plausibility.

The model at this point is likely to contain continuous covariates, and these should be examined carefully for linearity using the previously discussed methods. The next step is to see if there are any interactions which significantly improve the model. The procedure for stepwise selection is to use as candidate variables a list of plausible interactions among the main effects previously identified during the initial stepwise model building. One must begin with a model containing all the main effects, and the final model is selected using usual levels of statistical significance.

As an example of stepwise selection we consider covariates in the UIS. The list of candidate variables includes: age, Beck score, number of previous drug treatments, race, treatment, site, three design variables for previous heroin or cocaine use and two design variables for previous IV drug use, for a total of 11 covariates. The exact order of variable selection will depend on whether one uses the partial likelihood ratio test, the score test or the Wald test. The results presented in Table 5.12 were obtained using the partial likelihood ratio test. The variables selected are the same as those selected by the score test.

For illustrative purposes, the results presented in Table 5.12 were obtained using entry and removal p-values of $p_E = 0.5$ and $p_R = 0.8$. There were a total of 10 steps, counting step 0. At step 0, the variable with the smallest p-value was the number of previous drug treatments, NDRUGTX, with $p = 0.004$. Since this value is smaller than $p_E = 0.5$, the variable enters the model at step 1. At step 1 AGE has the smallest p-value with $p = 0.0064$ and it is smaller than $p_E = 0.5$, so AGE enters the model at step 2. At step 2, both AGE and NDRUGTX have p-values to remove which are less than $p_R = 0.8$ and thus remain in the model. Among the variables not in the model, the design variable for IV drug use at level 3, IVHX_3 (a recent user), has the smallest p-value and it is less than the criteria for entry into the model. Then the program goes to step 3, where the three-variable model is fit. All the p-values to remove are less than 0.8 and no variables are taken from the model. The variable with the smallest p-value for entry is treatment, TREAT, with $p = 0.0107$, which is less then 0.5. The program then goes to step 4 and fits the four-variable model.

This process of fitting, checking for continued significance, and selection continues until step 9. At this step, each of the nine variables in the model has a p-value to remove which is less than 0.8, and the p-values to enter for the two variables not in the model exceed 0.5. Therefore, the program terminates the selection process at step 9.

We use the results in Table 5.12 with a significance level of 0.15 to identify the preliminary main effects model by proceeding sequentially to the next step, as long as the smallest p-value for entry is less than 0.15. The first time it exceeds 0.15 is at step 6. At this step, the potential variable for inclusion is the design variable for previous IV drug use at level 2, IVHX_2 (previous user). Using the 15 percent rule, we would take as our model the one fit at step 6 which contains NDRUGTX, AGE, IVHX_3, TREAT, RACE and BECKTOTA. Inclusion of only IVHX_3 implies a recoding of previous IV drug use to 1 = recent, 0 = not recent. If we were performing variable selection for the first time, we might

Table 5.12   Results of Stepwise Selection of Covariates, p-Value for Entry to the Right and p-Value to Remove to the Left of the Solid Line in Each Row for the UIS (n = 575).   Columns are in Order of Entry.

Step	NDRUGTX	AGE	IVHX_3	TREAT	RACE	BECKTOT	IVHX_2	HERC_3	SITE	HERC_2	HERC_4
0	0.0004	0.0759	0.001	0.011	0.005	0.035	0.883	0.058	0.294	0.044	0.434
1	0.0004	0.0064	0.010	0.009	0.013	0.036	0.832	0.109	0.426	0.113	0.832
2	<.0001	0.0064	0.001	0.007	0.016	0.046	0.997	0.037	0.327	0.048	0.618
3	0.0008	0.0005	0.001	0.011	0.058	0.102	0.090	0.275	0.790	0.481	0.577
4	0.0005	0.0004	0.001	0.011	0.081	0.109	0.101	0.232	0.631	0.431	0.476
5	0.0010	0.0007	0.003	0.015	0.081	0.085	0.148	0.258	0.382	0.528	0.526
6	0.0009	0.0011	0.007	0.015	0.064	0.085	0.179	0.208	0.420	0.528	0.450
7	0.0016	0.0005	0.003	0.016	0.093	0.102	0.179	0.279	0.357	0.537	0.545
8	0.0013	0.0004	0.014	0.014	0.098	0.087	0.238	0.280	0.348	0.676	0.908
9	0.0014	0.0004	0.027	0.012	0.069	0.095	0.213	0.273	0.348	0.688	0.968

choose the model at step 7 to avoid having to recode this variable. If we do this, the next phase of the model building process would be the steps we went through during the purposeful selection of covariates. Recall that that process suggested the same set of covariates.

At this point, we examine the scale of the continuous covariates in the model following the same procedure illustrated in the previous section. This analysis yields the same model as in Table 5.8 if we add SITE for the same reasons it was included when we discussed purposeful selection. Stepwise selection of interactions proceeds using as candidate variables the interactions listed in Table 5.9. At step 0, the model contains all the main effects, the model in Table 5.8. Since all the stepwise selection programs perform single degree-of-freedom selection tests, there are a total of 27 individual interaction terms to choose from at step 0, since the process of checking the scale of the continuous covariates has led to transforming NDRUGTX into two nonlinear terms.

For illustrative purposes we discuss the variables selected using 0.15 as the significance level for entry. We use a smaller level of significance for entry as we wish to include in the model only those interactions that are significant, since confounding is not an issue when selecting interactions. Three interactions were identified as being important: At step 1, the RACE by SITE interaction entered the model with $p = 0.001$; at step 2, the AGE by NDRUGFP2 interaction entered the model with $p = 0.006$; and at step 3, the AGE by SITE interaction entered the model

Table 5.13   Estimated Coefficients, Standard Errors, z-Scores, Two-Tailed p-Values and 95% Confidence Intervals for the Preliminary Final Proportional Hazards Model for the UIS ($n = 575$)

Variable	Coeff.	Std. Err.	z	$P>|z|$	95% CIE
AGE	−0.054	0.012	−4.53	<0.001	−0.077, −0.031
BECKTOTA	0.010	0.005	2.07	0.038	0.001, 0.020
RACE	−0.483	0.135	−3.59	<0.001	−0.747, −0.219
TREAT	−0.222	0.094	−2.37	0.018	−0.406, −0.039
SITE	−0.278	0.122	−2.28	0.023	−0.517, −0.039
IVHX_3	0.234	0.108	2.17	0.030	0.023, 0.445
NDRUGFP1	−0.838	0.160	−5.25	<0.001	−1.151, −0.525
NDRUGFP2	−0.229	0.049	−4.70	<0.001	−0.325, −0.134
RACEXSITE	0.897	0.247	3.63	<0.001	0.412, 1.382
AGEXNDRUGFP1	0.007	0.003	2.77	0.006	0.002, 0.012

Log-likelihood = −2628.739.

with $p = 0.113$. Using the 5 percent level of significance, we would choose the model at step 2 containing all the main effects, the RACE by SITE interaction and, surprisingly, the AGE by NDRUGFP2 interaction.

At this point we have a somewhat complicated model to sort out. Subsequent analyses (details are not presented) reveal that: (1) there is a significant interaction of AGE with either one of the two fractional polynomial variables for number of previous drug treatments, (2) it does not seem to matter which of the two fractional polynomial variables AGE interacts with as both models have almost the same log partial likelihood and (3) the AGE by SITE interaction is significant only if the AGE by NDRUGFP1 or the AGE by NDRUGFP2 interaction is not included in the model (see Table 5.11).

Thus it appears that there are two possible models, each with the same eight main effect terms and two interaction terms: (1) the model in Table 5.11 containing the AGE $\times$ SITE and RACE $\times$ SITE interactions and (2) the model in Table 5.13 containing the RACE $\times$ SITE and AGE $\times$ NDRUGFP1 interactions. We defer deciding which of the two models to use for inferential purposes until after we have examined each for adherence to model assumptions, goodness-of-fit, and influential observations. From a practical point of view, we favor the model in Table 5.11 as it does not include any interaction terms involving fractional polynomials, making it easier to interpret. However, if nothing changes, the estimate of the effect of treatment is about the same in both models, so from that point of view we could use either model.

5.4 BEST SUBSETS SELECTION OF COVARIATES[1]

In the previous section we discussed stepwise selection of covariates. The advantage of stepwise selection is that most analysts are familiar with its use from other regression modeling settings and it is available in most major software packages. A disadvantage is that the procedure considers only a small number of the total possible models that can be formed from the covariates. The method of best subsets selection provides a computationally efficient way to screen all possible models.

The conceptual basis for best subsets selection of covariates in a proportional hazards regression is the same as in linear regression. The

---

[1] Implementation of the methods in this section requires matrix calculations not automatically performed by software packages. If one is familiar with simple matrix algebra and the software package has matrix capabilities, then they are relatively easy to perform.

procedure requires a criterion to judge a model. Given the criterion, the software screens all models containing q covariates and reports the covariates in the best, say 5, models for $q = 1, 2, \ldots, p$, where p denotes the total number of covariates.

Software to implement best subsets normal errors linear regression is readily available and has been used to provide best subsets selection capabilities for non-normal errors linear regression models such as logistic regression, see Hosmer and Lemeshow (1989, Chapter 4). There are three requirements to use the method described by Hosmer and Lemeshow: (1) It must be possible to obtain estimates of the coefficients of the model containing all p covariates from a weighted linear regression where the dependent variable is of the form

$$\mathbf{x}'\hat{\boldsymbol{\beta}} + \widehat{\text{weight}} \times (\widehat{\text{residual}}),$$

(2) the weight must be an easily computed function of the variance of the residual and (3) both weight and residual must be easily computed functions of the estimated coefficients and covariates. Only requirement 1 is satisfied by the proportional hazards regression model when it is fit using the partial likelihood. The difficulty is that even though the partial likelihood, see (3.17), is a product of n terms, the terms are not independent of each other. Each "subject" may contribute information to more than one term in the product, that is, "subjects" appear in every risk set until they fail or are censored. Thus Hosmer and Lemeshow's method may not be used to perform best subsets proportional hazards regression. We do not want to dwell on this point, but feel that it is important to explain why this well-known and easily used approach is not appropriate in this setting.

Kuk (1984) described how best subsets selection in a proportional hazards regression model may be performed with a normal errors linear regression best subsets program if the program allows input of the data via a covariance matrix. Kuk's method is related to a general method described by Lawless and Singhal (1978), which requires special software. We will illustrate Kuk's method using BMDP9R, but any best subsets linear regression program which permits a covariance matrix as data input can be used.

The computational steps one must perform to use Kuk's method are as follows:

(1) Fit the proportional hazards model containing all p covariates. This model must contain all the design variables for nominal scale

covariates coded at more than two levels. As was the case in stepwise selection, these related design variables will be considered as distinct binary variables in the best subsets selection method.

(2) Obtain the estimated covariance matrix of the estimated coefficients, denoted as $\widehat{\text{Var}}(\hat{\beta})$, and obtain its inverse, denoted

$$\mathbf{I}(\hat{\beta}) = \left[\widehat{\text{Var}}(\hat{\beta})\right]^{-1}.$$

This matrix is the observed information matrix. If the program does not provide the observed information matrix, then one must compute its value.

(3) Compute the $p \times 1$ matrix $\mathbf{B} = \mathbf{I}(\hat{\beta})'\hat{\beta}$ and the 1×1 matrix $\mathbf{C} = \hat{\beta}'\mathbf{I}(\hat{\beta})\hat{\beta}$.

(4) Use the matrices computed in steps 2 and 3 to form the $(p+1) \times (p+1)$ matrix

$$\mathbf{A} = \begin{bmatrix} \mathbf{I}(\hat{\beta}) & \mathbf{B} \\ \mathbf{B}' & (n-p)+\mathbf{C} \end{bmatrix}.$$

[There is a small mistake in Kuk (1984) in that he adds $(n-p-1)$ to $\mathbf{C}$.]

(5) Verify that the matrix $\mathbf{A}$ is correct. This may be done by performing linear regression with covariance matrix input, $\mathbf{A}$, declaring the $(p+1)$st variable as the dependent variable and assigning the names used in fitting the proportional hazards regression model in step 1 to the first p variables. The estimated coefficients and estimated standard errors of the estimated coefficients obtained from the linear regression output should be equal to those computed in step 1 from the proportional hazards regression model. The mean residual sum-of-squares should be equal to $n-1$. [Another small mistake in Kuk (1984) is that he states that this mean square is equal to 1.0.]

(6) Use a best subsets linear regression program with data set up as in step 5. Select Mallow's C [Mallows (1973)] as the criterion for best subsets [see Ryan (1997, Chapter 7) for a discussion of the use of Mallow's C in normal errors linear regression modeling].

The result of using these computational tricks is that best subsets are chosen using the values of multivariable Wald tests obtained after fitting the full p variable proportional hazards model, that is, the model fit in step 1. We will apply best subsets selection to the 11 possible main ef-

Table 5.14 Five Best Models Identified Using Mallow's C. Model Covariates, Mallow's C, the Wald Test for the Excluded Covariates, Its Degrees-of-Freedom and p-Value for the UIS $(n = 575)$

Model	Model Covariates	C	W	df	p
1	AGE, BECK, NDRUGTX, IVHX_3, RACE, TREAT	5.06	4.06	5	0.541
2	AGE, BECK, NDRUGTX, IVHX_2, IVHX_3, RACE, TREAT	5.21	2.21	4	0.697
3	AGE, BECK, NDRUGTX, HER_3, IVHX_3, RACE, TREAT	5.48	2.48	4	0.648
4	AGE, BECK, NDRUGTX, IVHX_2, IVHX_3, TREAT	5.93	4.93	5	0.424
5	AGE, NDRUGTX, IVHX_2, IVHX_3, RACE, TREAT	5.94	4.94	5	0.423

fect variables used in the UIS. For sake of illustration, consider a possible model that excludes 4 variables: the three design variables for heroin or cocaine use and SITE. The significance of the excluded variables may be assessed by the partial likelihood ratio test comparing the full 11-variable model to the 7-variable model containing AGE, BECKTOTA, NDRUGTX, IVHX_2, IVHX_3, RACE and TREAT. An equivalent test is the multivariable Wald test for the coefficients of the 4 excluded variables obtained following the fit of the full 11-variable model. The value of this 4 degrees-of-freedom Wald test is 2.41 and the value of Mallow's C for this 7-variable model is $5.21 = 2.21 + (11 - 2 \times 4)$.

In order to establish the relationship between Mallow's C and the Wald statistic in general, denote the value of the multivariable Wald test for q variables excluded from the full p variable model by W_q. The multivariable Wald test was described in Chapter 3 and is distributed as chi-square with degrees-of-freedom equal to the number of coefficients hypothesized to be equal to zero. The value of Mallow's C reported by BMDP9R, in step 6 above, is

$$C = W_q + (p - 2q) . \tag{5.7}$$

As in linear regression, good models will be ones with small values of C. Under the hypothesis that the coefficients for the q variables excluded from the model are zero, the mean of the Wald test is approximately q. Thus, the mean of Mallow's C is approximately $p - q$, the number of variables in the model. This is the same reference standard for Mallow's C used in normal errors linear regression.

Table 5.14 reports a summary of the five best models obtained by performing the six-step procedure with the UIS data. The best model contains the same six covariates identified using both purposeful and stepwise selection methods. The second best model is the same as one alternative model identified by the stepwise method. The remaining three models suggest other possible sets of covariates. The values of Mallow's C are relatively homogeneous across the five models. This is also the case for the Wald test p-values. The covariates selected for these models suggest that any good model is going to contain age, number of previous drug treatments, a binary variable for recent IV drug use, and treatment. There are four other covariates suggested.

Since no one model appears to be superior to the other four, one possible strategy is to fit a multivariable model containing all eight covariates used in the five models in Table 5.14. Following the fit of this model, we would proceed as in purposeful selection to try and reduce the size of the model. Based on models fit in the section on purposeful selection of covariates, this process would return us to model 1 in Table 5.14. The next steps in model development are the same as those described and illustrated in the section on purposeful selection of covariates: assessment of the scale of continuous covariates and identification of interactions.

An alternative method for best subset selection is to mimic the approach used in stepwise selection and choose as best models those in which the covariates in the model are significant. Selection of covariates thus proceeds by inclusion rather than exclusion. The best models containing $p - q$ covariates are those with the largest values of a test of the significance of the model. Theoretically, one could use any one of the three equivalent tests: partial likelihood ratio, Wald or score test. The SAS package, PROC PHREG, has implemented this selection method using the score test. Models identified are, for each fixed number of covariates, the ones with the largest value of the score test.

The problem with using the score test for model significance is that it is difficult to compare models of different sizes since the score test tends to increase with the number of covariates in the model. One possible solution is to use the values of the score test to approximate the value of Mallow's C in (5.7). Let the score test for the model containing all p covariates be denoted S_p and the score test for the model containing a particular set of $p - q$ covariates be denoted S_{p-q}. The value of the score test for the exclusion of the q covariates from the full p variable model is approximately $S_q = S_p - S_{p-q}$. Since the Wald and score tests are equivalent, this suggests that an approximation to Mallow's C for a fitted model containing $p - q$ covariates is

$$C = S_q + (p - 2q). \tag{5.8}$$

We note that if covariate selection had been based on the partial likelihood ratio test instead of the Wald and score test, the value of C in (5.7) would be equal to the value in (5.8).

As an example, consider model 1 in Table 5.14, with $p - q = 11 - 5 = 6$ covariates in the model. The value of the score test for the significance of the 11-covariate model is $S_{11} = 49.35$, and the value of the score test for the significance of the 6-covariate model is $S_6 = 45.52$. The approximation to the score test for the addition of the 5 covariates to the 6-covariate model is

$$S_5 \cong S_{11} - S_6 = 49.35 - 45.52 = 3.83.$$

The value of the approximation to Mallow's C is

$$C = 3.83 + (11 - 2 \times 5) = 4.83$$

and the correct value from Table 5.14 is 5.06. The approximation is close, but certainly not perfect. Of more practical interest is what models would be selected as best using (5.8) in conjunction with the values of the score tests provided by SAS in PROC PHREG. The best five models using this approach are summarized in Table 5.15.

The results in Table 5.15 are quite similar to those in Table 5.14. The three best models are the same and the fourth best model in Table 5.15 is the same as the fifth best model in Table 5.14. Thus it appears that the approximation in (5.8) provides a useful way to rank order models containing different numbers of covariates when models have

Table 5.15 Five Best Models Identified Using the Score Test Approximation to Mallow's C. Model Covariates, Approximate Mallow's C and the Approximate Score Test for the Excluded Covariates for the UIS ($n = 575$)

Model	Model Covariates	C	S_q
1	AGE, BECK, NDRUGTX, IVHX_3, RACE, TREAT	4.83	3.83
2	AGE, BECK, NDRUGTX, IVHX_2, IVHX_3, RACE, TREAT	5.00	2.00
3	AGE, BECK, NDRUGTX, HER_3, IVHX_3, RACE, TREAT	5.20	2.20
4	AGE, NDRUGTX, IVHX_2, IVHX_3, RACE, TREAT	5.43	4.43
5	AGE, NDRUGTX, IVHX_3, RACE, TREAT	5.94	6.52

been selected using the score test for model significance.

One point that must be kept firmly in mind when using procedures such as stepwise or best subsets selection to identify possible model covariates is that the results should be taken as suggestions for models to be examined in more detail. One cannot rule out the possibility that these methods may reveal new and interesting associations, but the collection of covariates must make clinical sense to the researchers. The statistical selection procedures suggest, but do not dictate, what the model might be.

5.5 NUMERICAL PROBLEMS

The software available in the major statistical packages for fitting the proportional hazards model is easy to use and, for the most part, contains checks and balances that warn the user of impending numerical disasters. However, there are certain configurations of data that cause numerical difficulties that may not produce a suitable warning to the user. The problem of *monotone likelihood* described by Bryson and Johnson (1981) is one such problem. This problem in a survival analysis is similar to the occurrence of a zero frequency cell in a two by two contingency table or when the distributions of a continuous covariate is completely separated by the binary outcome variable in logistic regression. The problem occurs in a proportional hazards regression when the rank ordering of the covariate and the survival times are the same. That is, at each observed survival time the subject who fails has the largest (smallest) value of one of the covariates among the subjects in the risk set.

To illustrate the problem, we created a hypothetical data set containing 100 observations of survival time in days, truncated at one year with approximately 30 percent of the observations censored. We created a dichotomous covariate whose value is equal to one if the observed

Table 5.16   Estimated Coefficient, Standard Error, z-Score, Two-Tailed p-Value and 95% Confidence Intervals for a Proportional Hazards Model Containing a Monotone Likelihood Covariate ($n = 100$)

Variable	Coeff.	Std. Err.	z	$P>|z|$	95% CIE
x	37.08	9.7E6	0.00	1.00	−1.92E7, 1.92E7

Log-likelihood = −209.74.

Table 5.17 Estimated Coefficients, Standard Errors, z-Scores, Two-Tailed p-Values and 95% Confidence Intervals for a Proportional Hazards Model Containing Two Highly Correlated Continuous Covariates ($n = 100$)

Variable	Coeff.	Std. Err.	z	$P>|z|$	95% CIE
x1	18.00	41.44	0.43	0.66	−63.2, 99.2
x2	−17.72	41.44	−0.43	0.66	−98.9, 63.5

Log-likelihood = −228.19.

survival time was less than the median and zero otherwise. The results of fitting the proportional hazards model are shown in Table 5.16, where the notation "9.7E6" means 9.7×10^6.

The estimated coefficient and its standard error are unreasonably large. The software also required 25 iterations to obtain this value. As in the case of logistic regression, any implausibly large coefficient and standard error is a clear indication of numerical difficulties. In this case, a graph of the covariate versus time would indicate the problem.

The example in Table 5.16 is a simple one since it involves a single covariate. In practice, the situation is likely to be more complex, with a combination of multiple covariates inducing the same effect. Bryson and Johnson (1981) show that certain types of linear combinations (e.g., a simple sum of the covariates) may yield monotone likelihood. In these situations the problem will manifest itself with unreasonably large coefficients and standard errors.

Extreme collinearity among the covariates is another possible problem. Most software packages contain diagnostic checks for highly correlated data, but clinically implausible results may be produced before the program's diagnostic switch is tripped. The results of fitting a proportional hazards model when the relationship between the two covariates is $x_2 = x_1 + u$, where u is the value of a uniformly distributed random variable on the interval (0, 0.01), are shown in Table 5.17. The correlation between the covariates is effectively 1.0, yet the program prints a result. Similar results were obtained until $u \sim U(0, 0.0001)$, at which point one of the covariates was dropped from the model by the program.

The bottom line is that it is ultimately the user of the software who is responsible for the results of an analysis. Any analysis producing "large" effect(s) or standard error(s) should be treated as a "mistake" until the involved covariate(s) are examined critically.

EXERCISES

For all exercises in this section involving analyses from the WHAS, use survival time defined by LENFOL, censoring defined by FSTAT, and data from all cohorts (i.e., ignore YEAR).

1. An important step in any model building process is assessing the scale of continuous variables in the model. The two continuous variables, AGE and CPK, in the WHAS present a challenge. Use the methods discussed in this chapter to assess the scale of AGE when it is the only covariate in a proportional hazards model. Repeat this process for CPK. In this problem, pay particular attention to the effect that a few subjects with either small or large values of the covariate can have on the methods for assessing the scale of a covariate.

2. Using the methods for model building discussed in this chapter, find the best model for estimating the effect of the covariates on long-term survival following hospitalization for an acute myocardial infarction in the WHAS. This process should include the following steps: variable selection, assessment of the scale of continuous variables and selection of interactions.

3. Present the results of the model selected in problem 2 in a table or tables that are suitable for publication in an applied journal. This presentation should include estimates of hazard ratios, with confidence intervals.

Note: Save any work done for problems 2 and 3 as there is a problem in Chapter 6 dealing with the assessment of fit of this model.

CHAPTER 6

Assessment of Model Adequacy

6.1 INTRODUCTION

Model-based inferences depend completely on the fitted statistical model. For these inferences to be "valid" in any sense of the word, the fitted model must provide an adequate summary of the data upon which it is based. A complete and thorough examination of a model's fit and adherence to model assumptions is just as important as careful model development.

The goal of statistical model development is to obtain the model which best describes the "middle" of the data. The specific definition of "middle" depends on the particular type of statistical model, but the idea is basically the same for all statistical models. In the normal errors linear regression model setting, we can describe the relationship between the observed outcome variable and one of the covariates with a scatter-plot. This plot of points for two or more covariates is often described as the "cloud" of data. In model development we find the regression line, plane or hyperplane that best fits/splits the cloud. The notion of "best" in this setting means that we have equal distances from observed points to fitted points above and below the surface. A "generic" main effects model with some nominal covariates, which treats continuous covariates as linear, may not have enough tilts, bends or turns to fit/split the cloud. Each step in the model development process is designed to tailor the regression surface to the observed cloud of data.

In most, if not all, applied settings the results of the fitted model will be summarized for publication using point and interval estimates of clinically interpretable measures. Examples of summary measures include the mean difference in linear regression, the odds ratio in logistic regression and the hazard ratio for the proportional hazards regression model. Since any summary measure is only as good as the model it is based on, it is vital that one evaluate how well the fitted regression surface describes the data cloud. This process is generally referred to as *assessing the adequacy of the model*; like model development, it involves a number of steps. Performing these in a thorough and conscientious manner will assure that the inferential conclusions based on the fitted model are the best and most valid possible.

The methods for assessment of a fitted proportional hazards model are essentially the same as for other regression models, and we assume some experience with these, particularly with logistic regression [see Hosmer and Lemeshow (1989, Chapter 5)]. Requirements for model assessment are: (1) methods for testing the assumption of proportional hazards, (2) subject-specific diagnostic statistics that extend the notions of leverage and influence to the proportional hazards model and (3) overall summary measures of goodness-of-fit.

6.2 RESIDUALS

Central to the evaluation of model adequacy in any setting is an appropriate definition of a residual. As we discussed in Chapter 1, the fact that the outcome variable is time to some event and the observed values may be incomplete or censored is what sets a regression analysis of survival time apart from other regression models. In earlier chapters we suggested that the semiparametric proportional hazards model is a useful model for data of this type and we described why and how it may be fit using the partial likelihood. This combination of data, model and likelihood make definition of a residual much more difficult in modeling survival time than is the case with other statistical models.

Consider a logistic regression analysis of a binary outcome variable. In this setting, values of the outcome variable are "present" ($y = 1$) or "absent" ($y = 0$) for all subjects. The fitted model provides estimates of the probability that the outcome is present (i.e., the mean of Y). Thus, a natural definition of the residual is the difference between the observed value of the outcome variable and that predicted by the model. This form of the residual also follows as a natural consequence of characterizing the observed value of the outcome as the sum of a systematic component and an error component. The two key assumptions in this definition of a residual are: (1) the value of the outcome is known and (2) the fitted model provides an estimate of the "mean of the dependent variable" or systematic component of the model. Since assumption 2 and, more than likely, assumption 1 are not true when using the partial likelihood to fit the proportional hazards model to censored survival data, there is no obvious analog to the usual "observed minus predicted" residual used with other regression models.

The absence of an obvious residual has lead to the development of several different residuals, each of which plays an important role in examining some aspect of the fit of the proportional hazards model. Most software packages provide access to at least one of these residuals. Only two packages, SAS and S-PLUS, have full residual analysis capabilities at this time. This situation is likely to change as other packages update and modify their proportional hazards routines.

We assume, for the time being, that there are p covariates and that the n independent observations of time, covariates and censoring indicator are denoted by the triplet $(t_i, \mathbf{x}_i, c_i)$, $i = 1, 2, ..., n$, where $c_i = 1$ for uncensored observations and is zero otherwise. Schoenfeld (1982) proposed the first set of residuals for use with a fitted proportional hazards model and packages providing them refer to them as the "Schoenfeld residuals." These are based on the individual contributions to the derivative of the log partial likelihood. This derivative for the kth covariate is shown in (3.21) and is repeated here as

$$\frac{\partial L_p(\boldsymbol{\beta})}{\partial \beta_k} = \sum_{i=1}^{n} c_i \left\{ x_{ik} - \frac{\sum_{j \in R(t_i)} x_{jk} e^{\mathbf{x}_j'\boldsymbol{\beta}}}{\sum_{j \in R(t_i)} e^{\mathbf{x}_j'\boldsymbol{\beta}}} \right\}$$

$$= \sum_{i=1}^{n} c_i \left\{ x_{ik} - \bar{x}_{w_i k} \right\}, \tag{6.1}$$

where

$$\bar{x}_{w_i k} = \frac{\sum_{j \in R(t_i)} x_{jk} e^{\mathbf{x}_j'\boldsymbol{\beta}}}{\sum_{j \in R(t_i)} e^{\mathbf{x}_j'\boldsymbol{\beta}}}. \tag{6.2}$$

The estimator of the Schoenfeld residual for the ith subject on the kth covariate is obtained from (6.1) by substituting the partial likelihood estimator of the coefficient, $\hat{\boldsymbol{\beta}}$, and is

$$\hat{r}_{ik} = c_i \left(x_{ik} - \hat{\bar{x}}_{w_i k} \right), \tag{6.3}$$

where

$$\hat{\bar{x}}_{w_i k} = \frac{\displaystyle\sum_{j \in R(t_i)} x_{jk} e^{x'_j \hat{\beta}}}{\displaystyle\sum_{j \in R(t_i)} e^{x'_j \hat{\beta}}}$$

is the estimator of the risk set conditional mean of the covariate. Since the partial likelihood estimator of the coefficient, $\hat{\beta}$, is the solution to the equations obtained by setting (6.1) equal to zero, the sum of the Schoenfeld residuals is zero. Software packages set the value of the estimate of the Schoenfeld residual to missing for subjects whose observed survival time is censored.

Grambsch and Therneau (1994) suggest that scaling the Schoenfeld residuals by an estimator of its variance yields a residual with greater diagnostic power than the unscaled residuals. Let the vector of p Schoenfeld residuals for the ith subject be denoted as

$$\hat{\mathbf{r}}'_i = (\hat{r}_{i1}, \hat{r}_{i2}, \ldots, \hat{r}_{ip}) ,$$

where $\hat{r}_{ik}$ is the estimator in (6.3), with the convention that $\hat{r}_{ik}$ = missing if $c_i = 0$. Let the estimator of the $p \times p$ covariance matrix of the vector of residuals for the ith subject, as reported in Grambsch and Therneau (1994), be denoted by $\hat{\text{Var}}(\hat{\mathbf{r}}_i)$, and the estimator is missing if $c_i = 0$. The vector of scaled Schoenfeld residuals is the product of the inverse of the covariance matrix times the vector of residuals, namely

$$\hat{\mathbf{r}}^*_i = \left[\hat{\text{Var}}(\hat{\mathbf{r}}_i) \right]^{-1} \hat{\mathbf{r}}_i . \tag{6.4}$$

The elements in the covariance matrix $\hat{\text{Var}}(\hat{\mathbf{r}}_i)$ are, in the current setting, a weighted version of the usual sum-of-squares matrix computed using the data in the risk set. For the ith subject, the diagonal elements in this matrix are

$$\hat{\text{Var}}(\hat{\mathbf{r}}_i)_{kk} = \sum_{j \in R(t_i)} \hat{w}_{ij} \left(x_{jk} - \hat{\bar{x}}_{w_i k} \right)^2 ,$$

and the off-diagonal elements are

$$\widehat{\mathrm{Var}}\left(\hat{\mathbf{r}}_i\right)_{kl} = \sum_{j \in R(t_i)} \hat{w}_{ij}\left(x_{jk} - \hat{\bar{x}}_{w_i k}\right)\left(x_{jl} - \hat{\bar{x}}_{w_i l}\right)$$

where

$$\hat{w}_{ij} = \frac{e^{x'_j \hat{\beta}}}{\sum_{l \in R(t_i)} e^{x'_l \hat{\beta}}} .$$

Grambsch and Therneau (1994) suggest use of an easily computed approximation for the scaled Schoenfeld residuals. This suggestion is based on their experience that the matrix, $\widehat{\mathrm{Var}}\left(\hat{\mathbf{r}}_i\right)$, tends to be fairly constant over time. If this matrix is constant, its inverse may be approximated by multiplying the estimator of the covariance matrix of the estimated coefficients by the number of events (i.e., the observed number of uncensored survival times m),

$$\left[\widehat{\mathrm{Var}}\left(\hat{\mathbf{r}}_i\right)\right]^{-1} = m\widehat{\mathrm{Var}}\left(\hat{\beta}\right).$$

The approximate scaled Schoenfeld residuals are the ones computed by software packages, namely

$$\hat{\mathbf{r}}_i^* = m\widehat{\mathrm{Var}}\left(\hat{\beta}\right)\hat{\mathbf{r}}_i. \tag{6.5}$$

Subsequent references to the scaled Schoenfeld residuals, $\hat{\mathbf{r}}_i^*$, will mean the approximation in (6.5), not the true scaled residual in (6.4).

The counting process approach is an extremely useful and powerful tool for studying the proportional hazards model. Most descriptions of it, however, including those in statistical software manuals, are difficult to understand without knowledge of calculus. In this section and those that follow we try to present the counting process results in an intuitive and easily understood manner. An expanded introduction to the counting process approach is presented in Appendix 2. A complete development of the theory as well as applications to other settings may be found in Fleming and Harrington (1991) and Andersen, Borgan, Gill and Keiding (1993).

Assume that we follow a single subject with covariates denoted by $\mathbf{x}$ from time "zero" and that the event of interest is death. We could use as the outcome any other event that can occur only once or the first occurrence of an event such as drug use. The counting process representation of the proportional hazards model is a linear-like model that "counts" whether the event occurs (e.g., the subject dies) at time t. The basic model is

$$N(t) = \Lambda(t, \mathbf{x}, \boldsymbol{\beta}) + M(t),\qquad (6.6)$$

where the function $N(t)$ is the "count" that represents the observed part of the model, the function $\Lambda(t, \mathbf{x}, \boldsymbol{\beta})$ is the "systematic component" of the model, and the function $M(t)$ is the "error component."

The function $N(t)$ is defined to be equal to zero until the exact time the event occurs and is equal to one thereafter. If the total length of follow-up is one year, and our subject dies on day 200, then

$$N(t) = \begin{cases} 0 \text{ for } t < 200, \\ 1 \text{ for } t \geq 200. \end{cases}$$

If the subject does not die during the one year of follow-up, then the count is always zero, $N(t) = 0$. Hence, the maximum value of the count function occurs at the end of follow-up of the subject and is equal to the value of the censoring indicator variable.

The systematic component of the model is, as we show in Appendix 2, equal to the cumulative hazard at time t under the proportional hazards model,

$$\Lambda(t, \mathbf{x}, \boldsymbol{\beta}) = H(t, \mathbf{x}, \boldsymbol{\beta}),$$

until follow-up ends on the subject and it is equal to zero thereafter. Thus, the value of the function for a subject who either dies or is censored on day 200 is

$$\Lambda(t, \mathbf{x}, \boldsymbol{\beta}) = \begin{cases} e^{\mathbf{x}'\boldsymbol{\beta}} H_0(t) & \text{for } t < 200, \\ e^{\mathbf{x}'\boldsymbol{\beta}} H_0(200) & \text{for } t \geq 200. \end{cases}$$

where $H_0(t)$ is the cumulative baseline hazard function. It follows that the maximum value for the systematic component also occurs at the end of follow-up, regardless of whether the event occurred. The function $M(t)$ in (6.6) is, under suitable mathematical assumptions, called a martingale and plays the role of the error component. It has many of the same properties that error components in other models have, in particular its mean is zero under the correct model. If we rearrange (6.6), $M(t)$ may be expressed in the form of a "residual" as

$$M(t) = N(t) - \Lambda(t, \mathbf{x}, \boldsymbol{\beta}). \tag{6.7}$$

The quantity in (6.7) is called the *martingale residual*. In theory it has a value at each time t, but the most useful choice of time at which to compute the residual is the end of follow-up, yielding a value for the ith subject of

$$M(t_i) = N(t_i) - H(t_i, \mathbf{x}, \boldsymbol{\beta})$$
$$= c_i - H(t_i, \mathbf{x}, \boldsymbol{\beta}), \tag{6.8}$$

since $\Lambda(t_i, \mathbf{x}, \boldsymbol{\beta}) = H(t_i, \mathbf{x}, \boldsymbol{\beta})$. For ease of notation, let $M_i = M(t_i)$. The estimator obtained by substituting the value of the partial likelihood estimator of the coefficients, $\hat{\boldsymbol{\beta}}$, is

$$\hat{M}_i = c_i - \hat{H}\left(t_i, \mathbf{x}, \hat{\boldsymbol{\beta}}\right). \tag{6.9}$$

The estimator $\hat{H}\left(t_i, \mathbf{x}, \hat{\boldsymbol{\beta}}\right)$ is defined in (3.41). We used this residual, (6.9), in Chapter 5 for the graphical methods to assess the scale of a continuous covariate.

The residual in (6.9) has also been called the Cox–Snell or modified Cox–Snell residual, see Cox and Snell (1968) and Collett (1994). This terminology is due to the work of Cox and Snell, who showed that the values of $\hat{H}\left(t_i, \mathbf{x}, \hat{\boldsymbol{\beta}}\right)$ may be thought of as observations from a censored sample with an exponential distribution and parameter equal to 1.0. Unfortunately, this distribution theory has not proven to be as useful for model evaluation as the theory derived from the counting process approach.

Using the counting process approach, the expressions in (6.7) and (6.8) are a completely natural way to define a residual. To see why it also makes sense to consider (6.8) as a residual in the proportional hazards regression model, assume for ease of notation that there are no ties and that the value of the baseline hazard at time t_i may be expressed as

$$h_0(t_i) = \frac{c_i}{\sum_{j \in R(t)} e^{x'_j \beta}},$$ (6.10)

and the expression for the cumulative baseline hazard is

$$H_0(t_i) = \sum_{t_j \le t_i} h_0(t_j).$$ (6.11)

The Breslow estimator of the cumulative baseline hazard is obtained from (6.11) by substituting the value of the partial likelihood estimator of the coefficients, $\hat{\beta}$. Under these assumptions, the derivative in (6.1) may be expressed as

$$\frac{\partial L_p(\beta)}{\partial \beta_k} = \sum_{i=1}^{n} x_{ik} \left[c_i - H(t_i, \mathbf{x}, \beta) \right].$$ (6.12)

The expression in (6.12) is similar to the equations obtained for other models, such as linear and logistic regression, in that it expresses the partial derivative as a sum of the value of the covariate times an "observed minus expected" residual.

The p equations obtained by setting (6.1) equal to zero for each covariate are called the *score equations* and some authors, for example, Collett (1994), call the Schoenfeld residual in (6.3) the *score residual*. However, an entirely different residual of the same name is the one currently calculated by software packages. This residual is obtained by expressing the martingale residual representation shown in (6.12) in a slightly different form. The score equation for the kth covariate may be expressed as

$$\frac{\partial L_p(\beta)}{\partial \beta_k} = \sum_{i=1}^{n} L_{ik}.$$ (6.13)

The expression for L_{ik} is somewhat complex. Readers who are willing to accept without further elaboration that the estimator of L_{ik} is the score *process* residual provided by software packages may skip the next paragraph where we describe L_{ik} in more detail.

The score process residual for the ith subject on the kth covariate in (6.13) may be expressed as

$$L_{ik} = \sum_{j=1}^{n} \left(x_{ik} - \bar{x}_{w_j k} \right) dM_i(t_j). \tag{6.14}$$

The mean in the expression, $\bar{x}_{w_j k}$, is the value of (6.2) computed at t_j. The quantity $dM_i(t_j)$ is the change in the martingale residual for ith subject at time t_j and is

$$dM_i(t_j) = dN_i(t_j) - Y_i(t_j) e^{x_i' \beta} h_0(t_j). \tag{6.15}$$

The first part of (6.15), $dN_i(t_j)$, is the change in the count function for the ith subject at time t_j. This will be always equal to zero for censored subjects. For noncensored subjects, it will be equal to zero except at the actual observed survival time, when it will be equal to one. That is, $dN_i(t_i) = 1$ for noncensored subjects. In the second part of (6.15), the function $Y_i(t_j)$ is called the *at risk process* and is defined as follows:

$$Y_i(t_j) = \begin{cases} 1 \text{ if } t_i \geq t_j \\ 0 \text{ if } t_i < t_j \end{cases}$$

and $h_0(t_j)$ is the value of (6.10) evaluated at t_j. An expanded computational formula yields the estimator

$$\hat{L}_{ik} = c_i \times \left(x_{ik} - \hat{\bar{x}}_{w_j k} \right) - x_{ik} \times \hat{H}\left(t_i, \mathbf{x}, \hat{\boldsymbol{\beta}} \right) + e^{x_i' \hat{\beta}} \sum_{t_j \leq t_i} \hat{\bar{x}}_{w_j k} \frac{c_j}{\sum_{l \in R_j} e^{x_i' \hat{\beta}}}. \tag{6.16}$$

Let the vector of p score process residuals for the ith subject be denoted as

$$\hat{\mathbf{L}}_i' = \left(\hat{L}_{i1}, \hat{L}_{i2}, \ldots, \hat{L}_{ip} \right). \tag{6.17}$$

Before moving on, we provide a brief summary of residuals. The martingale residual, $\hat{M}_i$ in (6.9) has the form typically expected of a residual in that it resembles the difference between an observed outcome and a predicted outcome. The other three residuals (score process, Schoenfeld and scaled Schoenfeld) are covariate-specific. Every subject has a value of the score process residual for the kth covariate, $\hat{L}_{ik}$ in (6.16), but the Schoenfeld residual in (6.3) and the scaled Schoenfeld residual in (6.5) are defined only at the observed survival times. Thus, there will be m subjects with values for these residuals. Each of these residuals provides a useful tool for examining one or more aspects of model adequacy. [Barlow and Prentice (1988) consider variations in the martingale residual obtained by including other functions of time in (6.8). As of this time, none of their generalized residuals have been added to software packages.]

We now consider methods for verifying the proportional hazards assumption.

6.3 METHODS FOR ASSESSING THE PROPORTIONAL HAZARDS ASSUMPTION

The proportional hazards assumption is vital to the interpretation and use of a fitted proportional hazards model, as discussed in detail in Chapter 4. Specifically, the proportional hazards model has a log-hazard function of the form

$$\ln[h(t, \mathbf{x}, \boldsymbol{\beta})] = \ln[h_0(t)] + \mathbf{x}'\boldsymbol{\beta}. \tag{6.18}$$

This function has two parts, the log of the baseline hazard function, $\ln[h_0(t)]$, and the linear predictor, $\mathbf{x}'\boldsymbol{\beta}$. Methods for building the linear predictor part of the model are discussed in detail in Chapter 5. The proportional hazards assumption characterizes the model as a function of time, not of the covariates per se. Assume for the moment that the model contains a single dichotomous covariate. A graph of the log-hazard, (6.18), over time would produce two curves, one for $x = 0$, $\ln[h_0(t)]$, and one for $x = 1$, $\ln[h_0(t)] + \beta$. Regardless of how simple or complicated the baseline hazard function is, the vertical distance between these two curves at any point in time is β. This fact is the reason that the hazard ratio, $\exp(\beta)$, has such a simple and useful interpretation.

As a second example, suppose age is the only covariate in the model and that it is scaled linearly. Consider the graphs of the log-hazard function for age a and age $a+10$. If the coefficient, β, is positive, the upper to lower vertical distance between the two curves will be 10β at every point in time. Assessing the proportional hazards assumption is an examination of the extent to which the two curves are equidistant over time.

There are, effectively, an infinite number of ways the model in (6.18) can be changed to yield non-proportional hazard functions or log-hazard functions that are not equidistant. As a result, a large number of tests and procedures have been proposed. However, recent developmental work by Grambsch and Therneau (1994) and simulation comparisons by Ng'andu (1997) have shown that one easily performed test and an associated graph yield a powerful and effective method for examining this critical assumption.

Grambsch and Therneau (1994) consider an alternative to the model in (6.18), originally proposed by Schoenfeld (1982), that has the following specific form of time-varying coefficient:

$$\beta_j(t) = \beta_j + \gamma_j g_j(t), \tag{6.19}$$

where $g_j(t)$ is a specified function of time. The rationale behind this model is that the effect of a covariate may change over the period of follow-up. For example, the baseline value of a specific test may lose its relevance over time. The opposite could also occur, where a baseline measure is more predictive of survival later in follow-up. Under this model, Grambsch and Therneau show that the scaled Schoenfeld residuals in (6.4), and their approximation in (6.5), have, for the jth covariate, a mean at time t of approximately

$$E\left[r_j^*(t)\right] \cong \gamma_j g_j(t). \tag{6.20}$$

The result in (6.20) suggests that a plot of the scaled Schoenfeld residuals over time may be used to visually assess whether the coefficient γ_j is equal to zero and, if not, what the nature of the time dependence, $g_j(t)$, may be. Grambsch and Therneau derive a generalized least squares estimator of the coefficients and a score test of the hypothesis that they are equal to zero, given specific choices for the functions $g_j(t)$. In addition, they show that specific choices for the function yield previously proposed tests. For example, use of $g(t) = \ln(t)$ yields a model first suggested by Cox (1972) and a test by Gill and Schumacher (1987) discussed by Chappell (1992). With this function, the model in (6.19) is

$$\beta_j(t) = \beta_j + \gamma_j \ln(t),$$

and the linear predictor portion of the model in (6.19) is

$$\beta_j x_j + \gamma_j x_j \ln(t). \tag{6.21}$$

The form of the linear predictor in (6.21) suggests that another way to test the hypothesis that $\gamma_j = 0$ is via the partial likelihood ratio test, score test or Wald test obtained when the interaction $x_j \ln(t)$ is added to the proportional hazards model. The advantage of this approach over the generalized least squares score test proposed by Grambsch and Therneau is that it may be done using the model fitting software in many statistical software packages. One should note that when the interaction term, $x_j \ln(t)$, is included in the model the partial likelihood becomes much more complicated. Since the interaction is a function of time, its value must be recomputed for each term in the risk set at each observed survival time. The interaction term is not simply the product of the covariate and the subject's observed value of time.

Other functions of time have been suggested. Quantin, et al. (1996) propose using $g(t) = \ln[H_0(t)]$. Based on simulations reported in their paper, this test appears to have good power, but it is not as easy to compute as the test based on $g(t) = \ln(t)$. This is because the Breslow estimator of $H_0(t)$ must be computed and must be accessible at each observed survival time. Based on the simulations in Quantin et al. (1996) and Ng'andu (1997), the test with $g(t) = \ln(t)$ has power nearly as high as or higher than all other commonly used tests to detect reasonable alternatives to proportional hazards. For this reason, we consider only the model in (6.21). These same simulations show that the performance of the partial likelihood score test and the Grambsch and Therneau generalized least squares score test are essentially the same. Thus, we will use the more easily computed model-based forms of the test.

Before evaluating the fitted models from the UIS developed in Chapter 5 for proportional hazards, we consider the methods in some simpler models. In the case of models containing nominal scale covariates, a purely graphical bivariate assessment may be obtained from the plots of the "log-negative-log" of the within-group Kaplan-Meier estimator of the survivorship functions versus log-time. If the hazard functions are proportional, this plot should have parallel lines and the vertical distance between each line and that of the reference group should be approximately equal to the coefficients from the fit of the proportional hazards model. One disadvantage of this graphical approach is that it is a univariate method that may be used only for nominal scale covariates or grouped continuous covariates. In addition it is difficult to visually assess whether the plotted lines deviate significantly from being parallel, particularly when sample sizes are small.

As an example of a multivariable model, we consider a situation in which the model contains one dichotomous covariate, denoted d, and one continuous covariate, denoted x. This data setting is used in several examples. In each example, two models were fit: the main effects model containing d and x and a second model obtained by including the interactions of each covariate with time, $d\ln(t)$ and $x\ln(t)$. For numerical reasons it is preferable to center log-time about its mean and use $\left[\ln(t) - \overline{\ln}(t)\right]$ in the interaction. As noted above, models containing interactions with log-time are easily fit in many statistical software packages.

The data for the first example illustrate a model in which the hazard is proportional in both covariates. The results of fitting the main effects model and the model with interactions with log-time are shown in Table 6.1. The main effects model in Table 6.1 shows that both covariates are highly significant. The p-values for the Wald statistics for both interaction coefficients are not significant, suggesting that the hazard function may be proportional in the two covariates. The value of the partial likelihood ratio test for the addition of the two interaction variables is $G = 0.134$ and, with 2 degrees-of-freedom, the p-value is 0.94. This is confirmed by the graphs in Figure 6.1. The scaled residuals scatter in a

Table 6.1 Estimated Coefficients, Standard Errors, z-Scores, Two-Tailed p-Values and 95% Confidence Intervals for Models with a Proportional Hazard Function in Both Covariates (n = 100 with 30% Censoring)

Variable	Coeff.	Std. Err.	z	P>|z|	95% CIE
d	0.579	0.249	2.33	0.020	0.092, 1.066
x	0.180	0.032	5.66	<0.001	0.118, 0.243
d	0.573	0.253	2.27	0.023	0.078, 1.068
x	0.186	0.035	5.26	<0.001	0.116, 0.255
$d\times\ln(t)$	−0.002	0.163	0.01	0.988	−0.322, 0.317
$x\times\ln(t)$	0.007	0.020	0.36	0.716	−0.032, 0.047

nonsystematic way about the zero line, and the polygon connecting the values of the smoothed residuals has approximately a zero slope and crosses the zero line several times. The initial upward trend in the smoothed residuals for the dichotomous covariate in Figure 6.1a is due to a few large negative residuals among the shortest survival times. One should also note that there are two bands of residuals for the dichotomous covariate. The upper band corresponds to subjects with $d = 1$ and the bottom one to those with $d = 0$. There are 70 points in the graph, as this is the observed number of survival times.

In a second example, the model is nonproportional in the continuous covariate. Table 6.2 presents the results of fitting the two models. The main effects model in Table 6.2 shows that both covariates are highly significant. The model with the interactions with log-time shows that the model may be non-proportional in the continuous covariate, as $p = 0.024$ for the Wald test for the $x \times \ln(t)$ coefficient. The value of the partial likelihood ratio test for the addition of the two interaction variables is $G = 5.424$ and, with 2 degrees-of-freedom, the p-value is 0.066. The interactions model shows some numeric instability, as the estimated standard error for the $d \times \ln(t)$ term is quite large. Based on this, two models (each containing only one of the log-time interactions) were fit, and the results supported the observation of non-proportionality in x but not d. Graphs of the scaled Schoenfeld residuals are shown in Figure 6.2.

Under the time-varying coefficient model in (6.21), if the covariate has a proportional hazard, the plot of the scaled Schoenfeld residuals and its smooth should show no trend over time. This is observed in Figure 6.2a, where it can be seen that the smoothed residuals have essentially a slope of zero. The apparent initial positive slope is due to one or two large negative residuals. On the other hand, the polygon based on the smoothed scaled residuals for the continuous covariate, shown in Figure 6.2b, displays a consistent positive slope, suggesting that the importance of the covariate increases over time and thus has a nonproportional hazard.

As a third example, data were created such that the hazard function was nonproportional for the dichotomous covariate and was proportional in the continuous covariate. Table 6.3 presents the results of fitting the main effects and log-time interactions models. Graphs of the scaled Schoenfeld residuals are presented in Figure 6.3.

The results in Table 6.3 show evidence of nonproportional hazards in the dichotomous covariate, as $p = 0.001$ for the Wald test of the coef-

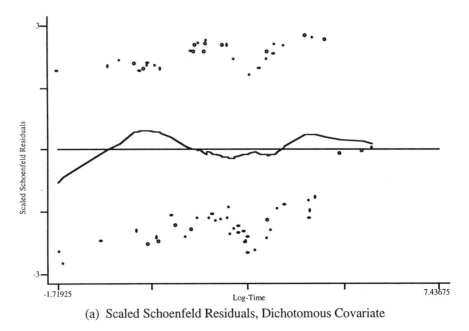

(a) Scaled Schoenfeld Residuals, Dichotomous Covariate

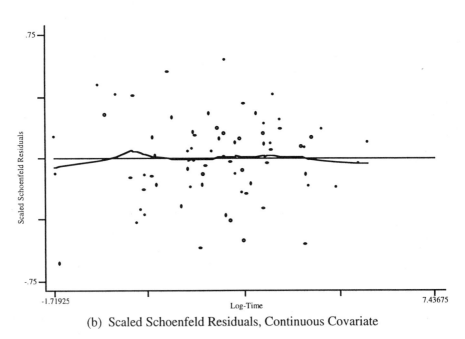

(b) Scaled Schoenfeld Residuals, Continuous Covariate

Figure 6.1 Graphs of the scaled Schoenfeld residuals and their lowess smooth obtained from main effects model in Table 6.1. Zero line is drawn for reference.

Table 6.2   Estimated Coefficients, Standard Errors, z-Scores, Two-Tailed p-Values and 95% Confidence Intervals for Models with a Nonproportional Hazard Function in the Continuous Covariate, x ($n = 100$ with 30% Censoring)

Variable	Coeff.	Std. Err.	z	$P>\lvert z\rvert$	95% CIE
d	0.561	0.256	2.19	0.028	0.059, 1.062
x	0.444	0.050	8.83	<0.001	0.343, 0.539
d	0.540	0.267	2.02	0.043	0.015, 1.063
x	0.539	0.070	7.68	<0.001	0.401, 0.676
$d\times\ln(t)$	0.498	5.149	0.10	0.923	−9.594, 10.590
$x\times\ln(t)$	2.337	1.037	2.25	0.024	0.304, 4.368

ficient for $d \times \ln(t)$. The Wald test for the coefficient for the $x \times \ln(t)$ term is not significant, suggesting that the model has a proportional hazard in the continuous covariate. The value of the partial likelihood ratio test for the addition of the two interaction variables is $G = 11.355$ and, with 2 degrees-of-freedom, the p-value is 0.003.   The polygon connecting the smoothed scaled Schoenfeld residuals in Figure 6.3a shows a strong initial positive slope that levels off.   The shape of this plot suggests that the dichotomous covariate may be an important de-terminant of survival initially, but not later in the follow-up period.   The polygon of the smoothed scaled Schoenfeld residuals for the continu-ous covariate essentially has a zero slope, supporting the lack of signifi-cance of the interaction with time that was seen in Table 6.3.

These examples demonstrate the utility of the two-step procedure for assessing the proportional hazards assumption: (1) add the covariate by log-time interactions to the model and assess their significance using the partial likelihood ratio test, score test or Wald test and (2) plot the scaled and smoothed scaled Schoenfeld residuals obtained from the model without the interactions terms.   The results of the two steps should support each other.   Procedures for modeling in the presence of nonproportional hazards are discussed in Chapter 7, when extensions of the proportional hazards model are considered.   We now turn to evalu-ating the model developed in Chapter 5 for the UIS, shown in Table 5.11.   We leave evaluation of the model in Table 5.13 as an exercise.

The model shown in Table 5.11 for the UIS is relatively complex in that it contains 10 terms, two of which are interactions and two of which model nonlinear effects of a continuous covariate.   As a first step in as-sessing the proportional hazards assumption, interactions of each main effect with log-time were added to the model, using only NDRUGFP1

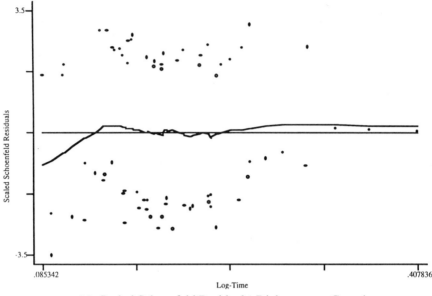

(a) Scaled Schoenfeld Residuals, Dichotomous Covariate

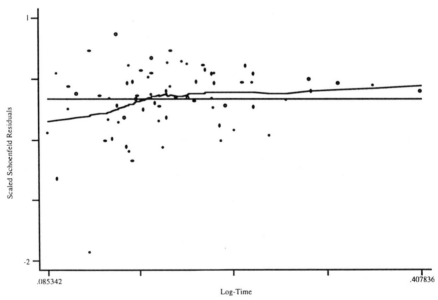

(b) Scaled Schoenfeld Residuals, Continuous Covariate

Figure 6.2 Graphs of the scaled Schoenfeld residuals and their lowess smooth obtained from the main effects model in Table 6.2. Zero line is drawn for reference.

Table 6.3   Estimated Coefficients, Standard Errors, z-Scores, Two-tailed p-Values and 95% Confidence Intervals for Models with a Nonproportional Hazard Function in the Dichotomous Covariate, d ($n = 100$ with 30% Censoring)

Variable	Coeff.	Std. Err.	z	$P>|z|$	95% CIE
d	4.238	0.608	6.98	<0.001	3.046, 5.430
x	0.171	0.032	5.40	<0.001	0.009, 0.133
d	8.977	1.884	4.77	<0.001	5.285, 12.669
x	0.185	0.034	5.54	<0.001	0.120, 0.251
$d \times \ln(t)$	2.709	0.837	3.24	0.001	1.069, 4.350
$x \times \ln(t)$	0.009	0.018	0.53	0.598	−0.025, 0.044

for number of drug treatments. Table 6.4 presents the estimated coefficients, standard errors, Wald statistics and p-values for the Wald statistics for the interactions with log-time. The value of the partial likelihood ratio test comparing the model in Table 5.11 to the 17 term model containing the seven interactions with log-time is $G = 5.538$ which, with 7 degrees-of-freedom, yields $p = 0.595$. These results suggests that the model may have proportional hazards in each of the seven covariates.

The next step is to examine a plot, similar to those in Figures 6.1–6.3, for each of the 10 terms in the model. The plots of the scaled Schoenfeld residuals and the lowess smooths shown in Figure 6.4 support the assumption of proportional hazards for each of the eight covariates shown. That is, each subplot in the figure has slope essentially equal to zero. The only possible exception is for the covariate

Table 6.4   Estimated Coefficients, Standard Errors, z-Scores and Two-Tailed p-Values for the Seven Interactions with Log-Time Added to the Model in Table 5.11 for the UIS ($n = 575$)

Variable	Coeff.	Std. Err.	z	$P>|z|$
AGE×ln(t)	0.002	0.009	0.20	0.838
BECKTOTA×ln(t)	−0.007	0.005	1.38	0.166
NDRUGFP1×ln(t)	−0.016	0.018	0.89	0.375
IVHX_3×ln(t)	−0.030	0.113	0.27	0.791
RACE×ln(t)	0.113	0.125	0.91	0.364
TREAT×ln(t)	0.128	0.100	1.28	0.201
SITE×ln(t)	−0.023	0.114	0.20	0.842

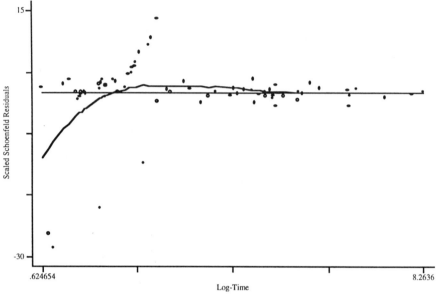

(a) Scaled Schoenfeld Residuals, Dichotomous Covariate

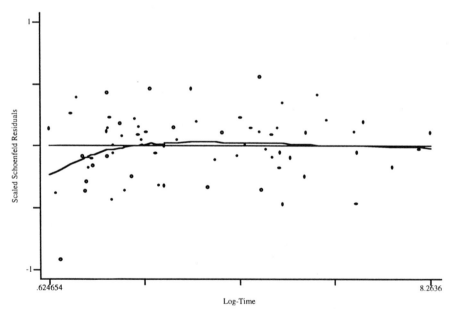

(b) Scaled Schoenfeld Residuals, Continuous Covariate

Figure 6.3 Graphs of the scaled Schoenfeld residuals and their lowess smooth obtained from the main effects model in Table 6.3. Zero line is drawn for reference.

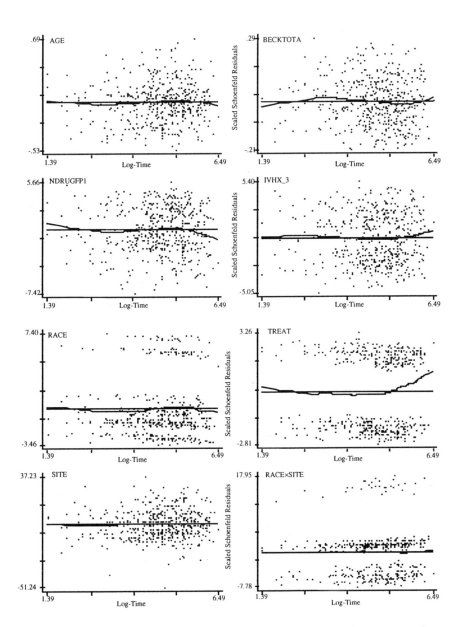

Figure 6.4 Graphs of the scaled Schoenfeld residuals and their lowess smooth obtained from the model in Table 5.11 for covariates (top left to bottom right): AGE, BECKTOTA, NDRUGFP1, IVHX_3, RACE, TREAT, SITE and RACE×SITE. The zero line is drawn for reference.

TREAT (third row, right plot). This plot could be interpreted to mean that the effect of the longer treatment (TREAT = 1) is most pronounced in the earlier and later periods of follow-up. However, we will not consider the possible departure from proportional hazards to be significant, since the Wald test of the treatment by log-time interaction is not significant, $p = 0.201$. We reexamine the effect of treatment in Chapter 7, when covariates that vary with time are discussed in detail. The plots for NDRUGFP2 and AGE×SITE (the two terms in the model in Table 5.11 that are not shown in Figure 6.4), also support the proportional hazards assumption.

The two-step procedure for assessing proportional hazards yields results that support this assumption for the 10-term model for the UIS shown in Table 5.11. We now consider the evaluation of the subject-specific diagnostic statistics for leverage and influence.

6.4 IDENTIFICATION OF INFLUENTIAL AND POORLY FIT SUBJECTS

Another important aspect of model evaluation is a thorough examination of regression diagnostic statistics to identify which, if any, subjects: (1) have an unusual configuration of covariates, (2) exert an undue influence on the estimates of the parameters, and/or (3) have an undue influence on the fit of the model. Statistics similar to those used in linear and logistic regression are available to perform these tasks with a fitted proportional hazards model. There are some differences in the types of statistics used in linear and logistic regression and proportional hazards regression, but the essential ideas are the same in all three settings.

Leverage is a diagnostic statistic that measures how "unusual" the values of the covariates are for an individual. In some sense it is a residual in the covariates. In linear and logistic regression leverage [see Hosmer and Lemeshow (1989), Kleinbaum, Kupper, Muller and Nizam (1998), and Ryan (1997)] is calculated as the distance of the value of the covariates for a subject to the overall mean of the covariates. It is proportional to $(x - \bar{x})^2$. The leverage values in these settings have nice properties in that they are always positive and sum over the sample to the number of parameters in the model. While it is technically possible to break the leverage into values for each covariate, this is rarely done in linear and logistic regression. Leverage is not quite so easily defined nor does it have the same nice properties in proportional hazards regression. This is due to the fact that subjects may appear in multiple risk sets and thus may be present in multiple terms in the partial likelihood.

The score residuals defined in (6.16) and (6.17) form the nucleus of the proportional hazards diagnostics. The score residual for the ith subject on the kth covariate, see (6.14), is a weighted average of the distance of the value, x_{ik}, to the risk set means, $x_{w_j k}$, where the weights are the change in the martingale residual, $dM_i(t_j)$. The net effect is that, for continuous covariates, the score residuals have the linear regression leverage property that the further the value is from the mean the larger the score residual is, but "large" may be either positive or negative. Thus, the score residuals are sometimes referred to as the leverage or partial leverage residuals.

The graphs of the score residuals for the covariates AGE, BECKTOTA, NDRUGFP1 and the AGE × SITE interaction obtained from the fitted model in Table 5.11 are shown in Figure 6.5. These four terms were chosen because they are the continuous variables in the fitted model and are therefore most amenable to having their score residuals examined graphically. The graphs for the dichotomous covariates are less interesting in that all the values fall on two vertical bands at zero and one, the two covariate values.

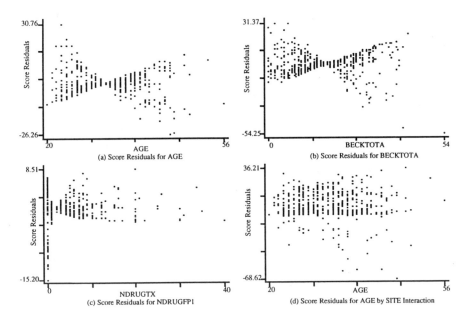

Figure 6.5 Graphs of the score residuals computed from the model in Table 5.11 for (a) AGE, (b) BECKTOTA, (c) NDRUGFP1 and (d) AGE × SITE Interaction.

The score residuals for AGE in Figure 6.5a display the fan shape expected, being smallest near the mean age of 32 and increasing in absolute value for ages increasingly older or younger than 32. The purpose of the plot is to see whether there are subjects whose ages yield unexpectedly large values. This would be seen in the graph as a point lying well away from the others in the plot. In Figure 6.5a there is one point in the top left and there are two in the bottom right that fall a bit away from the rest of the points. However, the distance between these points and the others is not striking. The two oldest subjects, ages 53 and 56, have score residuals that are well within the observed range of values. Thus, we conclude that there are no high leverage values for age.

The score residual values for BECKTOTA are plotted in Figure 6.5b. Recall that when we examined the scale of this covariate in Chapter 5 there was evidence in one of the plots, Figure 5.3, of some nonlinearity, but it was attributed to a few high values. These same values appear in the bottom right corner of Figure 6.5b as high leverage points that fall well away from all the other points. For the moment we do nothing more than note this fact.

The covariate, NDRUGTX, entered the model non-linearly with two terms. The plot of the score residuals for the first term, NDRUGFP1, is shown in Figure 6.5c. The plot of the score residuals for the second term, NDRUGFP2, is nearly identical in appearance to Figure 6.5c and is not presented. The fan shape is not quite as apparent because the mean number of drug treatments is about 5 while the maximum number is 40. We chose to plot the score residuals versus the number of drug treatments, rather than the transformation, NDRUGFP1, in order to more easily identify values associated with large residuals. The plot of the residuals versus NDRUGFP1 is essentially the mirror image of this plot since the transformation is the inverse of the variable. The vertical line of values at the left of the plot corresponds to the residuals for subjects with zero previous treatments. The only possible high leverage point is the one on the bottom left for a subject with zero previous treatments, but this value is not too distant from the other values. Thus, we conclude that none of the score residuals for NDRUGFP1 are abnormally large.

The score residuals for the AGE × SITE interaction covariate are plotted versus AGE in Figure 6.5d. The plot does not have the fan shape seen in Figure 6.5a since, at any age, there is a mix of subjects from the two sites. There are a few points in the bottom of the plot that fall a bit away from the others. However, the plot tends to drift down with no distinct break, so we conclude that none of the points have large residuals.

In summary, the plots in Figure 6.5 have shown that, except for the two subjects with the highest values for BECKTOTA, there are no strikingly large score residuals. Graphs and histograms, not shown, of the score residuals for the dichotomous covariates in the model did not yield any strikingly large values.

In linear and logistic regression, high leverage is not necessarily something to be concerned about. How high leverage contributes to a measure of the influence that a covariate value has on the estimate of a coefficient is of concern. The same is true in proportional hazards regression. To examine influence in the proportional hazards setting, we need statistics analogous to Cook's distance in linear regression. The purpose of Cook's distance is to obtain an easily computed statistic that approximates the change in the value of the estimated coefficients if a subject is deleted from the data. This is denoted as

$$\Delta \hat{\beta}_{ki} = \hat{\beta}_k - \hat{\beta}_{k(-i)}, \tag{6.22}$$

where $\hat{\beta}_k$ denotes the partial likelihood estimator of the coefficient computed using the entire sample of size n and $\hat{\beta}_{k(-i)}$ denotes the value of the estimator if the ith subject is removed. Cain and Lange (1984) show that an approximate estimator of (6.22) is the kth element of the vector of coefficient changes

$$\Delta \hat{\beta}_i = \left(\hat{\beta} - \hat{\beta}_{(-i)} \right) = \hat{Var}(\hat{\beta}) \hat{L}_i, \tag{6.23}$$

where $\hat{L}_i$ is the vector of score residuals, (6.17), and $\hat{Var}(\hat{\beta})$ is the estimator of the covariance matrix of the estimated coefficients. These are commonly referred to as the *scaled score residuals* and their values may be obtained from some software packages, for example, SAS and S-PLUS.

Graphs of the scaled score residuals, (6.23), are presented in Figure 6.6 for the covariates whose score residuals were graphed in Figure 6.5. The plots in Figures 6.5 and 6.6 are quite similar in appearance, but the scaling has enhanced the fan shape. The points seen in the top left of Figure 6.5a and the bottom right of Figure 6.5b are more noticeable in Figures 6.6a and 6.6b, confirming that they may exert an undue influence on the estimates of the coefficients.

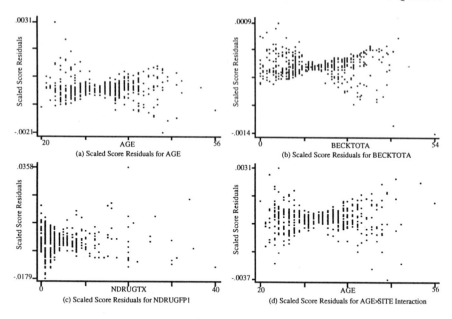

Figure 6.6 Graphs of the scaled score residuals computed from the model in Table 5.11 for (a) AGE, (b) BECKTOTA, (c) NDRUGFP1 and (d) AGE × SITE interaction.

One point in the top and middle of Figure 6.6c lies well away from the other scaled score residuals for subjects with the same number of drug treatments. This subject has some potential for influence on the coefficient for NDRUGFP1. We examine the effect this subject, as well as others, have on the model later in this section.

The plot in Figure 6.6d is much more distinctly fan shaped than its counterpart in Figure 6.5d, and none of the points seem to fall well away from the others. There are no distinct breaks in the points, they just slowly drift out, so nothing is noted in this plot for further examination. Plots of the scaled score residuals for the dichotomous covariates in the model revealed no points of potential high influence and thus are not shown.

Cook's distance in linear and logistic regression may be used to provide a single overall summary statistic of the influence a subject has on the estimators of all the coefficients. The overall measure of influence is

$$\left(\hat{\beta}-\hat{\beta}_{(-i)}\right)'\left[\widehat{\text{Var}}\left(\hat{\beta}\right)\right]^{-1}\left(\hat{\beta}-\hat{\beta}_{(-i)}\right),$$

and using (6.23) it may be approximated using

$$ld_i = \left(\Delta\hat{\boldsymbol{\beta}}_i\right)'\left[\widehat{\text{Var}}\left(\hat{\boldsymbol{\beta}}\right)\right]^{-1}\left(\Delta\hat{\boldsymbol{\beta}}_i\right)$$

$$= \left(\hat{\mathbf{L}}_i\right)'\left[\widehat{\text{Var}}\left(\hat{\boldsymbol{\beta}}\right)\right]\left[\widehat{\text{Var}}\left(\hat{\boldsymbol{\beta}}\right)\right]^{-1}\left[\widehat{\text{Var}}\left(\hat{\boldsymbol{\beta}}\right)\right]\left(\hat{\mathbf{L}}_i\right)$$

so

$$ld_i = \left(\hat{\mathbf{L}}_i\right)'\left[\widehat{\text{Var}}\left(\hat{\boldsymbol{\beta}}\right)\right]\left(\hat{\mathbf{L}}_i\right). \tag{6.24}$$

The statistic in (6.24) has been shown by Pettitt and Bin Daud (1989) to be an approximation to the amount of change in the log partial likelihood when the ith subject is deleted. In this context the statistic is called the likelihood displacement statistic, hence the rationale for labeling it ld in (6.24). Thus

$$ld_i \cong 2\left[L_p\left(\hat{\boldsymbol{\beta}}\right) - L_p\left(\hat{\boldsymbol{\beta}}_{(-i)}\right)\right]. \tag{6.25}$$

Another form of the likelihood displacement statistic is obtained from a matrix form of (6.24). In particular let $\hat{\mathbf{L}}$ denote the n by p matrix whose ith row is $\hat{\mathbf{L}}_i'$, see (6.17), and let the n by n matrix of scaled score residuals be

$$\hat{\mathbf{L}}\left[\widehat{\text{Var}}\left(\hat{\boldsymbol{\beta}}\right)\right]\hat{\mathbf{L}}'. \tag{6.26}$$

When the matrix in (6.26) is broken into its eigenvalues and associated eigenvectors, the n elements in the eigenvector associated with the largest eigenvalue are called the "l-max" statistics and are denoted lm_i for the ith subject. Since both ld_i and lm_i are overall summary statistics, we feel it makes the most sense to plot them versus another summary statistic. The one we like to use is the martingale residual. Other possible choices are the estimated survival probability or estimated cumulative hazard function. We feel that plots of these values against something like a study identification code or case number may be somewhat useful in locating large values in small data sets but provide little additional information about the subject. An additional enhancement that aids in the interpretation of the plot is to use a different symbol for the two values of the censoring variable.

Plots of the values of the likelihood displacement and l-max statistics versus the martingale residuals are shown in Figures 6.7a and 6.7b, respectively. Both plots have the same asymmetric "cup" shape with the bottom of the cup at zero. In linear and logistic regression, the influence diagnostic, Cook's distance, is a product of a residual measure and leverage. While the same concise representation does not hold in proportional hazards regression, it is approximately true in the sense that an influential subject will have a large residual and/or leverage. Thus, the largest values of both the likelihood displacement and l-max statistic form the sides of the cup and correspond to poorly fit subjects (ones with either large negative or positive martingale residuals).

Examining the plots in Figure 6.7a, we find that there is a group of four points lying well away from the others in the top left corner of the plot, with two other points slightly below this cluster. Five of the six subjects have censored survival times and all have martingale residuals less than about –2.0. Figure 6.7b is not quite as cup-shaped as Figure 6.7a, but the four points in the top left of each figure correspond to the same subjects. The principal difference in the two figures is that the two subjects on the right edge of Figure 6.7b correspond to a cluster with the next largest values of the l-max statistics, whereas in Figure 6.7a these same two subjects do not have large values of the likelihood displacement statistic. Thus the two statistics, while similar for the extreme values, do identify different subjects in the mid range. From this point of view it makes sense, in an applied setting, to look at both statistics to locate those subjects with large values on both or only one of the statistics.

In summary, use of the plots of the diagnostic statistics for change in individual coefficients identified four possible subjects whose effect on the model should be checked in more detail: one for AGE in Figure 6.6a, two for BECKTOTA in Figure 6.6b and one for NDRUGFP1 in Figure 6.6c. Four to six subjects were identified in Figures 6.7a and 6.7b as having extreme values for the summary change statistics, likelihood displacement and l-max. These subjects are possibly different from the ones previously identified in Figure 6.6. We emphasize "possibly different" because the summary measures take into account values of all the covariates. A subject extreme on only one covariate may be near enough to the middle for the others that an extreme value of the likelihood displacement or l-max statistic would not be generated.

The next step in the modeling process is to identify explicitly the subjects with the extreme values, refit the model deleting these subjects, and calculate the change in the individual coefficients. The final decision on the continued use of a subject's data to fit the model will depend on the observed percent change in the coefficients that results from deleting the subject's data and, more importantly, the clinical plausibility of that subject's data.

Deleting the subjects with extreme values in the change in coefficient diagnostic for AGE and NDRUGFP1 individually did not produce marked changes in the coefficients. However, deletion of the two subjects with the extreme values of the diagnostic for change in the BECKTOTA coefficient yielded a model in which the BECKTOTA coefficient was 33.5 percent larger than the coefficient in Table 5.11.

Through a process of deleting and refitting models, we determined that deletion of the four subjects with the most extreme values of the likelihood displacement or of the l-max statistics yielded models with important changes in several coefficients. The deletion of the two subjects with the next largest values of either statistic did not produce additional important changes in the coefficients.

Further examination of the data showed that the two subjects with the extreme values for change in the BECKTOTA coefficient were not among the four with the extreme values of either summary change measure. The model obtained by deleting six subjects (four [based on l-max] and two others [based on BECKTOTA]) had a coefficient for BECKTOTA that increased by 60.8 percent, a coefficient for the RACE ×SITE interaction that increased by 33.6 percent, a coefficient for the AGE × SITE interaction that increased by 20.3 percent, and a coefficient for SITE that increased by 14.9 percent. All other coefficients changed by less than 9 percent. In particular, the coefficient for TREAT increased by only 6.3 percent. At this point we reviewed the data for these six most influential subjects. We felt that only the value of 54 for BECKTOTA was a bit unusual, but not too extreme. In the end, therefore, we decided to keep all subjects in the data set.

We leave as an exercise the fitting of the models with the specified subjects deleted and computation of the reported percent change in coefficients. We remind the reader that we calculate the percent change in a coefficient as

$$\Delta \hat{\beta} \% = 100 \left(\hat{\beta}_{\text{reduced}} - \hat{\beta}_{\text{all}} \right) / \hat{\beta}_{\text{all}} \, ,$$

where $\hat{\beta}_{\text{all}}$ stands for the estimate of the coefficient from the model with no subjects deleted and $\hat{\beta}_{\text{reduced}}$ stands for the estimate of the coefficient from the model with subjects deleted.

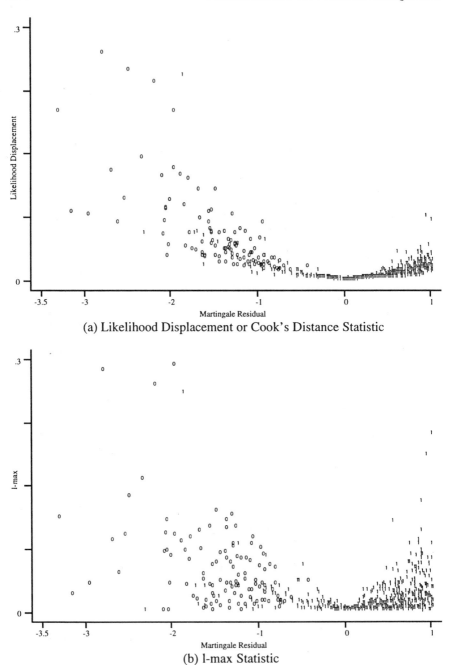

(a) Likelihood Displacement or Cook's Distance Statistic

(b) l-max Statistic

Figure 6.7 Graphs of the likelihood displacement or Cook's distance statistic and maximum eigenvalue or l-max statistic computed from the model in Table 5.11 versus the martingale residual (0 = censored, 1 = uncensored).

In summary, we feel that it is important to exa｜
score residuals, scaled score residuals, the likelihood d｜
tic and the l-max statistic. The first two statistics are
fying subjects with high leverage or who influence the
coefficient. The latter two provide useful information
fluence on the vector of coefficients. Each statistic po
tant aspect of the effect a particular subject has on the fitted model.
One always hopes that major problems are not uncovered. However, if
the model does display abnormal sensitivity to the subjects deleted, this
is a clear indication of fundamental problems in the model and we recommend going back to "square-one" and redoing each step in the
modeling process, perhaps with these subjects deleted.

The next step in the modeling process is to compute an overall
goodness-of-fit test.

6.5 OVERALL GOODNESS-OF-FIT TESTS AND MEASURES

Until quite recently, all of the proposed tests for the overall goodness-
of-fit of a proportional hazards model were difficult to compute in most
software packages. For example, the test proposed by Schoenfeld
(1980) compares the observed number of events to a proportional hazards regression model-based estimate of the expected number of events
in each of G groups that are formed by partitioning the time axis and
covariate space. Unfortunately, the covariance matrix required to form
a test statistic comparing the observed to expected number of events is
quite complex to compute. The test proposed by Lin, Wei and Ying
(1993) is based on the maximum absolute value of partial sums of
martingale residuals. This test requires complex and time consuming
simulations to obtain a significance level. Other tests [e.g., O'Quigley
and Pessione (1989) and Pettitt and Bin Daud (1990)] require that the
time axis be partitioned and interactions between covariates and interval-
specific, time-dependent covariates be added to the model. Overall
goodness-of-fit is based on a significance test of the coefficients for the
added variables.

Grønnesby and Borgan (1996) propose a test similar to the Hosmer-
Lemeshow test [Hosmer and Lemeshow (1989)] used in logistic regression. They suggest partitioning the data into G groups based on the
ranked values of the estimated risk score, $x'\hat{\beta}$. The test is based on the
sum of the martingale residuals within each group, and it compares the

rved number of events in each group to the model-based estimate the expected number of events. Using the counting process approach, they derive an expression for the covariance matrix of the vector of G sums. They show that their quadratic form test statistic has a chi-square distribution with $G-1$ degrees-of-freedom when the fitted model is the correct model and the sample is large enough that the estimated expected number of events in each group is large. As presented in their paper, the calculations of Grønnesby and Borgan (1996) are not a trivial matter.

May and Hosmer (1998), following the method used by Tsiatis (1980) to derive a goodness-of-fit test in logistic regression, prove that Grønnesby and Borgan's test is the score test for the addition of $G-1$ design variables, based on the G groups, to the fitted proportional hazards model. Thus, the test statistic may be calculated in any package that performs score tests. Using the asymptotic equivalence of score tests and likelihood ratio tests, one may approximate the score test with the partial likelihood ratio test, which may be done in any package.

One may be tempted to define groups based on the subject-specific estimated survival probabilities,

$$\hat{S}\left(t_i, \mathbf{x}_i, \hat{\boldsymbol{\beta}}\right) = \left[\hat{S}_0\left(t_i\right)\right]^{\exp\left(\mathbf{x}_i'\hat{\boldsymbol{\beta}}\right)}.$$

This should not be done as the values of time differ for each subject. If groups are to be based on the survival probability scale, they should be computed using the risk score and a fixed value of time for each subject. For example, in the UIS we could use the estimated one-year survival probability

$$\hat{S}\left(365, \mathbf{x}_i, \hat{\boldsymbol{\beta}}\right) = \left[\hat{S}_0\left(365\right)\right]^{\exp\left(\mathbf{x}_i'\hat{\boldsymbol{\beta}}\right)}.$$

Since the choice of a time is arbitrary, one cannot interpret the probability as a prediction of the number of events in each decile of risk. It merely provides another way to express the risk score.

The value of the score test for the inclusion of the nine decile-of-risk design variables to the model in Table 5.11 is 7.86 which, with 9 degrees-of-freedom, has a p-value of 0.549. The partial likelihood ratio test comparing the model in Table 5.11 to the one including the nine design variables is $G = 7.56$ which, with 9 degrees-of-freedom, has a p-value of 0.579. The two test statistics have nearly the same value and neither is significant, suggesting that there is no evidence that the model does not fit.

May and Hosmer's (1998) result not only greatly simplifies the calculation of the test, but it also suggests that a two by ten table presenting the observed and expected numbers of events in each group is a useful way to summarize the model fit. The individual observed and expected values in the table may be compared by appealing to counting process theory. Under this theory, the counting function is approximately a Poisson variate with mean equal to the cumulative hazard function. Sums of independent count functions will be approximately Poisson distributed, with mean equal to the sum of the cumulative hazard function. This suggests considering the observed counts within each decile of risk to be distributed approximately Poisson, with mean equal to the estimated expected number of counts. Furthermore, the fact that the Poisson distribution may be approximated by the normal for large values of the mean suggests that an easy way to compare the observed and expected counts is to form a z-score by dividing their difference by the square root of the expected. The two-tailed p-value is obtained from the standard normal distribution. There are obvious dependencies in the counts due to the fact that the same estimated parameter vector is used to calculate the individual expected values and some dependency due to grouping subjects into deciles. The effect of these dependencies has not been studied, but it is likely to smooth the counts toward the expected counts. Thus, the proposed cell-wise z-score comparisons should, if anything, be a bit conservative.

Table 6.5 presents the observed and estimated expected numbers of events, the z-score and two-tailed p-value within each decile of risk for the fitted model in Table 5.11. The numbers in Table 6.5 are large enough that we feel comfortable using the normal approximation to the Poisson distribution. With a p-value equal to 0.049, only the sixth decile has a possibly significant difference between the observed and model-based expected count. If we use Bonferroni's method to adjust the 5 percent level of significance for multiple testing to 0.005, then none of the deciles has a significant difference between the observed and expected counts. Thus, we conclude that there is agreement between observed and expected number of events within each of the 10 deciles of risk.

Arjas (1988) suggests plotting the cumulative observed versus the cumulative estimated expected number of events for subjects with observed, not censored, survival times within partitions of the data to assess model fit. If the model is the correct one, the points should follow a 45

Table 6.5 Observed Number of Events, Estimated Number of Events, z-Scores and Two-Tailed p-Values within Each Decile of Risk Based on the Model in Table 5.11

Decile of Risk	Observed Number of Events	Estimated Number of Events	z	p-Value
1	34	33.96	0.007	0.994
2	43	36.35	1.103	0.270
3	37	44.88	−1.176	0.240
4	44	45.77	−0.262	0.744
5	46	52.82	−0.939	0.348
6	51	38.73	1.972	0.049
7	49	49.28	−0.041	0.968
8	53	52.73	0.037	0.971
9	52	53.19	−0.164	0.870
10	55	56.3	−0.173	0.863
Total	464	464		

degree line beginning at the origin. Arjas suggests forming groups based on covariate values, such as the treatment variable for the model in Table 5.11. Rather than using groups based on only a few covariates, we feel that a partition based on the risk score is a convenient way to incorporate all study covariates into the grouping strategy. The Arjas plot for each of the deciles of risk in Table 6.5 provides graphical support for the conclusion that the model fits within each decile of risk. These plots do, in fact, support model fit. For illustrative purposes we demonstrate the Arjas plots in Figure 6.8 using quartiles of risk instead of deciles of risk.

The plots in Figure 6.8 are obtained as follows: first, create groups based on quartiles of risk and sort on risk score within each group; second, compute the cumulative sum of the zero-one censoring variable and the cumulative sum of the estimated cumulative hazard function within each group; third, plot the pairs of cumulative sums within each group only for subjects with an observed survival time.

In Figures 6.8a–6.8d the polygons connecting the points are each close to the 45 degree line. They display small departures that do not necessarily indicate a poorly fitting model. Thus, these Arjas plots do not contradict earlier conclusions regarding the model's fit to the data.

As in all regression analyses, some measure analogous to R^2 may be of interest as a measure of model performance. As shown in a detailed study by Schemper and Stare (1996), there is not a single, simple, easy

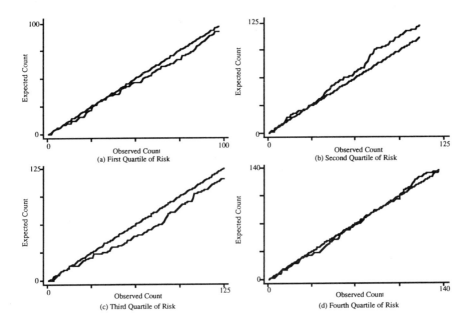

Figure 6.8 Plots of the cumulative estimated expected count versus the cumulative observed count within each quartile of risk based on the fitted model in Table 5.11, for subjects with an observed survival time.

to calculate, useful, easy to interpret measure for a proportional hazards regression model. In particular, all measures depend on the proportion of values that are censored. A perfectly adequate model may have what, at face value, seems like a terribly low R^2 due to a high percent of censored data. In our opinion, further work needs to be done before we can recommend one measure over another. However, if one must compute such a measure, then

$$R_p^2 = 1 - \left\{ \exp\left[\frac{2}{n} \left(L_0 - L_p \right) \right] \right\}$$

is perhaps the easiest and best one to use, where L_p is the log partial likelihood for the fitted model with p covariates, and L_0 is the log partial likelihood for model zero, the model with no covariates. For the fitted model in Table 5.11, the value is

$$R_p^2 = 1 - \left(\exp\left\{ \left[\frac{2}{575} \right] \times \left[(-2663.985) - (-2630.418) \right] \right\} \right) = 0.11.$$

The model displayed in Table 5.11 has passed all the tests for a good fitting model. We are now in a position to discuss the interpretation of this model and how to best present the results to the audience of interest.

6.6 INTERPRETATION AND PRESENTATION OF THE FINAL MODEL

The model fit to the UIS data, shown in Table 5.11, is reported again in Table 6.6. It is an excellent model for teaching purposes, as it contains an example of just about every possible covariate one is likely to encounter in practice. The model contains two simple dichotomous covariates (treatment and recent IV drug use), a continuous linear covariate (Beck score), a continuous non-linear covariate (number of prior drug treatments), an interaction between a continuous and a dichotomous covariate (age and site) and and interaction between two dichotomous covariates (race and site). In this section, when we refer to "the model" we are referring to the one in Table 6.6.

Table 6.6   Estimated Coefficients, Standard Errors, z-Scores, Two-Tailed p-Values and 95% Confidence Intervals for the Final Proportional Hazards Model for the UIS ($n = 575$)

Variable	Coeff.	Std. Err.	z	$P>|z|$	95% CIE
AGE	−0.041	0.010	−4.18	<0.001	−0.061, −0.022
BECKTOTA	0.009	0.005	1.76	0.078	−0.001, 0.018
NDRUGFP1	−0.574	0.125	−4.59	<0.001	−0.820, −0.329
NDRUGFP2	−0.215	0.049	−4.42	<0.001	−0.310, −0.119
IVHX_3	0.228	0.109	2.10	0.036	0.015, 0.441
RACE	−0.467	0.135	−3.47	0.001	−0.731, −0.203
TREAT	−0.247	0.094	−2.62	0.009	−0.432, −0.062
SITE	−1.317	0.531	−2.48	0.013	−2.359, −0.275
AGE×SITE	0.032	0.016	2.02	0.044	0.001, 0.064
RACE×SITE	0.850	0.248	3.43	0.001	0.365, 1.336

Log-likelihood = −2630.418

We begin by discussing how to prepare point and interval estimates of hazard ratios for the covariates. We wish to call attention to the fact that we have assiduously avoided including any exponentiated coefficients in tables of estimated coefficients in Chapters 5 and 6. While most software packages automatically provide these quantities, they probably will be useful summary statistics for only a few model covariates. We feel it is best not to even attempt estimating any hazard ratios until one has completed all steps in both model development and model checking: i.e., the model fits, satisfies the proportional hazards assumption, and any and all highly influential subjects have been dealt with in a scientifically appropriate manner.

Only the covariates for Beck score, recent IV drug use and treatment have hazard ratios that may be estimated by exponentiating their estimated coefficients. This is because the other covariates are either involved in interactions or are nonlinear in the model. It is convenient to display these estimated hazard ratios and their confidence intervals in a table similar to Table 6.7.

The estimated hazard ratio for a 10-point increase in the Beck score is $1.09 = \exp(10 \times 0.009)$, which shows a slight increase in the rate of return to drug use. The interpretation is that subjects with the 10-point higher score are returning to drug use at a rate that is 9 percent higher than for subjects at the lower score. The 95 percent confidence interval suggests that an increased rate of return to drug use as high as 20 percent or even a decreased rate of 1 percent is consistent with the data. Since the model is linear in the Beck score, this interpretation holds over the observed range of Beck scores.

The estimated hazard ratio for recent IV drug use is 1.26. The interpretation of this is that subjects who have a recent history of IV drug use are returning to drug use at a rate that is 26 percent higher than for subjects who are not recent IV drug users. The confidence interval indicates that the rate could actually be as much as 55 percent higher or as little as 2 percent higher.

The hazard ratio for treatment of 0.78 means that subjects in the longer or extended treatment program are returning to drug use at a rate that is 22 percent lower than for subjects with the shorter treatment. The 95 percent confidence interval suggests that the rate could be as much as 35 percent lower to only 6 percent lower. The estimated hazard ratio points to a significant benefit for the longer of the two treatments, controlling for all other model covariates. In studies in which there is a single covariate of primary interest, such as a treatment covariate, one may encounter tables of results in which a summary statistic for

**Table 6.7   Estimated Hazard Ratios and 95%
Confidence Intervals for Beck Score, Recent IV
Drug Use and Treatment for the UIS ($n = 575$)**

Variable	Hazard Ratio	95% CIE
BECKTOTA*	1.09	0.99, 1.20
IVHX_3	1.26	1.02, 1.55
TREAT	0.78	0.65, 0.94

* Hazard ratio for a 10-point increase.

this covariate only is presented, with the other covariates in the model relegated to footnote status. We feel that this is not good statistical or scientific practice. With such an oversimplified summary, the reader has no way of evaluating whether an appropriate model building and model checking paradigm has been followed or what the actual fitted model contains. We feel that the full model should be presented in a table similar to Table 6.6 at some point in the results section.

The number of previous drug treatments is modeled with two non-linear terms, so any hazard ratio will depend on the values of the number of previous drug treatments being compared. The graph in Figure 5.1d of the log hazard using the two non-linear terms shows an initial decrease in risk followed by a progressive nonlinear increase. One possible strategy would be to compare the hazard ratio for an increase of one in the number of previous drug treatments (i.e., hazard ratios for 0 vs. 1, 1 vs. 2, 2 vs. 3, etc.). These hazard ratios could either be tabulated or presented graphically, along with their confidence limits. We will do the latter.

One must proceed carefully when calculating hazard ratios for non-linear functions of a covariate. The first step is to write down the expression for the log-hazard function, keeping all the other covariates constant. For ease of presentation, let the log-hazard function computed at a particular value of the number of previous drug treatments, NDRUG, holding all other covariates fixed and denoted as z, be

$$g(\text{NDRUGTX}, z) = \beta_1 \text{NDRUGFP1} + \beta_2 \text{NDRUGFP2} + \beta'z,$$

where

$$\text{NDRUGFP1} = \left[10/(\text{NDRUGTX} + 1) \right]$$

and

$$\text{NDRUGFP2} = \text{NDRUGFP1} \times \ln\left[(\text{NDRUGTX} + 1)/10 \right].$$

The next step is to write down the equation for the difference of interest in the log hazard function. In this case it is, for a one-unit increase,

$$g(\text{NDRUGTX}+1,\mathbf{z})-g(\text{NDRUGTX},\mathbf{z}).$$

Note that the first term in the difference in the log-hazard functions is the log-hazard evaluated at a one-unit increase in the previous number of drug treatments, not a one-unit increase in the nonlinear transformation. If we denote the values of the transformed variables at the increased value as

$$\text{NDRUGFP11}=\left[10/(\text{NDRUGTX}+2)\right]$$

and

$$\text{NDRUGFP21}=\text{NDRUGFP11}\times\ln\left[(\text{NDRUGTX}+2)/10\right],$$

then the difference in the log-hazard functions is

$$g(\text{NDRUGTX}+1,\mathbf{z})-g(\text{NDRUGTX},\mathbf{z})=a\beta_1+b\beta_2, \qquad (6.27)$$

where

$$a=\text{NDRUGFP11}-\text{NDRUGFP1}$$
$$=\left[10/(\text{NDRUGTX}+2)\right]-\left[10/(\text{NDRUGTX}+1)\right]$$

and

$$b=\text{NDRUGFP21}-\text{NDRUGFP2}.$$

The estimated difference in the log-hazard function is obtained by evaluating (6.27) using the values of the coefficients from Table 6.6, $\hat{\beta}_1=-0.574$ and $\hat{\beta}_2=-0.215$, and NDRUGTX = 0, 1, 2, 3, etc. The estimated hazard ratios are then obtained by exponentiating the estimated differences in the log-hazard functions,

$$\hat{\text{HR}}(\text{NDRUGTX}+1,\text{NDRUGTX},\mathbf{z})=\exp\!\left(a\hat{\beta}_1+b\hat{\beta}_2\right). \qquad (6.28)$$

The estimator of the endpoints of the $100(1-\alpha)$ percent confidence interval for the difference in the log-hazard functions is

$$\left(a\hat{\beta}_1+b\hat{\beta}_2\right)\pm z_{1-\alpha/2}\hat{\text{SE}}\!\left(a\hat{\beta}_1+b\hat{\beta}_2\right), \qquad (6.29)$$

where

$$\hat{\text{SE}}\!\left(a\hat{\beta}_1+b\hat{\beta}_2\right)=\left[a^2\hat{\text{Var}}\!\left(\hat{\beta}_1\right)+b^2\hat{\text{Var}}\!\left(\hat{\beta}_2\right)+2ab\hat{\text{Cov}}\!\left(\hat{\beta}_1,\hat{\beta}_2\right)\right]^{0.5}. \qquad (6.30)$$

The estimators of the variances and covariance in (6.30) may be obtained from output of the covariance matrix of the estimated coefficients from software packages. In the current example these values are $\widehat{Var}\left(\hat{\beta}_1\right) = 0.015672$, $\widehat{Var}\left(\hat{\beta}_2\right) = 0.002361$ and $\widehat{Cov}\left(\hat{\beta}_1, \hat{\beta}_2\right) = 0.006022$. The endpoints of the confidence interval estimator for the hazard ratio are obtained by exponentiating the estimators in (6.29).

We calculated the value of the hazard ratio in (6.28) for the entire range of values for number of previous drug treatments, 0–40, and observed that, after about 10 previous treatments, there was not much change in the hazard ratio for a one-unit increase. Thus, we present in Figure 6.9 the graph of the hazard ratio and its confidence interval for up to 10 previous drug treatments.

The point estimate of the hazard ratio at 0 previous drug treatments in Figure 6.9 is 0.70. The interpretation is that subjects who have had one previous drug treatment are returning to drug use at a rate that is 30 percent lower than subjects who have had no previous drug treatments.

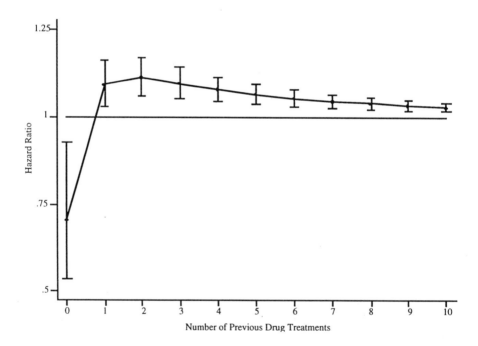

Figure 6.9 Graph of the estimated hazard ratio and associated 95 percent confidence interval for a one unit increase in the number of drug treatments from the labeled value.

The point estimate of the hazard ratio at 1 in Figure 6.9 is 1.09. This means that subjects with two previous treatments are returning to drug use at a rate that is 9 percent higher than subjects with one previous drug treatment. The estimate of the hazard ratio at 2 is 1.11, and then it slowly falls to 1.03 at 10. Since the estimated hazard ratios in Figure 6.9 exceed 1 at one or more previous drug treatments, the graph indicates a continuing increase in risk as the number of treatments increase; but the increase is progressively less, that is, the change in rate from 1 to 2 is much greater than the change from 9 to 10. We note that, since none of the confidence intervals include 1.0, the increase in rate of return to drugs is significant at all values.

An alternative presentation is obtained if we define one previous treatment as a common reference value. The resulting estimated hazard ratios could be either graphed or tabulated. We present, in Table 6.8, the estimated hazard ratios and corresponding confidence intervals comparing NDRUGTX = 0, 2, 5, 10 to NDRUGTX = 1. These results are obtained by using (6.27)–(6.30), with a change in the values being compared. The results in Table 6.8, in the column labeled "0," are the reciprocals of the values shown in Figure 6.9 for 0 versus 1 previous drug treatments, while the values for column "2" are the same as those in Figure 6.9 for 1 versus 2 previous drug treatments. The values in the columns for 5 and 10 previous treatments cannot be obtained from Figure 6.9. The results in Table 6.8 demonstrate in a more direct manner the increase in hazard rate relative to the modeled minimum at 1. We note that the rates at 0 and 5 previous treatments are about 40 percent higher than the rate at 1. Although not shown, the rates increase progressively, so that the point estimate for 40 versus 1 is about 2.5 (1.70, 3.74).

Age and site are present in the model, with both main effects and their interaction. Since site is at two levels and is fixed by design of the study, we present hazard ratios for age at each site rather than for site at each age. The process is essentially the same as the one used to compute hazard ratios for the number of previous drug treatments, but it is a

Table 6.8 Estimated Hazard Ratios and 95% Confidence Intervals for the Stated Number of Previous Drug Treatments versus One Treatment.

	0	2	5	10
HR	1.42	1.09	1.44	1.82
95% CIE	(1.08, 1.87)	(1.03, 1.16)	(1.22, 1.71)	(1.40, 2.36)

bit simpler as we don't have to deal with nonlinear scaling. Again, the first step is to write down the equation for the hazard ratio as a function of the variables of interest, holding all the others fixed,

$$g(\text{AGE},\text{SITE},\mathbf{z}) = \beta_1 \text{AGE} + \beta_2 \text{SITE} + \beta_3 \text{AGE} \times \text{SITE} + \boldsymbol{\beta}'\mathbf{z}.$$

The second step is to write down the expression for the difference of interest, in this case an increase of c years of age holding SITE fixed:

$$g(\text{AGE}+c,\text{SITE},\mathbf{z}) - g(\text{AGE},\text{SITE},\mathbf{z})$$
$$= \left\{ \beta_1(\text{AGE}+c) + \beta_2\text{SITE} + \beta_3(\text{AGE}+c)\times\text{SITE} + \boldsymbol{\beta}'\mathbf{z} \right\}$$
$$- \left\{ \beta_1(\text{AGE}) + \beta_2\text{SITE} + \beta_3(\text{AGE})\times\text{SITE} + \boldsymbol{\beta}'\mathbf{z} \right\}$$
$$= \beta_1 c + \beta_3 c \times \text{SITE}. \tag{6.31}$$

The next step is to choose a value for c, say 5 years, and to estimate the value of (6.31) using the estimated coefficient of AGE from Table 6.6, $\hat{\beta}_1 = -0.041$, and the estimated coefficient of the interaction of AGE and SITE, $\hat{\beta}_3 = 0.032$. The estimated hazard ratio for an increase of 5 years of AGE at SITE $= 0$ is

$$\hat{\text{HR}}(\text{AGE}+5,\text{AGE},\text{SITE}=0) = e^{5\times(-0.041)} = 0.815,$$

and at SITE $= 1$ it is

$$\hat{\text{HR}}(\text{AGE}+5,\text{AGE},\text{SITE}=1) = e^{5\times(-0.041)+5\times0.032} = 0.956.$$

The endpoints of the $100(1-\alpha)$ percent confidence interval estimator of the hazard ratio are computed by exponentiating the endpoints of the confidence interval of the estimator of (6.32), which are

$$\hat{\beta}_1 c + \hat{\beta}_3 c \times \text{SITE} \pm z_{1-\alpha/2}\hat{\text{SE}}\left(\hat{\beta}_1 c + \hat{\beta}_3 c \times \text{SITE}\right), \tag{6.32}$$

where

$$\hat{\text{SE}}\left(\hat{\beta}_1 c + \hat{\beta}_3 c \times \text{SITE}\right) = \left[\begin{array}{c} c^2 \times \hat{\text{Var}}\left(\hat{\beta}_1\right) + c^2 \times \text{SITE}^2 \times \hat{\text{Var}}\left(\hat{\beta}_3\right) \\ + 2c^2 \times \text{SITE} \times \hat{\text{Cov}}\left(\hat{\beta}_1,\hat{\beta}_3\right) \end{array} \right]^{0.5}, \tag{6.33}$$

and since SITE is coded zero or one, $SITE^2 = SITE$. Again the values of the variances and covariance needed to compute the standard error are available from software packages. In this example, these are

$$\widehat{Var}\left(\hat{\beta}_1\right) = 0.000098,$$

$$\widehat{Var}\left(\hat{\beta}_3\right) = 0.000259$$

and

$$\widehat{Cov}\left(\hat{\beta}_1, \hat{\beta}_3\right) = -0.00009.$$

Using these values, the 95 percent confidence interval at $SITE = 0$ is (0.739, 0.898) and at $SITE = 1$ is (0.839, 1.089). The fact that age is linear within site means that these results hold at all ages.

The interpretation is that being older by 5 years at $SITE = 0$ reduces the rate of return to drug use by about 18 percent, and the fact that 1.0 is not contained in the confidence interval points to a significant age effect at this site. At the other site, $SITE = 1$, the rate is only 5 percent lower and is not significant.

The remaining hazard ratio involves race and site. We present the hazard ratio and its confidence interval for race at each site. These may be obtained by using (6.27)–(6.33) with $c = 1$, reflecting the fact that race is dichotomous and has been recoded zero and one. The estimated hazard ratio and 95 percent confidence interval for $RACE =$ other versus $RACE =$ white at $SITE = 0$ is

$$\hat{HR}(\text{other}, \text{white}, SITE = 0) = e^{-0.467} = 0.627$$

and (0.481, 0.816) while at $SITE = 1$ they are

$$\hat{HR}(\text{other}, \text{white}, SITE = 1) = e^{-0.467+0.850} = 1.467$$

and (0.972, 2.214). The interpretation is that non-whites are returning to drug use at a rate that is about 37 percent lower than whites at $SITE = 0$. The confidence interval for the ratio suggests that the rate could be as much as 52 percent lower to only about 18 percent lower. The reverse seems to be the case at $SITE = 1$, where the non-whites are returning to drug use at much higher rate than whites (about 46 percent higher), and the confidence interval suggests that this could be as high as 121 percent or even slightly negative. In any event, the results point not only to important racial differences, but to differences between the two sites. This was discussed with the study team and deemed to be an appropriate interpretation.

In this section we have had to emphasize both the calculation and interpretation of the estimated hazard ratios. In practice, the estimated hazard ratios and their confidence intervals would likely be tabulated with no computational details presented and thus lend themselves to a discussion with more continuity than was possible here. However, for a complicated nonlinear variable like the number of previous drug treatments, inclusion of an appendix providing an outline of how the graphed (or tabulated) hazard ratios and their confidence intervals have been computed can be a helpful addition to a paper.

We conclude our presentation of the fitted model in Table 6.6 with graphs of the covariate-adjusted survivorship functions for the two levels of treatment. Since the model is complicated, it is not clear what we could use for a mean or median subject, so we use the modified risk score method discussed in Section 4.3 and illustrated in (4.29) and (4.30). The modified risk score is calculated for each subject as $r\hat{m}_i = \hat{r}_i - (-0.247)\text{TREAT}_i$ and the median is $r\hat{m}_{50} = -2.088$. The plotted points for the covariate-adjusted survivorship function for the shorter treatment are

$$\hat{S}(t_i, r\hat{m}_{50}) = \left[\hat{S}_0(t_i)\right]^{\exp(-2.088)}$$

and for the longer treatment are

$$\hat{S}(t_i, r\hat{m}_{50}) = \left[\hat{S}_0(t_i)\right]^{\exp(-2.088-0.2468)}.$$

Graphs of these two functions, at all observed values of time, are shown in Figure 6.10.

The figure shows that, at all times, the covariate-adjusted proportion of subjects who have not returned to drug use is higher for the longer of the two treatments. The covariate-adjusted estimated median times to return to drug may be obtained from the graphs in Figure 6.10 or from a time-sorted list of the functions. The median times are 157 days for the shorter treatment and 190 days for the longer treatment. As noted in Section 4.5, there is no easily computed confidence interval for the estimator of the median time from a modified risk-score-adjusted survi-

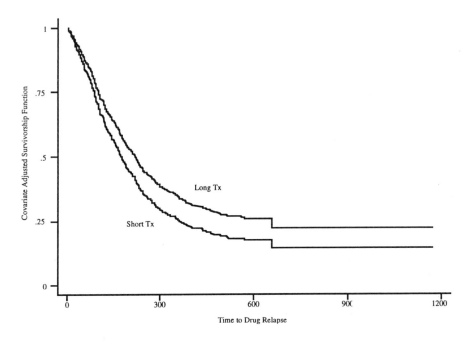

Figure 6.10 Graphs of the covariate-adjusted survivorship functions for the longer and shorter treatments computed using the model in Table 6.6.

vorship function.

In conclusion, the fitted model, shown in Table 6.6, has allowed description of a number of interesting relationships between time to return to drug use and study factors. The notable results include, besides a significant treatment effect, a differential effect of race within treatment site, the nonlinear effect of the number of previous drug treatments and significant effects due to Beck score and recent IV drug use.

In the next chapter we consider alternative methods for modeling study covariates. These methods are of interest in themselves, but they often also provide alternatives to models that are not adequate due to poor fit or violations of the proportional hazards assumption.

EXERCISES

1. Using data from the HMO-HIV+ study, assess the fit the proportional hazards model containing AGE and DRUG. This assessment of fit should include the following steps: evaluation of the proportional haz-

ards assumption for each of the two covariates, examination of diagnostic statistics, and an overall test of fit. If the model does not fit or adhere to the proportional hazards assumption what would you do next? Note: the goal is to obtain a model to estimate the effect of AGE and DRUG on the survivorship experience.

2. Using the model obtained at the conclusion of problem 1, present a table of estimated hazard ratios, with confidence intervals. Present graphs of the age-adjusted, at the mean age, estimated survivorship functions for the two drug use groups. Use the estimated survivorship functions to estimate the age-adjusted median survival time for each of the two drug use groups.

3. In Section 6.4 diagnostic statistics were plotted and a few subjects were identified as being possibly influential. Fit the model shown in Table 6.6 deleting these subjects one at a time and then, collectively, calculate the percent change in all coefficients with each deletion. Do you agree or disagree with the conclusion in Section 6.4 to keep all subjects in the analysis? Explain the rationale for your decision.

4. A considerable amount of the material presented in this chapter dealt with the evaluation of fit, and the presentation and interpretation of the fitted model shown in Table 5.11 (and repeated in Table 6.6). Repeat the entire process for the fitted model shown in Table 5.13. This model contains an interaction b tween AGE and NDRUGFP1 and, as a result, estimation and presentation of hazard ratios for age, controlling for the number of previous drug treatments and for the number of previous drug treatments controlling for age, is a major challenge.

5. Repeat the full model evaluation and presentation process using the fitted model developed for the WHAS in problems 3 of Chapter 5.

CHAPTER 7

Extensions of the Proportional Hazards Model

7.1 INTRODUCTION

Up to this point we have made several simplifying assumptions in developing and interpreting proportional hazards models. We have used a proportional hazards model with a common unspecified baseline hazard function where all the study covariates had values that remained fixed over the follow-up period. Additionally, we have assumed that the observations of the time variable were continuous and subject only to right censoring. In some settings one or more of these assumptions may not be appropriate.

We may have data from a study in which subjects were randomized within study sites. If we account for site by including it as a covariate, the model forces the baseline hazards to be proportional across study sites. This may not be justified and, if it isn't, a careful analysis of the proportional hazards assumption (as discussed in Chapter 6) for site should reveal the problem. One possible solution is to use site as a stratification variable, whereby each site would have a separate baseline hazard function.

When study subjects are observed on a regular basis during the follow-up period, the course of some covariates over time may be more predictive of survival experience than the original baseline values. For example, continued survival of intensive care unit patients may depend more on changes in their physiologic condition since admission than on their absolute state at admission. Covariates whose values change over time are commonly called *time-varying* or *time-dependent* covariates. These may include measurements on individual subjects or measurements that record study conditions and apply to a number of study subjects.

Suppose that in the process of model checking we determine that the hazard function is nonproportional in one or more of the covariates. One approach is to include in the model an explicit function of the covariate and time, which is essentially the test for nonproportionality described in Chapter 6. In some settings this may provide a useful solution to the problem. A similar approach is to partition the time axis and to fit different functional forms for the covariate within time intervals. We could use the smoothed plots of the scaled Schoenfeld residuals to identify the intervals where the "slope" was the same. For example, suppose in the UIS that the model was nonproportional in the Beck score and that the plot showed an early linear effect up to 100 days. We could model Beck score linearly in the first 100 days and use a different parametric form for the rest of the follow-up period. In general, we wouldn't attempt such an analysis unless there was strong evidence for it from subject matter considerations, statistical tests and plots of the type discussed in Chapter 6.

While right censoring is by far the most frequently encountered censoring mechanism, there may be settings in which incomplete observations arise from other types of censoring. Thus it may be necessary to model the data taking into account a variety of different patterns of censoring.

In settings in which subjects are not under continual observation, but rather their follow-up status is determined at "fixed" intervals, another problem may arise. For example, if we are studying survival of patients admitted to a coronary intensive care unit who were discharged alive, we may contact subjects at 3-month intervals. All that may be known on the subjects is their vital status at these interval points, and the recorded values of time would be 3, 6, 12, and so on. Such data are highly discrete and, in this setting, we should consider using an extension of the proportional hazards model designed to handle interval censored data.

These and other topics that extend the basic proportional hazards model will be discussed in some detail in this chapter. We begin with the stratified proportional hazards model.

7.2 THE STRATIFIED PROPORTIONAL HAZARDS MODEL

The rationale for creating a stratified proportional hazards model is the same as that for other stratified analyses. We assume that there are variables which are known to affect the outcome, but we consider obtaining estimates of their effects to be of secondary importance to those of other covariates. These covariates might be fixed by the design of the study or they might have been identified in earlier analyses. For example, in the UIS, subjects were randomized to treatment length within study site. Thus study site was fixed by design. In previous chapters, we treated site as if it were not fixed for purposes of demonstration; however, site is most appropriately considered as a stratification variable. The stratified proportional hazards model is also sometimes used to accommodate nonproportional hazards in a nominal scale covariate.[1]

The model is, in spirit, quite similar to the one used in a matched logistic regression analysis [see Hosmer and Lemeshow (1989, Chapter 7)]. The effect of all covariates whose values are constant within each stratum is incorporated into a stratum-specific baseline hazard function. The effects of other covariates may be modeled either with a constant slope across strata or with different slopes. For example, in the model developed for the UIS in Chapters 5 and 6, we included age both as a main effect and as part of an interaction term with site. This interaction term is used to model the difference in the effect of age between the two sites. In general, this modeling decision, equal or different slopes, may be made based on subject matter considerations or by using the methods for selecting interactions discussed in Chapter 6. We now describe the stratified proportional hazards model using constant slopes, but note that the non-constant slopes model may be handled simply by specifying one of the covariates as an interaction. The proportional hazard function for stratum s is

$$h_s(t, \mathbf{x}, \boldsymbol{\beta}) = h_{s0}(t)e^{\mathbf{x}'\boldsymbol{\beta}}, \qquad (7.1)$$

---

[1]Note that, if the hazard function for a continuous covariate has been identified as being nonproportional using the methods in Chapter 6, one could create a grouped or categorized version of the covariate and treat it as if it were a nominal scale covariate.

where we assume there are $s = 1, 2, \ldots, S$ strata. In the UIS, there are $S = 2$ strata indicating the two study sites.

Hazard ratios are computed within each stratum. For example, in a stratified model, we may find that the stratum-specific hazard ratio for a dichotomous covariate is 2. The interpretation of this hazard ratio is the same as in the nonstratified model, that is, the hazard rate in the $x = 1$ group is twice the hazard rate in the $x = 0$ group, and this interpretation applies to each stratum.

The form of the partial likelihood for the sth stratum is identical to the partial likelihood used in earlier chapters, see (3.17), but it includes an additional subscript, s, indicating the stratum. The contribution to the partial likelihood for the sth stratum is

$$l_{sp}(\beta) = \prod_{i=1}^{n_s} \left[\frac{e^{x'_{si}\beta}}{\sum_{j \in R(t_{si})} e^{x'_{sj}\beta}} \right]^{c_{si}}, \qquad (7.2)$$

where n_s denotes the number of observations in the sth stratum, t_{si} denotes the ith observed value of time in the sth stratum, c_{si} is the value of the 0/1 censoring variable associated with t_{si}, $R(t_{si})$ denotes the subjects in stratum s in the risk set at time t_{si} and $\mathbf{x}_{si}$ is the vector of p covariates. The full stratified partial likelihood is obtained by multiplying the contributions to the likelihood, namely

$$l_{Sp}(\beta) = \prod_{s=1}^{S} l_{sp}(\beta). \qquad (7.3)$$

The subscript S is used in (7.3) to differentiate the stratified partial likelihood from the nonstratified partial likelihood, $l_p(\beta)$, used in previous chapters. The maximum stratified partial likelihood estimator of the parameter vector, β, is obtained by solving the p equations obtained by differentiating the log of (7.3) with respect to the p unknown parameters and setting the derivatives equal to zero. We do not provide these equations since they are quite similar in form to those in (3.30). The only difference is that the weighted risk set covariate means are based on the data within each stratum. The estimator of the covariance matrix of the estimated coefficients is obtained from the inverse of the observed information matrix in a manner similar to the nonstratified setting, see (3.31)–(3.33).

The general steps in model building and assessment are the same for the stratified model as for the nonstratified model, the only difference being that the stratified analysis is based on the partial likelihood in (7.3). The full stratified analysis of the UIS data using site as a stratification variable would require that we repeat all the steps in Chapters 5 and 6 relevant to model building and assessment. Since repeating these steps would not demonstrate anything new, we begin our discussion of the application of the stratified proportional hazards model in the UIS assuming that the stratified model variables are the same as those shown in Table 5.11, but with SITE excluded from the model and defined as the stratification variable. The results of fitting the stratified model are shown in Table 7.1

The results in Table 7.1 are nearly identical to those in Table 5.11. This is not too surprising since the nonstratified model that was fit in Chapter 5 was shown to have proportional hazards in SITE in Chapter 6. The interpretation of the estimates of hazard ratios and their confidence intervals would not differ substantially from those presented and discussed in Section 6.6 and thus are not repeated here.

After a stratified model has been fit, the methods described in Section 3.5 may be used to estimate stratum-specific baseline survivorship functions. These functions may then be used to estimate covariate-adjusted stratum-specific survivorship functions using the methods described in Section 4.5.

Table 7.1 Estimated Coefficients, Standard Errors, z-Scores, Two-Tailed p-Values and 95% Confidence Intervals for the Proportional Hazards Model Stratified by Site of the Intervention for the UIS ($n = 575$)

Variable	Coeff.	Std. Err.	z	$P>|z|$	95% CIE
AGE	−0.041	0.010	−4.17	<0.001	−0.061, −0.022
BECKTOTA	0.009	0.005	1.75	0.079	−0.001, 0.018
NDRUGFP1	−0.573	0.125	−4.57	<0.001	−0.818, −0.327
NDRUGFP2	−0.214	0.049	−4.39	<0.001	−0.309, −0.118
IVHX_3	0.231	0.109	2.13	0.033	0.018, 0.444
RACE	−0.464	0.135	−3.44	0.001	−0.728, −0.199
TREAT	−0.249	0.094	−2.63	0.008	−0.434, −0.064
AGEXSITE	0.033	0.016	2.04	0.042	0.001, 0.064
RACEXSITE	0.847	0.248	3.42	0.001	0.361, 1.333

Log-likelihood = −2349.776.

In the current example, we denote the estimators of the stratum-specific baseline survivorship function as $\hat{S}_{s0}(t), s = 0,1$ since SITE was coded zero and one. The fitted model in Table 7.1 is complicated, so we choose to use the modified risk score approach to obtain graphs to describe the effect of treatment within each stratum, see (4.26)–(4.30). Since this estimate is to be made within strata, and since the distribution of the risk score could be different across strata, we favor using stratum-specific median modified risk scores. The median values based on the fitted model in Table 7.1 are $\hat{rm}_{050} = -2.0079$ for SITE $= 0$ and $\hat{rm}_{150} = -0.8903$ for SITE $= 1$. It follows that the estimators for the stratum-specific modified risk-score-adjusted survivorship functions for the shorter treatment are described by the equations

$$\hat{S}(t, \text{SITE} = 0, \text{TREAT} = 0, \hat{rm}_{050}) = \left[\hat{S}_{00}(t)\right]^{\exp(-2.0079)} \tag{7.4}$$

and

$$\hat{S}(t, \text{SITE} = 1, \text{TREAT} = 0, \hat{rm}_{150}) = \left[\hat{S}_{10}(t)\right]^{\exp(-0.8903)}. \tag{7.5}$$

The estimators for the longer treatment are described by the equations

$$\hat{S}(t, \text{SITE} = 0, \text{TREAT} = 1, \hat{rm}_{050}) = \left[\hat{S}_{00}(t)\right]^{\exp(-2.0079+(-0.2486))} \tag{7.6}$$

and

$$\hat{S}(t, \text{SITE} = 1, \text{TREAT} = 1, \hat{rm}_{150}) = \left[\hat{S}_{10}(t)\right]^{\exp(-0.8903+(-0.2486))}, \tag{7.7}$$

where -0.2486 is the estimated coefficient of treatment from Table 7.1.

Figure 7.1 presents the graphs of the four modified risk-score-adjusted survivorship functions. The range of time has been restricted to 880 days to emphasize the differences between the stratum-specific functions. The lowest of the four functions corresponds to the modified risk-score-adjusted survivorship function for the short treatment at SITE $= 0$. The uppermost curve corresponds to the modified risk-score-adjusted survivorship function for the long treatment at SITE $= 1$. The middle two curves are essentially indistinguishable from each other and correspond to the modified risk-score-adjusted survivorship function for the long treatment at SITE $= 0$ and the short treatment at SITE $= 1$. The distance between the two curves for each stratum is such that the estimated hazard ratio for treatment 1 versus treatment 0 is

$$\hat{HR} = \exp(-0.2486) = 0.78$$

at each value of time.

The modified risk-adjusted survivorship functions may be used in the same manner illustrated in Section 6.6 to obtain estimates of adjusted median time to return to drug use, and these are presented in Table 7.2. The difference in the medians is 31 days at SITE = 0 and 44 days at SITE = 1. These differences are consistent with the difference of 33 days from the nonstratified analysis presented in Figure 6.10. As discussed in Sections 4.5 and 4.6, there is not an easily obtained confidence interval estimator for the estimator of median time to response when the modified risk score is used.

The stratified proportional hazards model is the correct model to use in a setting in which covariates have values fixed by the design of the study. In addition, using stratified analysis provides the possibility for a simple solution to nonproportional hazards in a nominal scale covariate. When the hazard function is nonproportional over the levels of the stratification variable, there may be more substantial differences

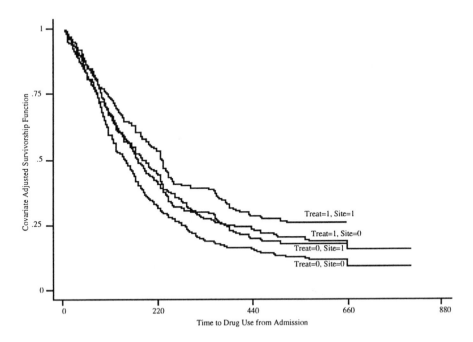

Figure 7.1 Graphs of the modified risk-score-adjusted stratum-specific survivorship functions for treatment.

Table 7.2 Modified Risk-Score-Adjusted Estimated Median Time to Return to Drug Use for Each Level of Treatment Within Site

SITE = 0		SITE = 1	
Treatment		Treatment	
Short	Long	Short	Long
144	175	184	228

between the results of fitting stratified and nonstratified proportional hazards models than those seen in the UIS. In all cases, one must pay attention to the number of observations and survival times within each stratum. Small numbers within any stratum will result in an estimated baseline survivorship function with greater variance than the estimates from strata with more data. However, the variance of the estimates of the coefficients in the model is a function of the total sample size and total number of survival times, even in a stratified model.

7.3 TIME-VARYING COVARIATES

Until now we have assumed that the values of all covariates were determined at the point when follow-up began on each subject, time zero, and that these values did not change over the period of observation. There may, however, be situations in which the values of the covariates change over time, and in which the value of the hazard function depends more on the current value of the covariate than on the value at time zero. As will be shown, it is not difficult to generalize the proportional hazards regression model and its partial likelihood to include time varying covariates. However, from a conceptual point of view, the model becomes much more complicated, and one should give serious consideration to the nature of any time-varying covariate before including it in the model. Specifically, one must pay close attention to the definition of the zero value of time, since for different subjects, the zero value may occur at different calendar times. However, the value of any time-varying covariate must depend only on study time, not on calendar time. Another concern is the potential to overfit a model when using time-varying covariates. In all instances, inclusion of time-varying covariates should be based on strong clinical evidence. One of the earliest applications of the use of time-varying covariates in a biomedical setting may be found in Crowley and Hu (1977). Andersen (1992), in a summary paper, illustrates the use of time-varying covariates in several examples and discusses some of the problems one may encounter when including them in a model. Collett (1994) and Marubini and Valsecchi (1995) discuss using time-varying covariates at a level comparable to this text, while Fleming and Harrington (1991) and Andersen, Borgan, Gill and Keiding (1993) present the topic from the counting process point of view.

Time-varying (or time-dependent) covariates are usually classified as being either *internal* or *external*. An internal time-varying covariate is one whose value is subject-specific and requires that the subject be under periodic observation. For example, consider a clinical trial for a new cancer treatment in which the endpoint is death due to the cancer. If the current value of a measure of physiologic status is related to the progression of the disease, measurement of this covariate depends on continued observation of the subject. In contrast, an external time-varying covariate is one whose value at a particular time does not require subjects to be under direct observation. Typically, these covariates are study or environmental factors that apply to all subjects under observation. For example, consider a study of a new medication for relief from symptoms of hay fever in which the time variable for the outcome records the number of hours until self-perceived relief from symptoms. An external covariate might be the average hourly pollen count. A subject-specific external covariate is the subject's age. If we follow subjects for a long enough period of time, their current age may have more of an effect on survival than their age when the study began. However, once we know a subject's birth date, age may be computed at any point in time, regardless of whether the subject is still under observation. Another important external time-varying covariate is time itself. We made extensive use of this time-varying covariate in Chapter 6, where it was used to test the assumption of proportional hazards in fixed covariates via the inclusion of the interaction $x \ln(t)$ in the model. In the remainder of this section we will not differentiate between internal and external time-varying covariates.

We need to generalize the notation to include time-varying covariates in the proportional hazards regression model. Let $x(t)$ denote the value of the covariate x measured at time t. Assume, as we have up to this point, that we have adjusted each subject's follow-up time to begin at 0 as opposed to using real or calendar time. Thus we assume that t determines the value of $x(t)$ even though the same t may arise from different calendar times. For example, suppose we define a time-varying covariate as the length of time on study. Subject 1 may begin on May 1 and be on the study 60 days. Subject 2 may begin the study on July 1 and also be on the study 60 days. What is important is that both subjects were on the study 60 days, not that they began at different times. In order to include such a covariate in the partial likelihood, (3.17) or (3.18) and their multivariable equivalents, we need to account for the subject as well as the specific time. Let $x_l(t_i)$ denote the value of the covariate for subject l at time t_i. To allow for multiple covariates, we let $x_{lk}(t_i)$, $k = 1, 2, \ldots, p$ denote the value of the kth covariate for subject l at time t_i and denote the vector of covariates as

$$\mathbf{x}'_l(t_i) = \left[x_{l1}(t_i), x_{l2}(t_i), \ldots, x_{lp}(t_i) \right].$$ (7.8)

The notation in (7.8) is completely general in the sense that, if a particular covariate, x_k, is fixed (i.e., not time-varying) then

$$x_{lk}(t_i) = x_{lk}(t = 0) = x_{lk},$$

and this has lead some authors, for example, Andersen, Borgan, Gill and Keiding (1993), to use the time-dependent notation in (7.8) exclusively. The generalization of the proportional hazards regression function (3.7) to include possibly multiple time-varying covariates is

$$h(t, \mathbf{x}(t), \boldsymbol{\beta}) = h_0(t) \exp[\mathbf{x}'(t)\boldsymbol{\beta}],$$ (7.9)

and the generalization of the partial likelihood function (3.17) is

$$l_p(\boldsymbol{\beta}) = \prod_{i=1}^{n} \left[\frac{e^{\mathbf{x}'_i(t_{(i)})\boldsymbol{\beta}}}{\sum_{l \in R(t_{(i)})} e^{\mathbf{x}'_i(t_{(i)})\boldsymbol{\beta}}} \right]^{c_i}.$$ (7.10)

An important assumption of the model in (7.9) is that the effect for the time-varying covariate, as measured by its coefficient, does not depend on time. Models with time-varying coefficients are complex and are discussed in the context of additive models in Chapter 9.

The estimators of the coefficients and their associated standard errors are obtained in a manner identical to the one described in Chapter 3, using (7.10) in place of (3.17) and (3.18). Often the biggest problem in practice is the data management necessary to describe to the software the values of the time-varying covariates. We strongly suggest that, before data collection actually begins, one consult with experts in data management for whatever software package is going to be used for model development. There is significant variability among software packages with respect to the ease with which time-varying covariates are handled.

We illustrate modeling a time-varying covariate with data from the UIS. The analyses of the UIS data in this text have used survival time defined as the length of time from admission to the treatment program to subsequent return to drug use. The actual length of treatment (LOT), as defined in Table 1.3, is an internal time-varying covariate recorded for each subject. The two treatment protocols under study were of different durations. It is possible that the significant treatment effect seen in the fitted model (shown in Tables 5.11 and 6.6 and discussed in Section 6.6) may be due to the fact that a subject's risk of returning to drug use was reduced while under treatment. Even though subjects were free to leave the treatment program at any time and possibly return to drug use, those on the longer treatment may have had a tendency to stay longer and thus appear to remain drug free for a longer period after admission. In order to explore this possibility we created the dichotomous internal time-varying covariate OFF_TRT(t), where

$$OFF\_TRT(t) = \begin{cases} 0 \text{ if } t \le LOT, \\ 1 \text{ if } t > LOT, \end{cases}$$

where LOT stands for the length of treatment. For example, consider one of the terms in (7.10) and suppose the survival time indexing the risk set is 200 days. Subjects in the risk set would have OFF_TRT(200) = 0 if their value of LOT was greater than 200, meaning that after 200 days of follow-up they were still in a treatment facility. Once the length of follow-up exceeds the duration of treatment, a subject is "off treatment" and their value of OFF_TRT changes to one. There is a considerable amount of computation involved when fitting a model with time-varying covariates. Not only must the composition of each risk set be determined, the actual values of the time-varying covariates also need to be computed. Table 7.3 presents the results obtained when OFF_TRT is added to the final UIS model developed in Chapters 5 and 6.

The results in Table 7.3 demonstrate a rather dramatic effect due to the time-varying covariate, OFF_TRT. Not only is its coefficient highly significant, but its inclusion in the model induces a substantial change in the coefficient for TREAT. In addition, each of the three coefficients involving SITE, while still significant, has also changed. The estimated hazard ratio comparing a subject who is off treatment to one who is on treatment at any time t is

$$\hat{HR}(t) = e^{2.571} = 13.08.$$

This estimated hazard ratio, although quite large, is still plausible. The interpretation is that subjects who are not actively involved in a treatment program are at substantially increased risk of returning to drug use, regardless of the site or type of treatment. Estimation and interpretation of hazard ratios for the other covariates in the model would be done as shown in Section 6.6.

In Chapter 4 we discussed presentation and interpretation of covariate-adjusted survival curves. These methods were used in Section 6.6 and in the previous section. For models containing time-varying covariates, it is possible to estimate the baseline survivorship function because it does not depend on the covariates. However, it is not feasible to present individual covariate-adjusted survivorship functions, since these are explicit functions of time, which is changing.

Time-varying covariates can provide a useful and powerful adjunct

Table 7.3   Estimated Coefficients, Standard Errors, z-Scores, Two-Tailed p-Values and 95% Confidence Intervals for the Model Adding the Time-Varying Covariate OFF_TRT to Model Shown in Table 5.11 for the UIS ($n = 575$)

Variable	Coeff.	Std. Err.	z	$P>|z|$	95% CIE
AGE	−0.038	0.010	−3.77	<0.001	−0.058, −0.018
BECKTOTA	0.008	0.005	1.62	0.105	−0.002, 0.018
NDRUGFP1	−0.609	0.128	−4.74	<0.001	−0.860, −0.357
NDRUGFP2	−0.226	0.050	−4.55	<0.001	−0.323, −0.128
IVHX_3	0.275	0.109	2.52	0.012	0.061, 0.488
RACE	−0.517	0.135	−3.84	<0.001	−0.781, −0.253
TREAT	0.019	0.096	0.20	0.840	−0.169, 0.208
SITE	−0.969	0.516	−1.88	0.060	−1.980, 0.042
AGEXSITE	0.036	0.016	2.30	0.021	0.005, 0.067
RACEXSITE	0.511	0.257	1.99	0.047	0.007, 1.014
OFF_TRT	2.571	0.157	16.40	<0.001	2.264, 2.878

Log-likelihood = −2461.657.

to the covariate composition of a survival time regression model. However, one should carefully consider all the implications of adding them to the model before doing so.

7.4 TRUNCATED, LEFT CENSORED, AND INTERVAL CENSORED DATA

Another simplifying assumption that we have made up to this point is that the data are subject only to right censoring. That is, each observation of time begins at a well-defined and known zero value and follow-up continues until the event of interest occurs or until observation terminates for a reason unrelated to the event of interest (e.g., the subject moves away or the study ends). In practice this is the most frequently encountered type of survival data; however, other reasons for incomplete observation of survival time do occur. In this section, we consider some of these reasons, as well as methods for extending the proportional hazards model to address them.

Imagine the time scale for a study as a horizontal line. Incomplete observation of time can occur anywhere, but it is most common at either the beginning (i.e., on the left) or at the end of the time scale (i.e., on the right). The two most common causes of incomplete observation are censoring and truncation. Thus when we consider an observation of time, we must evaluate whether the observation is subject to left censoring or left truncation and/or right censoring or right truncation. In practice, it is unlikely that an observation will be both censored and truncated on the same end of the time scale; however, an observation can be subject to incomplete observation on both sides (e.g., left truncated and right censored).

We begin with left truncation as it is, after right censoring, the next most common source of incomplete observation of survival time. To illustrate, suppose that in the UIS we begin following subjects after they have completed the treatment program, but we define the drug free period (i.e., survival time) as beginning at the time the subject entered the treatment program, that is, at randomization. In this case, there is a selection process taking place in that only those subjects who completed the treatment program are eligible to be included in the analysis. As a result, their minimum survival time would be the length of their particular treatment program. These observations of time are left truncated. In other words, all follow-up times must exceed some fixed value and those subjects whose follow-up times do not exceed this value are excluded from the analysis.

In the UIS example the truncation point was fixed at the lengths of the two treatment programs but in other settings it may be different for each subject. For example, we might be interested in modeling survival time among subjects in the WHAS who were discharged alive from the hospital. Thus subjects who died in the hospital are not included in the analysis. Survival time, defined by length of follow-up, includes the length of time spent in the hospital. Subjects who were discharged alive have a follow-up that is at least as long as their length of stay in the hospital.

Another term used to describe left truncation is *delayed entry*. Recall that the hazard rate at any time is, in the Nelson–Aalen sense, estimated by the ratio of the number events to the number at risk. The survival experience of subjects with left truncated survival times does not contribute to the analysis until time is at least as large as the left-truncated value. Their entry into the risk sets is delayed. Once entered they remain at risk until they have the event or are right censored.

From a model fitting point of view, left truncation or delayed entry is difficult to handle unless the statistical software package allows counting process type data. In this data structure each subject's follow-up time is described by a beginning value of time, that need not be zero, an end time and a right-censoring indicator variable. Currently, most of the major statistical software packages have this capability. In practice a particular study may contain follow-up times that represent all four possible combinations of left truncation and right censoring, and these are handled by the counting process style of data description. Once accounted for in the data setup, the analysis proceeds exactly as described in the previous chapters.

We use the UIS to provide an example of analysis with left truncation. Of the 628 subjects, 546 remained drug free for the duration of their treatment program. If we analyze drug free time from completion of the treatment program, then the observations are left truncated with delayed entry time equal to their respective lengths of stay. This analysis was performed in STATA, but any package allowing for delayed entry, counting process type data, could just as well have been used. Rather than going through a full model building and evaluation process, we fit the model obtained in Chapter 5 and evaluated in Chapter 6 (see Table 5.11). The results of fitting this model are presented in Table 7.4. Due to missing data in the covariates, the model is based on 504 of the 546 subjects.

The overall model is significant with the likelihood ratio test of $G = 53.76$ which, with 10 degrees-of-freedom, has $p < 0.001$. We note that the estimated hazard ratio for TREAT is

$$\hat{HR} = e^{0.140} = 1.15$$

suggesting a benefit for the short treatment. However, this effect is not significant. The Wald test for the coefficient for TREAT has a significance level $p = 0.184$, and the 95 percent confidence interval for the hazard ratio, $(0.94, 1.41)$, includes 1.0. This analysis is in agreement with that presented in the previous section where length of stay was modeled via a binary time-varying covariate. Both analyses suggest that the treatment effect described in Chapter 5 may be solely due to the difference in the length of the treatment programs. The point here is to not reanalyze the UIS data but to illustrate how to use the proportional hazards model with left-truncated or delayed entry data that is also subject to right censoring. Once delayed entry is accounted for, there are no substantive changes in the methods used for model development, assessment and interpretation.

In summary, the key elements defining a left-truncated observation of survival time are: (1) the observation must, by design of the study, exceed some minimum value, that may be the same or different for all subjects and (2) the beginning or zero value of time is known for each observation. Andersen, Borgan, Gill and Keiding (1993) present, in

Table 7.4 **Estimated Coefficients, Standard Errors, z-Scores, Two-Tailed p-Values and 95% Confidence Intervals Based on a Model Using Delayed Entry of Subjects for the UIS ($n = 504$)**

Variable	Coeff.	Std. Err.	z	$P>|z|$	95% CIE
AGE	−0.033	0.011	−3.05	0.002	−0.055, −0.012
BECKTOTA	0.005	0.005	0.85	0.395	−0.006, 0.015
NDRUGFP1	−0.546	0.143	−3.83	<0.001	−0.826, −0.266
NDRUGFP2	−0.204	0.055	−3.72	<0.001	−0.311, −0.096
IVHX_3	0.215	0.119	1.82	0.070	−0.017, 0.447
RACE	−0.494	0.142	−3.47	0.001	−0.774, −0.215
TREAT	0.140	0.105	1.33	0.184	−0.066, 0.345
SITE	−0.960	0.556	−1.73	0.084	−2.050, 0.129
AGEXSITE	0.040	0.017	2.33	0.020	0.006, 0.073
RACEXSITE	0.220	0.290	0.76	0.447	−0.347, 0.788

Log likelihood = −1910.161.

Chapter III, the mathematical details as well as insightful examples involving left truncation of survival time.

Left censoring of survival time is different from left truncation in that it occurs randomly at the individual subject level while left truncation involves a selection process that applies to all subjects. A follow-up time is left censored if we know that the event of interest took place at an unknown time prior to the actual observed time. Examples of left-censored data often involve age as the time variable and a life course event. For example, in a study modeling the age at which "regular" smoking starts, the data may come from interviews of 12-year olds. A 12-year-old study subject may report that he is a regular smoker but that he does not remember when he started smoking regularly. In this case we know the observed time, 12, is larger than the age when the subject became a regular smoker, the survival time.

The defining characteristic of left-censored data is that the event is known to have occurred and the observed time is larger than the survival time. In a sense, left censoring is the opposite of right censoring, where we know that the event of interest has not occurred and that the observed time is less than the survival time.

Ware and DeMets (1976) proposed one solution for the analysis of left-censored data. They suggest turning the time scale around and treating the data as if they were right censored. This method works if the data are only subject to left censoring. In practice, however, if left censoring can occur then right censoring is also likely to occur. In the example of age at first regular smoking many study subjects will not be regular smokers at the time they are interviewed and these observations of time are right censored. Alioum and Commenges (1996) present a method for fitting the proportional hazards model to arbitrarily censored and truncated data. We will not discuss their method as it is quite complex and requires computational skills greater than those assumed in the rest of this text. Klein and Moeschberger (1997) discuss general mechanisms for incomplete observation of survival time and necessary modifications in estimation methods. A somewhat simpler but less flexible approach is presented later in this section within the context of interval-censored data.

Right truncation occurs when, by design of the study, there is a selection process such that data are available only on subjects who have experienced the event. This typically occurs in settings where data come from a registry containing information on confirmed cases of a disease. For example, all subjects in a cancer registry, by definition, have cancer. Thus any analysis of a time variable that uses confirmed diagnosis of cancer as the event of interest will involve right truncation. Extensions of the methods for left-truncated data or right-censored data for the analysis of right-truncated data are not especially straightforward. We will not consider analysis of right-truncated data further in this text. Instead we refer the reader to Klein and Moeschberger (1997) for a general discussion of likelihood construction and to Alioum and Commenges (1996) for methods for the proportional hazards model.

Another type of incomplete observation of time can occur if we do not know the zero value when we begin observing a subject. The observation may also be subject to right censoring. For example, suppose in a study of survival time of patients with AIDS, some subjects enter the study with active, confirmed disease, but no precise information can be obtained as to when they converted from HIV+ to AIDS. Thus, for these patients, the zero value of time, that is, when they actually developed AIDS, is unknown. Statistical methods have been developed to handle this setting, see DeGruttola and Lagakos (1989) who describe this type of data as being doubly censored.

Interval-censored data is another form of incomplete observation of survival time that can involve left and right censoring as well as truncation. Interval censoring is used to describe a situation where a subject's survival time is known only to lie between two values. Data of this type typically arise in studies where follow-up is done at fixed intervals. For example, in the WHAS the aim is to model the survival time among patients admitted to a hospital for a myocardial infarction. Suppose that patients who were discharged alive from the hospital were contacted every 3 months to ascertain their vital status. Patients who die before the first contact at 3 months have survival times that are left censored. We know the event took place prior to 3 months, yet we are unsure of the exact time. All that is known is that these survival times are at most 3 months. For subjects who die between two contacts, all that is known is that survival time is at least as long as the time of the earlier contact and is no longer than the time of the most recent contact (e.g., between 9 and 12 months). For subjects still alive at their last follow-up, all we know is that their survival time is at least as long as the time associated with their last contact (e.g., alive at 18 months and then lost to follow-up). These observations are right censored.

Recently, Carstensen (1996) and Farrington (1996) have considered regression models for arbitrarily interval-censored survival time that extend methods developed by Finkelstein (1986) specifically for the proportional hazards model as well as earlier developmental work by Prentice and Gloecker (1978). Carstensen and Farrington show how arbitrarily interval-censored data may be fit by casting the problem as a binary outcome regression problem. We present this approach using data from the UIS. For purposes of illustration, the interval censoring is restricted to a few values. Collett (1994) presents methods for interval-censored data similar to those presented here.

Assume that all we know about the observed time for the ith subject is that it is bounded between two known values, denoted $a_i < T \le b_i$. In addition, we know whether the event of interest occurred. The outcome is indicated by the usual censoring variable, $c_i = 1$ if the event occurred and $c_i = 0$ otherwise. Observations that are left censored have $a_i = 0$ and $c_i = 1$. Observations that are right censored have $b_i = \infty$ and $c_i = 0$.

Let the survivorship function at time t for a subject with covariate vector $\mathbf{x}_i$ and associated parameter vector $\boldsymbol{\beta}$ be denoted $S(t, \mathbf{x}_i, \boldsymbol{\beta})$. The probability of the observed interval for the ith subject is

$$\left[S(a_i, \mathbf{x}_i, \boldsymbol{\beta})\right]^{1-c_i} \left[S(a_i, \mathbf{x}_i, \boldsymbol{\beta}) - S(b_i, \mathbf{x}_i, \boldsymbol{\beta})\right]^{c_i}. \qquad (7.11)$$

The expression in (7.11) yields $\left[1 - S(b_i, \mathbf{x}_i, \boldsymbol{\beta})\right]$ for left-censored observations, $\left[S(a_i, \mathbf{x}_i, \boldsymbol{\beta}) - S(b_i, \mathbf{x}_i, \boldsymbol{\beta})\right]$ for noncensored observations and $S(a_i, \mathbf{x}_i, \boldsymbol{\beta})$ for right-censored observations.

The first steps in obtaining estimators of parameters in a regression model are to construct the likelihood function and then to evaluate it with the chosen model. We will use the proportional hazards model, but the method is general enough that other models could be used, including the parametric models discussed in Chapter 8. The likelihood is the product of the terms obtained by evaluating (7.11) for each subject, namely

$$l(\boldsymbol{\beta}) = \prod_{i=1}^{n} \left[S(a_i, \mathbf{x}_i, \boldsymbol{\beta})\right]^{1-c_i} \left[S(a_i, \mathbf{x}_i, \boldsymbol{\beta}) - S(b_i, \mathbf{x}_i, \boldsymbol{\beta})\right]^{c_i}. \qquad (7.12)$$

The computations and model fitting procedure are simplified if only a few values are possible for a and b. In this case, it is easier to refer to the interval-censored values by intervals on the time scale common to all subjects. Assume that we have $J+1$ such intervals denoted $\left(t_{j-1}, t_j\right]$ for $j = 1, 2, \ldots, J+1$ with $t_0 = 0$ and $t_{J+1} = \infty$, and these intervals are the same for all subjects. If follow-up was every 3 months for 3 years in the WHAS, it would make sense to use the 13 intervals:

$$\{(0,3],(3,6],(6,9],\ldots,(33,36],(36,\infty]\}.$$

For ease of presentation, we let I_j denote the jth time interval $(t_{j-1},t_j]$. The binary variable indicating the specific time interval observed for the ith subject is defined as

$$y_{ij} = \begin{cases} 1 \text{ if } (a_i,b_i] = I_j, \\ 0 \text{ otherwise.} \end{cases}$$

In addition, we re-express the probability for the jth interval as

$$S(t_{j-1},\mathbf{x}_i,\boldsymbol{\beta}) - S(t_j,\mathbf{x}_i,\boldsymbol{\beta}) = S(t_{j-1},\mathbf{x}_i,\boldsymbol{\beta})\left[1 - \frac{S(t_j,\mathbf{x}_i,\boldsymbol{\beta})}{S(t_{j-1},\mathbf{x}_i,\boldsymbol{\beta})}\right]. \quad (7.13)$$

The right-most term in the square brackets in (7.13) is the conditional probability that the event occurs in the jth interval given that the subject was alive at the end of interval $j-1$, $\Pr(t_{j-1} < T \le t_j \mid T > t_{j-1})$. Under the proportional hazards model, the ratio of the survivorship function at successive interval endpoints can be simplified (algebraic steps not shown) to:

$$\frac{S(t_j,\mathbf{x}_i,\boldsymbol{\beta})}{S(t_{j-1},\mathbf{x}_i,\boldsymbol{\beta})} = \exp\left[-\exp(\mathbf{x}_i'\boldsymbol{\beta} + \tau_j)\right], \quad (7.14)$$

where

$$\tau_j = \ln\left\{-\ln\left[\frac{S_0(t_j)}{S_0(t_{j-1})}\right]\right\}.$$

In order to further simplify the notation, let us denote the conditional probability in (7.13) as

$$\theta_{ij} = 1 - \exp\left[-\exp(\mathbf{x}_i'\boldsymbol{\beta} + \tau_j)\right]. \quad (7.15)$$

Using the result in (7.15), it follows that the likelihood function in (7.12) may be expressed as follows:

$$l(\boldsymbol{\beta}) = \prod_{i=1}^{n}\prod_{j=1}^{J+1}\left\{\left[S\left(t_{j-1},\mathbf{x}_i,\boldsymbol{\beta}\right)^{1-c_i}\right]\left[S\left(t_{j-1},\mathbf{x}_i,\boldsymbol{\beta}\right)-S\left(t_j,\mathbf{x}_i,\boldsymbol{\beta}\right)\right]^{c_i}\right\}^{y_{ij}}$$

$$= \prod_{i=1}^{n}\prod_{j=1}^{J+1}\left\{\left[S\left(t_{j-1},\mathbf{x}_i,\boldsymbol{\beta}\right)\right]\left[1-\frac{S\left(t_j,\mathbf{x}_i,\boldsymbol{\beta}\right)}{S\left(t_{j-1},\mathbf{x}_i,\boldsymbol{\beta}\right)}\right]^{c_i}\right\}^{y_{ij}}$$

$$l = \prod_{i=1}^{n}\prod_{j=1}^{J+1}\left\{S\left(t_{j-1},\mathbf{x}_i,\boldsymbol{\beta}\right)\times\theta_{ij}^{c_i}\right\}^{y_{ij}}. \qquad (7.16)$$

The next step involves expressing the survivorship function at an interval endpoint as a product of successive conditional survival probabilities, a process similar to the one used to develop the Kaplan–Meier estimator in Chapter 2, but using the expression in (7.15). The algebraic details are not shown, but the result is

$$S\left(t_{j-1},\mathbf{x}_i,\boldsymbol{\beta}\right) = \prod_{l=1}^{j-1}\left(1-\theta_{il}\right). \qquad (7.17)$$

Substituting the expression in (7.17) into the function in (7.16) results in the following likelihood function:

$$l(\boldsymbol{\beta}) = \prod_{i=1}^{n}\prod_{j=1}^{J+1}\left[\prod_{l=1}^{j-1}\left(1-\theta_{il}\right)\theta_{ij}^{c_i}\right]^{y_{ij}}. \qquad (7.18)$$

Let the observed interval for the ith subject be denoted k_i, that is, $I_{k_i} = \left(a_i, b_i\right]$. The first thing to note in (7.18) is that the only time the terms in the product over j differ from 1 is when $j = k_i$. Thus the expression (7.18) simplifies to

$$l(\boldsymbol{\beta}) = \prod_{i=1}^{n}\theta_{ik_i}^{c_i}\prod_{j=1}^{k_i-1}\left(1-\theta_{ij}\right). \qquad (7.19)$$

The likelihood function in (7.19) can be made to look like the likelihood function for a binary regression model. We define a pseudo binary outcome variable as $z_{ij} = y_{ij} \times c_i$ and use it to re-express (7.19) as

$$l(\beta) = \prod_{i=1}^{n} \prod_{j=1}^{k_i-1+c_i} \left(1-\theta_{ij}\right)^{1-z_{ij}} \theta_{ij}^{z_{ij}} . \qquad (7.20)$$

For each subject, i, (7.20) is the likelihood for $k_i - 1 + c_i$ independent binary observations with probabilities θ_{ij} and outcomes z_{ij}. This observation allows us to use standard statistical software to fit the interval-censored proportional hazards regression model.

The first step in the model fitting process is to expand the data for each subject $k_i - 1 + c_i$ times and create the values of j and z_{ij}. For example, in the WHAS, if a subject died in the third interval, between the second and third follow-up calls, then $k_i = 3$ and $c_i = 1$. This subject would contribute 3 lines to the expanded data file. The covariates would be the same in each line; the interval indicator, j, would take on the values 1, 2, and 3; and the binary outcome variable, z, would be zero for the first 2 lines and 1 for the third line of data. If the subject was known to be alive at the fourth follow-up and not found on the fifth call, then $k_i = 5$ and $c_i = 0$. This subject would contribute 4 lines of data, and the value of the binary outcome variable would be zero in all 4 lines. If a subject was still alive at the end of the study (i.e., at the twelfth follow-up call), then $k_i = 13$ and $c_i = 0$. The subject would be represented with 12 lines of data, and the binary outcome variable would be zero for each line.

The binary regression model defined in (7.15) has as its link function, or linearizing transformation, the complementary log-log function, that is, $\ln\left[-\ln\left(1-\theta_{ij}\right)\right] = x_i'\beta + \tau_j$. The design variable, represented by τ_j in the model, is a 0/1 indicator variable for each time interval. All J of these variables are included in the model, which requires that we force the usual constant term to be zero.

An important point to keep firmly in mind is that the manipulations of the likelihood in (7.11)–(7.20) were designed to cast the interval-censored data problem in a form that would allow likelihood analysis by existing software. The problem is not a binary regression problem in the usual sense of the primary outcome variable being a 0/1 variable; however, we manipulated the problem to make it look like one. This is exactly the same sort of manipulation that is done to make the likelihood function for the conditional logistic regression model for a matched-pairs case-control study look like an ordinary logistic regression model with a user-created outcome variable and transformed covariate values.

The likelihood in (7.20) is identical to that which would be obtained using the general methods in Carstensen (1996) and Farrington (1996) under the restriction of a few intervals common to all subjects. If this assumption does not hold, then the more general method must be used. This method also yields a likelihood that may be analyzed using a binary regression model, but it requires a two-step fitting procedure. In addition, a second set of calculations is required to obtain estimates of standard errors of estimated model coefficients. For these reasons we have chosen not to present the general regression method.

The creation of the expanded data set and the corresponding model fitting is more easily demonstrated using an example. Suppose that instead of recording follow-up time in days in the UIS it was ascertained every 6 months. The survival time for each subject is recorded as the 6-month period in which they returned to drug use or as the last 6 month period when they were known to be drug free. There are 7 intervals, $I_j = (6 \times (j-1), 6 \times j]$ for $j = 1, 2, \ldots, 6$, and the seventh interval is $(36, \infty]$. In the UIS data set of 628 subjects, there is only one subject in the seventh interval, there are no subjects in the sixth interval, and only 4 in the fifth interval. The observations of time for these 5 subjects are all censored values. The thin data in the last few intervals highlights a limitation inherent in modeling interval-censored data in that the number of intervals that can be modeled is often limited by the data. In order to be able to estimate the interval-specific parameters, τ_j, each interval must contain at least one subject with a noncensored survival time. Before any model fitting is done using the expanded data set, one should check the cross tabulation of the interval variable by the pseudo binary outcome to be sure that there are no zero frequency cells. When this was done in the UIS using the 7 intervals described above, intervals 5 and 7 generated cells with a zero frequency. We chose to pool the data in intervals 4, 5, 6 and 7. The smallest frequency in the cross tabulation for the pooled data was 3 for the last interval. This indicates that, among the 78 subjects who had a follow-up time categorized as being between 18 and 36 months, only 3 were noncensored observations. While a frequency of 3 is a bit small, we chose to continue with four intervals rather than reduce the data further. This data limitation is essentially the same one encountered in logistic regression when modeling a categorical variable that generates a cell frequency of 0 in the cross tabulation of outcome by the categorical variable [see Hosmer and Lemeshow (1989, Chapter 4)].

Table 7.5 presents a few specific examples from the expanded data set. The first two lines of data in this table represent two subjects who returned to drug use during the first 6-month period. This is indicated by month = 6, interval = 1, censor = 1 and the binary outcome variable $z = 1$. The covariate chosen, for demonstration purposes only, is AGE.

The second block of four lines represents two subjects who returned to drug use between 6 and 12 months. This is noted by month = 12. Each of these subjects contributes two lines of data to the expanded data set. The first line corresponds to the fact that they had not returned to drug use at 6 months. For this line the interval = 1 and $z = 0$. The second line is for the second 6-month period, denoted by interval = 2, and return to drug use in this interval is denoted by $z = 1$. The value of age is the same in both lines.

The third block of six lines represents two subjects: one who returned to drug use between 12 and 18 months and another who, when last contacted at 18 months, had not returned to drug use. Each of these subjects contributes three lines of data, one for each of the three inter-

Table 7.5 Examples of the Expanded Data Set Required to Fit the Binary Regression Model in Equation (7.20) for the UIS

ID	Month	Interval	Censor	z	Age
2	6	1	1	1	30
4	6	1	1	1	29
1	12	1	1	0	36
1	12	2	1	1	36
3	12	1	1	0	30
3	12	2	1	1	30
7	18	1	1	0	36
7	18	2	1	0	36
7	18	3	1	1	36
31	18	1	0	0	36
31	18	2	0	0	36
31	18	3	0	0	36
5	24	1	0	0	21
5	24	2	0	0	21
5	24	3	0	0	21
5	24	4	0	0	21
388	24	1	1	0	40
388	24	2	1	0	40
388	24	3	1	0	40
388	24	4	1	1	40

vals. During the first two intervals of follow-up, neither had returned to drug use so, for each subject, $z = 0$. The value is changed to $z = 1$ in the third line for the subject who returned to drug use and is kept at $z = 0$ for the one who had not returned to drug use.

The last block of eight lines represents two subjects who were last contacted at 24 months or the fourth interval. Subject 5 was still drug free at 24 months, so $z = 0$ for all four lines. Subject 388 had returned to drug use between 18 and 24 months, so $z = 1$ on the fourth line. In all cases, the value of age is repeated on all lines. The same would be true of any other covariates being considered in the model.

The technical details of expanding the data set will vary by software package, but most can perform the expansion without too much trouble. Analyses presented here were performed in STATA, where the data expansion is especially easy to perform.

In order to illustrate the model fitting we chose to use the same model developed in Chapter 5 and used in the earlier sections of this chapter. Table 7.6 presents the results of fitting the interval-censored proportional hazards model using the likelihood in (7.20) with the expanded data set.

It is interesting to note that the results in Table 7.6 are in general agreement with those presented in Table 5.11. This should not be taken

Table 7.6 Estimated Coefficients, Standard Errors, z-Scores, Two-Tailed p-Values and 95% Confidence Intervals Based on the Interval Censored Proportional Hazards Model for the UIS (n = 575)

Variable	Coeff.	Std. Err.	z	P>|z|	95% CIE
AGE	−0.040	0.010	−3.91	<0.001	−0.060, −0.020
BECKTOTA	0.006	0.005	1.19	0.236	−0.004, 0.016
NDRUGFP1	−0.531	0.130	−4.08	<0.001	−0.786, 0.276
NDRUGFP2	−0.197	0.050	−3.91	<0.001	−0.295, −0.098
IVHX_3	0.236	0.111	2.12	0.034	0.018, 0.453
RACE	−0.444	0.137	−3.25	0.001	−0.712, −0.176
TREAT	−0.235	0.097	−2.43	0.015	−0.425, −0.046
SITE	−1.220	0.548	−2.23	0.025	−2.293, −0.146
AGEXSITE	0.029	0.017	1.74	0.081	−0.004, 0.062
RACEXSITE	0.825	0.255	3.24	0.001	0.326, 1.324
INT_1	1.827	0.432	4.23	<0.001	0.980, 2.675
INT_2	1.745	0.446	3.92	<0.001	0.872, 2.619
INT_3	0.904	0.475	1.90	0.057	−0.027, 1.835
INT_4	−0.816	0.728	−1.12	0.262	−2.242, 0.611

as an endorsement for collecting survival time data at broad intervals. Rather, it shows that, if one must use interval-censored data, one can obtain reasonable estimates of covariate effects.

The presentation and interpretation of the estimated coefficients for study covariates in Table 7.6 would proceed in exactly the same manner as illustrated in Chapter 6 for the fitted, noninterval-censored, proportional hazards model. In this regard, it is important to remember that even though a binary regression program was used to obtain coefficient estimates, the model generating the likelihood is the proportional hazards model.

We can use the results of the fitted model and (7.17) to obtain estimates of the covariate–adjusted survivorship function. Since there are only four time intervals, the survivorship function is not likely to be of great practical value. In other settings with more intervals, it might, however, and for this reason we present and illustrate the method.

As in the noninterval-censored setting, we begin by obtaining an estimator of the baseline survivorship function. In this case the values are obtained at endpoints of each interval, using the interval-specific parameter estimates. It follows from the definition of τ_j, and the fact that $S_0(t_0 = 0) = 1$, that the estimator of the baseline survivorship function at the end of the first interval is

$$\hat{S}_0(t_1) = \exp\left[-\exp(\hat{\tau}_1)\right].$$

The estimator of the baseline survivorship function at the end of the second interval is

$$\hat{S}_0(t_2) = \hat{S}_0(t_1) \times \exp\left[-\exp(\hat{\tau}_2)\right].$$

In general, the estimator at the end of the jth interval, t_j, is

$$\hat{S}_0(t_j) = \hat{S}_0(t_{j-1}) \times \exp\left[-\exp(\hat{\tau}_j)\right] \tag{7.21}$$

for $j = 1, 2, \ldots, J$. As in the continuous time setting, the actual values of these estimators will not be particularly useful unless the data have been centered in such a way that having covariates equal to zero corresponds to a clinically plausible subject. We obtain the covariate-adjusted estimator of the survivorship function from the estimator of the baseline survivorship function in the same way as in the noninterval-censored case, namely

$$\hat{S}\left(t_j, \mathbf{x}, \hat{\boldsymbol{\beta}}\right) = \left[\hat{S}_0\left(t_j\right)\right]^{\exp\left(\mathbf{x}'\hat{\boldsymbol{\beta}}\right)}. \tag{7.22}$$

The covariate–adjusted survivorship function can also be used to provide a graphical, but model–based, comparison of survivorship experience in groups, such as the two treatments in the UIS. In the current example, since the model is quite complicated, we use the modified risk– score approach, shown in Chapters 4 and 6, to compare the survivorship functions for the two treatments. In doing so, one must not perform the calculations in the expanded data set as it has more than one observation or data line per subject for those whose follow–up goes beyond the first interval. One must either delete the extra lines from the expanded data set or return to the original data set to perform the calculations.

We return to the original data. The first step is to calculate the modified risk score for each subject. This calculation uses the coefficients from the model in Table 7.6, excluding those for the intervals. The median value of the modified risk score among the 575 subjects who had complete data on the covariates used in the model is –2.027. The estimated survivorship functions for subjects in the short treatment are obtained by evaluating

$$\hat{S}\left(t_j, \text{TREAT} = 0, rm_{50} = -2.027\right) = \left[\hat{S}_0\left(t_j\right)\right]^{\exp(-2.027)}$$

for $j = 0, 1, \ldots, 5$. The estimated survivorship functions for subjects in the long treatment are obtained by evaluating

$$\hat{S}\left(t_j, \text{TREAT} = 1, rm_{50} = -2.027\right) = \left[\hat{S}_0\left(t_j\right)\right]^{\exp(-2.027 + (-0.235))}$$

for $j = 0, 1, \ldots, 5$, where –0.235 is the estimated coefficient for TREAT in Table 7.6. Figure 7.2 presents the graphs of the two estimated survivorship functions. The plotted points are connected by steps to emphasize that, in the interval censored data setting, we have no information about the baseline survivorship function between interval endpoints, and one must assume it is constant.

The graph itself is not particularly interesting in this example since only four intervals were used. However, in an analysis with more intervals, it could be used to provide estimates of quantiles of survival time using the methods discussed in Chapter 4 and illustrated in Chapter 6.

In summary, the modeling paradigm for interval-censored survival data is essentially the same as for noninterval censored data. Interpretation and presentation of the results of a fitted proportional hazards model is identical for the two types of data. The difference is that model building with interval-censored data uses the binary regression likelihood in (7.20). This means that model building details, such as variable selection, identification of the scale of continuous covariates, and inclusion of interactions, use techniques based on binary regression modeling. These are discussed in detail for the logistic regression model in Hosmer and Lemeshow (1989) and may be used without modification for the complimentary log-log model. The methods for assessing model adequacy and fit discussed in Hosmer and Lemeshow (1989) are for use with the logistic regression model only. McCullagh and Nelder (1989, Chapter 12) discuss these methods for generalized

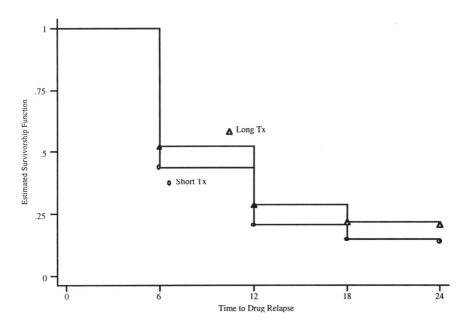

Figure 7.2 Graphs of the modified risk-score-adjusted survivorship functions for the two treatments based on the fitted model in Table 7.6 for the UIS ($n = 575$).

linear models. Details of how to obtain measures of leverage, residual and influence, as well as overall goodness-of-fit from existing software packages for the complementary log-log link function have not been described.

The complementary log-log binary regression model and the logistic regression model are quite similar for probabilities less than 0.15 [see McCullagh and Nelder (1989, page 109)]. This fact lead us to suggest [see Hosmer and Lemeshow (1989, Chapter 8)] that the logistic regression model could be used in place of the complementary log–log model in (7.20). In 1989, when these books were published, software to handle the complementary log-log link was not readily available and use of the logistic regression model was proposed as a practical solution to a computing problem. Given the capabilities of current software, there is no longer any reason to use a logistic regression approximation to the complementary log-log model. In fact, use of the logistic regression model in the current example yields coefficients (output not shown) that are quite different from those in Table 7.6.

It is possible, in some settings, that survival time could be thought of as being a discrete random variable. Kalbfleisch and Prentice (1980) discuss the discrete-time proportional hazards model and show that it may be obtained as a grouped–time version of the continuous-time proportional hazards model. The grouped-time model discussed by Kalbfleisch and Prentice is identical to the interval-censored model presented in this section, so analysis of a discrete-time proportional hazards model would use the methods described in this section.

Kalbfleisch and Prentice (1980) also present the discrete-time logistic model, the same one used by Hosmer and Lemeshow (1989, Chapter 8). They show, using a partial-likelihoodtype argument, that the interval–specific parameters can be eliminated from the model and only the slope coefficients estimated [see Kalbfleisch and Prentice (1980), page 78, equation (4.14)]. The capability of modeling data using this particular likelihood is available in SAS, as well as in an addendum to STATA (Stata Technical Bulletin #37). They note that, if discrete data arise from a grouped continuous-time model, the estimates of the parameters are not consistent, and they do not recommend using it in this case. The implication is that one should not use the discrete-time logistic model partial likelihood for the analysis of continuous-time interval-censored data. This is important to remember since, when one is modeling interval-censored data with many intervals, the temptation will be great to "sweep under the rug" the interval-specific parameters. As one might expect, use of the logistic partial likelihood in the current example yields parameter estimates that are quite close to ones obtained when using the logistic regression model in (7.20). As noted above, these estimates were quite different from those shown in Table 7.6, supporting the recommendation that the logistic model not be used as an approximation to the analysis of a proportional hazards model analysis of interval-censored data.

EXERCISES

1. Problem 3 at the end of Chapter 5 involved finding the best model in the WHAS for LENFOL as survival time and FSTAT as the censoring variable. The fit and adherence to model assumptions was assessed in problem 5 in Chapter 6. In this problem and section we call this model the "WHAS model."

(a) Treat grouped cohort, YRGRP, as a stratification variable and fit the WHAS model. Compare the estimated coefficients with those from the fit of the WHAS model obtained in problem 3 of Chapter 5. Are there any important differences, i.e., changes greater than 15 percent?

(b) If the WHAS model examined in problem 5 of Chapter 6 was not proportional in any covariates, examine the effectiveness of using these covariates to define stratification variables. (In this problem do not include YRGRP as a stratification variable.)

(c) It is possible that the effect of the covariates in the WHAS model are not constant across cohorts defined by YRGRP. Examine this via inclusion of interactions between YRGRP and model covariates in an extended WHAS model with YRGRP as a stratification variable.

2. In this chapter we demonstrated, via a time-varying dichotomous covariate, that much of the apparent treatment effect in the UIS might be due to differences in length of treatment, LOT. It is quite possible in the WHAS that differences in long-term survival in the grouped cohorts, YRGRP, could be due to different lengths of stay in the hospital. Examine this by creating a dichotomous time-varying covariate comparing LENSTAY and LENFOL and adding it to the WHAS model.

3. An alternative approach to the analysis of long-term survival in the WHAS is to study the survival experience of patients post-hospital discharge using LENFOL as survival time. Restrict the analysis to patients discharged alive, but account for differing lengths of stay by defining LENSTAY as the delayed entry time. That is, use counting process–

type input of the form (LENSTAY, LENFOL, FSTAL). Note that subjects with LENSTAY = LENFOL must be excluded as 0 is not an allowable value for survival time in any software package. Alternatively one could add a small constant (less than 1) to the value of LENFOL for these subjects. Examine the effect on the WHAS model of this alternative method of assessing long-term survival in the WHAS. The analysis in this problem could be performed at several levels. The simplest analysis is just a refit of the WHAS model from problem 3 of Chapter 4, but with delayed entry. The most elaborate analysis involves a complete repeat of all steps performed in the model building and assessment in Chapters 5 and 6.

4. The survival time variable in the WHAS, LENFOL, is calculated as the days between the hospital admission date, HOSPDAT, and the date of the last follow–up, FOLDAT. Follow–up of patients is not done continuously but at various intervals. Thus it is possible that a reported number of days of follow–up could be inaccurate. In order to explore this, create a new variable MNTHFOL = LENFOL/30.416667. Use MNTHFOL to create a discrete, in multiples 24 months, (e.g., 2 year intervals). For purposes of this problem, restrict analysis to those subjects with at most 120 months of follow–up.

   (a) Use the method for fitting interval-censored data presented in Section 7.4, fit the WHAS model from problem 3 in Chapter 5. Compare the results of the two fitted models. Is this comparison helpful or not in evaluating the described grouping strategy as a method for dealing with imprecisely measured survival times?

   (b) Use the fitted model in problem 4(a) and present covariate-adjusted survivorship functions comparing the survivorship experience of patients with and without complications due to cardiogenic shock.

   (c) Use the graphs in problem 4(b) to estimate the covariate-adjusted median survival time for the two levels of cardiogenic shock.

   (d) Describe a strategy for including grouped cohort, YRGRP, as a stratification variable in the interval censored data analysis in problem 4(b). If possible implement this strategy.

CHAPTER 8

Parametric Regression Models

8.1 INTRODUCTION

In the previous chapters, we have focused almost exclusively on the use of either nonparametric or semiparametric models for the analysis of censored survival time data. The rationale for using these techniques, in particular the semiparametric proportional hazards regression model, was to avoid having to specify the hazard function completely. The utility of the proportional hazards model stems from the fact that a reduced set of assumptions is needed to provide the hazard ratios formed from the coefficients that are easily interpreted and clinically meaningful. However, there may be settings in which the distribution of survival time, through previous research, has a known parametric form that justifies use of a fully parametric model to better address the goals of the analysis. These models have some advantages. In particular, (1) full maximum likelihood may be used to estimate the parameters, (2) the coefficients can be clinically meaningful and, for some models, are related to those from a proportional hazards model, (3) fitted values from the model can provide estimates of survival time and (4) residuals can be computed that are differences between observed and predicted values of time. The result is that an analysis using a fully parametric model can have the look and feel of a normal errors linear regression analysis.

Two parametric regression models, the exponential and Weibull, were used in Chapter 1 to compare and contrast the analysis of right-censored survival time data with the usual linear regression model with normally distributed errors. We expanded the discussion in Section 3.1 to introduce regression modeling of survival time. We encourage the reader to review Sections 1.1 and 3.1 before proceeding further in this chapter. In Sections 8.2, 8.3 and 8.4 we present more detailed discussions of these two parametric survival time models, as well as the log-logistic model, within the context of a class of models called *accelerated failure time* models. We conclude the chapter with a few comments and observations on the use of parametric survival time models.

There are a number of texts that discuss parametric survival time models; see, for example, Crowder, Kimber, Smith and Sweeting (1991), Cox and Oakes (1984), Elandt-Johnson and Johnson (1980), Gross and Clark (1975), Lawless (1982), Lee (1992) and Nelson (1982, 1990). Collett (1994) presents parametric models at a level comparable to the material in this chapter, while Klein and Moeschberger (1997) treat the topic at a slightly higher mathematical level. Andersen, Borgan, Gill and Keiding (1993) discuss theoretical aspects of parametric regression models.

In Chapter 1, we showed that a convenient and clinically plausible way to describe survival time, shown here for a model with a single covariate, is with the equation

$$T = e^{\beta_0 + \beta_1 x} \times \varepsilon. \tag{8.1}$$

We express survival time, T, in (8.1) as the product of the systematic component of the model, $\exp(\beta_0 + \beta_1 x)$, and the error component, ε. We noted in Chapter 1 that this model can be "linearized" by taking the natural log of each side of the equation, yielding

$$\ln(T) = \beta_0 + \beta_1 x + \varepsilon^*, \tag{8.2}$$

where $\varepsilon^* = \ln(\varepsilon)$. The error component, ε^*, follows the extreme minimum value distribution, denoted $G(0,\sigma)$. We noted that under this model survival time followed the exponential distribution when $\sigma = 1$ and the Weibull distribution when $\sigma \neq 1$. In Chapter 1, our purpose was to consider a survival time regression model that could be compared and contrasted to the usual linear regression model. We presented a brief discussion of the use of likelihood techniques to fit the model in (8.2), but there was no discussion of the interpretation of model coefficients.

Survival time models that can be linearized by taking logs are called *accelerated failure time* models. The reason for this terminology is that the effect of the covariate is multiplicative on the time scale. That is, the effect of the covariate is said to "accelerate" survival time. We have shown in the previous chapters that, with the proportional hazards model, the effect of the covariates is multiplicative on the hazard scale. Two other distributions that are sometimes used in (8.2) to describe the distribution of the errors are the log-logistic and log-normal. The texts by Collett (1994) and Klein and Moeschberger (1997) discuss the use of these distributions. A model based on a general linearizing transformation is discussed by Andersen, Borgan, Gill and Keiding (1993).

Wei (1992) discusses semi-parametric approaches to fitting accelerated failure time models. He suggests that, since the parameters in the accelerated failure time models are interpreted as effects on the time scale, they may be more easily understood than the hazard ratios by clinical investigators, especially those unfamiliar with survival time analyses. Fisher (1992), commenting on Wei's paper, notes that in his experience most research-oriented clinicians have little or no trouble understanding the proportional hazards model or the hazard ratio. The proportional hazards model has, in fact, been used by many investigators for years. It is now accepted as the standard method for regression analysis of survival times in many applied settings. It is so deeply embedded in the statistical practice of some fields that it is unlikely that it will be replaced by another model in the foreseeable future. The parametric accelerated failure time regression model can, in some instances, provide a concise and easily interpreted analysis of censored survival time data. As we will show in this chapter, there are two, related, parametric accelerated failure time models that also have proportional hazards. Thus, the accelerated failure time property of these two models can provide an alternative way of explaining covariate effects, even when the first choice may be a proportional hazards interpretation.

8.2 THE EXPONENTIAL REGRESSION MODEL

First we consider the single covariate model in (8.2), where the error distribution is log-exponential, that is, the extreme minimum value distribution denoted $G(0,1)$. The survivorship function may be obtained by expressing (1.7) in terms of time and is

$$S(t, x, \beta) = \exp\left(-t / e^{\beta_0 + \beta_1 x}\right). \tag{8.3}$$

To obtain the median survival time we set the right-hand side of (8.3) equal to 0.5 and solve the resulting equation, yielding the equation

$$t_{50}(x, \beta) = -e^{\beta_0 + \beta_1 x} \times \ln(0.5). \tag{8.4}$$

When the covariate in (8.4) is dichotomous, coded 0/1, the ratio of the median survival time for the group with $x = 1$ to the group with $x = 0$ denoted, $\mathrm{TR}(x = 1, x = 0)$, is

$$\mathrm{TR}(x = 1, x = 0) = \frac{t_{50}(x = 1, \boldsymbol{\beta})}{t_{50}(x = 0, \boldsymbol{\beta})} = \frac{-e^{\beta_0 + \beta_1} \times \ln(0.5)}{-e^{\beta_0} \times \ln(0.5)} = e^{\beta_1}. \qquad (8.5)$$

Alternatively, the relationship between the two median times is

$$t_{50}(x = 1, \boldsymbol{\beta}) = e^{\beta_1} t_{50}(x = 0, \boldsymbol{\beta}). \qquad (8.6)$$

If, for example, $\exp(\beta_1) = 2$, then the median survival time in the group with $x = 1$ is twice the median survival time in the group with $x = 0$. In a similar manner if $\exp(\beta_1) = 0.5$, then the median survival time in the group with $x = 1$ is one-half the median survival time in the group with $x = 0$. The multiplicative covariate effect on the time scale is clear and easy to understand. The result in (8.5) holds not only for the median but for all percentiles. The quantity $\exp(\beta_1)$ is commonly referred to as the *acceleration factor*, even though its effect can be either to "accelerate" or "decelerate" survival time.

An alternative way to present the multiplicative effect is via the survivorship function. It follows from (8.3) that the relationship between the survivorship functions for the two groups is

$$S(t, x = 1, \boldsymbol{\beta}) = S(te^{-\beta_1}, x = 0, \boldsymbol{\beta}). \qquad (8.7)$$

The interpretation of the result in (8.7) is that the value of the survivorship function at time t for the group with $x = 1$ may be obtained by evaluating the survivorship function for the group with $x = 0$ at time $t \exp(-\beta_1)$. The change in the sign of the coefficient is due to the fact that time percentiles and survival probabilities are inverse operations of one another.

The hazard function for the model in (8.3), as shown in (3.3), is

$$h(t, x, \boldsymbol{\beta}) = e^{-(\beta_0 + \beta_1 x)} \qquad (8.8)$$

and the hazard ratio for the dichotomous covariate is

$$HR(x = 1, x = 0) = e^{-\beta_1}. \qquad (8.9)$$

Thus, we see that the exponential regression model is an example of an accelerated failure time model that has proportional hazards. Cox and Oakes (1984) show that the only accelerated failure time models that have proportional hazards are the exponential and Weibull regression models. If an exponential regression model fits the data, one may express the effect of covariates either as a time ratio or as a hazard ratio.

Some software packages, including SAS and BMDP, fit the linearized or log-time form of the model and provide estimates of the accelerated failure time form of the coefficients. Other packages, including STATA, provide a choice of expressing the results in terms of either hazards or accelerated failure time coefficients. One should always consult the software manual to confirm the form of the model fit and the scale (time or hazard) on which the results are displayed. With the exponential regression model, it is easy to switch between the two forms of the coefficients because each is simply the negative of the other. As we show in the next section, the Weibull model is more complicated in this respect.

In this chapter, we assume that observations are subject only to right censoring, but the analysis may be extended to other types of censoring and truncation using the methods discussed in Chapter 7 for the proportional hazards model. Under the assumption that we have n independent observations of time, p covariates and a censoring indicator denoted $(t_i, \mathbf{x}_i, c_i)$, $i = 1, 2, \ldots, n$, the log-likelihood function is

$$L(\boldsymbol{\beta}) = \sum_{i=1}^{n} c_i z_i - e^{z_i}, \qquad (8.10)$$

where $z_i = y_i - \mathbf{x}_i'\boldsymbol{\beta}$, $y_i = \ln(t_i)$, $\mathbf{x}_i' = (x_{i0}, x_{i1}, \ldots, x_{ip})$ and $x_{i0} = 1$. The likelihood equations are obtained by differentiating the log-likelihood function with respect to the unknown parameters and setting the expressions equal to zero. This process yields the equation for the constant term of

$$\sum_{i=1}^{n} \left(c_i - e^{z_i} \right) = 0 \qquad (8.11)$$

and equations for the nonconstant covariates of

$$\sum_{i=1}^{n} x_{ij}\left(c_i - e^{z_i}\right) = 0, \tag{8.12}$$

for $j = 1, 2, \ldots, p$. If we denote the solutions to (8.11) and (8.12) as $\boldsymbol{\beta}' = \left(\hat{\beta}_0, \hat{\beta}_1, \ldots, \hat{\beta}_p\right)$, the model-based predicted or fitted values of time are computed as $\hat{t}_i = \exp\left(\mathbf{x}_i'\hat{\boldsymbol{\beta}}\right)$ for $i = 1, 2, \ldots, n$. In the usual linear regression model, the inclusion of the constant term in the model forces the mean of the observed and fitted values to be equal. However, in the exponential regression model, its effect is to force the sum of the ratio of observed to fitted values to be equal to the number of non-censored observations, that is,

$$\sum_{i=1}^{n} c_i = m = \sum_{i=1}^{n} \frac{t_i}{\hat{t}_i}. \tag{8.13}$$

The actual value of $\hat{\beta}_0$ depends to a large extent on m, the number of non-censored observations. Thus, the fitted values are predictions of values from a censored exponential distribution — not the true mean survival time of a subject with covariate value $\mathbf{x}_i$.

We obtain the estimator of the variances and covariances of the estimator of the coefficients, using standard theory of maximum likelihood estimation, from the second partial derivatives of the log-likelihood function. The general form of the second partial derivative of the log-likelihood function in (8.10) is

$$\frac{\partial^2 L(\boldsymbol{\beta})}{\partial \beta_j \partial \beta k} = -\sum_{i=1}^{n} x_{ij} x_{ik} e^{z_i}, \quad j, k = 0, 1, \ldots, p. \tag{8.14}$$

In this setting, estimators are based on the observed information matrix, denoted $\mathbf{I}(\hat{\boldsymbol{\beta}})$, which is the matrix with elements given by the negative of (8.14) evaluated at the estimator of the coefficients. The inverse of the observed information matrix provides the estimators of the variances and covariances, namely

$$\widehat{\mathrm{Var}}\,(\hat{\boldsymbol{\beta}}) = \mathbf{I}(\hat{\boldsymbol{\beta}})^{-1}. \tag{8.15}$$

Typically, software packages provide estimates of the standard errors of each of the coefficients in the model that are the square roots of the elements on the main diagonal of the matrix in (8.15). Most packages provide an option for obtaining all the elements of (8.15). The end-points of a $100(1-\alpha)$ percent Wald-statistic-based confidence interval for the jth coefficients are

$$\hat{\beta}_j \pm z_{1-\alpha/2}\hat{SE}(\hat{\beta}_j), \qquad (8.16)$$

where $\hat{SE}(\hat{\beta}_j)$ denotes the estimator of the standard error of the estimator of the coefficient.

As a first example, we consider an exponential regression model containing the dichotomous covariate for history of IV drug use (DRUG) in the HMO-HIV+ study. Table 8.1 presents the results of fitting this model.

The estimated coefficients shown in Table 8.1 are expressed on the log-time scale and, using (8.4), the estimated median time to death for a subject with no history of IV drug use is

$$\hat{t}_{50}(\text{DRUG} = 0) = -e^{3.024} \times \ln(0.5) = 14.26$$

months, and the estimate for a subject with a history of IV drug use is

$$\hat{t}_{50}(\text{DRUG} = 1) = -e^{3.024-1.056} \times \ln(0.5) = 4.96$$

months. The estimated time ratio, from (8.5), is

$$\hat{TR}(x = 1, x = 0) = e^{-1.056} = 0.35. \qquad (8.17)$$

That is, the median time to death in the group with a history of IV drug use is approximately one-third the median time to death in the group with no prior history.

The value of any estimated time percentile depends on the value of $\hat{\beta}_0$ which, in turn, depends on the observed number of actual survival times m. Thus the values of 14.26 and 4.96 are only useful for descriptive purposes in the 80 percent of the data with observed survival times. Since the model is an accelerated failure time model, the ratio of any two time percentiles depends only on the coefficient of the covariate of

Table 8.1   Estimated Coefficients, Standard Errors, z-Scores, Two-Tailed p-Values and 95% Confidence Intervals for the Exponential Regression Model in the HMO-HIV+ Study ($n = 100$)

Variable	Coeff.	Std. Err.	z	$P>|z|$	95% CIE
DRUG	−1.056	0.224	−4.72	<0.001	−1.494, −0.617
Constant	3.024	0.154	19.60	<0.001	2.721, 3.326

interest, $\hat{\beta}_1$. Hence, we feel that the ratio is a more useful summary measure of survival experience than estimates of percentiles that depend on the observed censoring rate.

In particular, the interpretation of the estimated time ratio in (8.16) is that the survival time for a subject with a history of IV drug use is estimated to be 35 percent of that of a subject without such a history. Another way to state the effect is that a history of IV drug use is estimated to shorten survival time by 65 percent. In a manner identical to that used in previous chapters for the hazard ratio, the confidence interval for the time ratio is obtained by exponentiating the endpoints of the confidence interval for the coefficient. In the example, the 95 percent confidence interval reported in Table 8.1 for $\hat{\beta}_1$ is obtained using (8.16). Thus the 95 percent confidence interval for the time ratio is

$$\left(e^{-1.494} \le \mathrm{TR} \le e^{-0.617}\right) = \left(0.22 \le \mathrm{TR} \le 0.54\right).$$

The interpretation of this confidence interval is that subjects with a history of IV drug use have survival times estimated to be between 22 percent and 54 percent of those for subjects with no history of drug use. Alternatively, survival is shortened by between 46 percent and 78 percent for subjects with a history of IV drug use.

Since the exponential regression model has proportional hazards, we can express the effect of covariates using hazard ratios. Using the negative of the coefficient for drug use from the results in Table 8.1 and (8.9), the estimated hazard ratio for history of IV drug use is

$$\widehat{\mathrm{HR}}(x = 1, x = 0) = e^{1.056} = 2.87.$$

The endpoints of a 95 percent confidence interval are the inverse of those of the time ratio, namely

$$\left(e^{0.617} \leq \mathrm{HR} \leq e^{1.494}\right) = \left(1.85 \leq \mathrm{TR} \leq 4.46\right).$$

The interpretation of these results is that subjects with a history of IV drug use are dying a rate that is estimated to be 2.87 times that of subjects without such a history, and it is consistent with the data that the hazard ratio could be as low as 1.85 or as high as 4.46.

The two descriptions of the effect are statistically equivalent, but one describes the effect on the time scale and the other with the hazard rate. For some investigators, time may be more easily understood than a rate expressed by the hazard function. For such an audience, results presented as time ratios might be preferable to ones based on hazard ratios.

We wish to emphasize that the dual interpretation of results discussed here is a result of the fact that the exponential regression model is both an accelerated failure time model and a proportional hazards model. One should never automatically invert hazard ratios estimated from a fitted proportional hazards model and interpret them as estimated time ratios. This is due to the fact that there are many models with proportional hazards but only two of them are accelerated failure time models.

The steps in building an exponential regression model are essentially identical to those described in detail in Chapter 5 for the proportional hazards model and, as such, are not repeated in this section. The only difference is that the log-likelihood function is used to compare models rather than the *partial* log-likelihood. Most software packages have routines for stepwise selection of covariates. As discussed in Chapter 5, issues of scale selection for continuous covariates, statistical adjustment, and selection of interactions are still vital steps in the model building process.

It is also possible to perform a best subsets variable selection by casting the problem as weighted linear regression. The approach is similar to one used for best subsets selection in logistic regression. However, this form of best subsets selection may be used only with the exponential regression model. Since it cannot be extended to the other parametric regression models in this chapter, we do not present it in detail.

Since most of the model building steps are the same as those presented in Chapter 5, we discuss only a few of the details. In particular the techniques for checking the scale of all continuous covariates are similar to those illustrated in Chapter 5. Recall that possible methods include the use of design variables, smoothed residual plots and the method of fractional polynomials. Each of these methods may be used in the same way with an exponential regression model, with a few modifications. The smoothed residual plots discussed in Chapter 5 used the martingale residuals. The martingale residuals for an exponential regression model are

$$\hat{M}_i = c_i - t_i e^{-x_i'\hat{\beta}}, \tag{8.18}$$

and one may plot them using the method described for the proportional hazard model in Section 5.2. We remind the reader that application of the Grambsch, Therneau and Fleming (1995) plot assumes that the coefficients are on the log-hazard scale. If the model has been fit on the log-time scale, the sign of the coefficients has to be reversed before being added to the log of the ratio of smoothed values.

Rather than repeat the detailed analysis of the data from the UIS presented in Chapters 5 and 6, we leave this as an exercise and instead illustrate the exponential regression model using the HMO-HIV+ study. Application of the modeling methods, left as an exercise, reveals that the model is linear in AGE and there is no interaction between AGE and DRUG. We regard the model in Table 8.2 as the preliminary final model until we check model adequacy and fit.

The methods and procedures for assessing the fit of an exponential regression model are nearly identical to those described in Chapter 6 for the proportional hazards model, with one exception. We check the hazard function for a specific parametric form rather than the more general proportional hazards assumption. The methods for identifying whether any subjects have an undue influence on the estimated parameters or are poorly fit use the same score function residuals used for the proportional hazards model. Not all software packages provide these score

Table 8.2   Estimated Coefficients, Standard Errors, z-Scores, Two-Tailed p-Values and 95% Confidence Intervals for the Preliminary Final Exponential Regression Model for the HMO-HIV+ Study ($n = 100$)

Variable	Coeff.	Std. Err.	z	$P>|z|$	95% CIE
AGE	−0.092	0.016	−5.69	<0.001	−0.124, −0.060
DRUG	−1.010	0.224	−4.51	<0.001	−1.449, −0.571
Constant	6.152	0.606	10.15	<0.001	4.964, 7.340

Log-likelihood = −130.397.

function residuals; however, they are not too difficult to calculate, as shown below.

Overall goodness-of-fit can be assessed using the test proposed by Grønnesby and Borgan (1996) and simplified by May and Hosmer (1998). This test was discussed in Section 6.5, and the regularity conditions for the proofs given in Grønnesby and Borgan's paper also hold for the right-censored data exponential regression model.

The basis for examining the influence of individual subjects on the values of the estimated parameters is the score residuals. These residuals are the individual contributions to score equations represented generally in (8.12). Specifically the score residual for subject i on covariate j is the individual component of the derivative of the log-likelihood in equation (8.10), namely

$$\hat{L}_{ij} = x_{ij}\left(t_i e^{-x_i'\beta} - c_i\right) = -x_{ij}\hat{M}_i . \tag{8.19}$$

We denote the vector containing the score residuals for all the covariates as

$$\hat{\mathbf{L}}_i' = \left(\hat{L}_{i0}, \hat{L}_{i1}, \hat{L}_{i2}, \ldots, \hat{L}_{ip}\right). \tag{8.20}$$

As was the case in (6.23) for the proportional hazards model, the statistic for assessing the influence of the ith subject on the coefficients is the components of the vector

$$\Delta\hat{\boldsymbol{\beta}}_i = \hat{\text{Var}}\left(\hat{\boldsymbol{\beta}}\right)\hat{\mathbf{L}}_i , \tag{8.21}$$

where $\hat{\text{Var}}\left(\hat{\boldsymbol{\beta}}\right)$ is the estimated covariance matrix of the estimated coefficients shown in (8.15). The covariate-specific score residuals in (8.19) and their respective individual components of (8.21) can be plotted against several measures. We feel that when a diagnostic statistic relates to a specific covariate, using the values of that covariate leads to the most informative plot. With this plot, we are able to identify large values of the diagnostic statistic, and to determine at approximately which covariate values they occur. Figure 8.1 presents an example of such a plot.

A measure of the overall influence of a subject on the vector of coefficients, similar to the result in (6.24), is the statistic

$$ld_i = \hat{\mathbf{L}}_i' \widehat{\text{Var}}\left(\hat{\boldsymbol{\beta}}\right)\hat{\mathbf{L}}_i. \tag{8.22}$$

This statistic was first proposed by Hall, Rogers and Pregibon (1982), who suggested using the scaled statistic

$$\frac{ld_i}{\left(1 - ld_i\right)^2}. \tag{8.23}$$

Either form, (8.22) or (8.23), could be used, but, to be in agreement with approaches described in Chapter 6, we use the unscaled form in (8.22). When referring to the statistic in (8.22), we use the same term(s) used in Chapter 6, namely likelihood displacement or Cook's distance statistic. The measure of overall influence depends on all covariates, so it makes sense to plot it against another summary statistic. We feel a plot using the martingale residuals with separate plotting symbols for censored and noncensored observations is most informative.

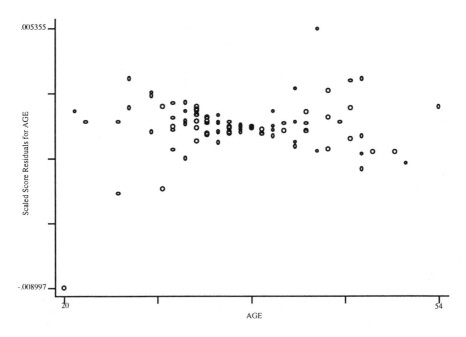

Figure 8.1 Graph of the scaled score residuals for AGE computed from the model in Table 8.2.

A second overall measure discussed in Chapter 6 was the l-max statistic obtained from a matrix version of (8.22). This statistic has not been used for parametric regression models, but it seems reasonable that the derivation of the statistic could be extended to cover the current setting. However, based on the similarity of the results seen in Figures 6.7a and Figure 6.7b, it is not clear whether it would provide any meaningful additional diagnostic information.

Unfortunately, most of the major software packages do not routinely calculate the values of the diagnostic statistics for parametric regression models, although this is likely to change in the future. At this time, only S-PLUS provides the diagnostic statistics. Users of other packages will need to write their own procedures to calculate the values of the diagnostic statistics. This requires some facility with creating matrices, performing matrix operations and saving the results as new variables. Collett (1994) provides macros for use with SAS. The values of the diagnostic statistics for the model in Table 8.3 were calculated in STATA using a program we wrote.

As was the case in the proportional hazards model, there is no distribution theory to provide critical values for significance tests of the diagnostic statistics. The best approach is to plot them and look for values that seem too large. Figure 8.1 presents the plot of scaled score residuals for AGE from (8.21).

The first thing we note is that the basic hourglass shape of the plot is similar to the plots shown in Figure 6.6 for the proportional hazards model. Specifically, the plot is narrow in the middle near the mean age of 36 years and fans out for younger and older values. One value in the bottom left corner stands out and represents a subject who could have an important influence on the magnitude of the coefficient for AGE.

A box plot of the scaled score residuals for DRUG is shown in Figure 8.2. In the figure, we see one negative value for a subject with no previous history and three positive values for subjects with a history that stand out as possible subjects with an influence on the magnitude of the coefficient for DRUG.

The graph of the overall likelihood displacement or Cook's distance statistic is shown in Figure 8.3, where the plotting symbol is the value of the censoring variable. The plot shows two subjects with large values of the statistic. Both of these subjects have large negative martingale residuals, indicating that the number of observed events was smaller than expected.

The largest value of the likelihood displacement statistic in Figure 8.3 corresponds to the 20-year-old subject in Figure 8.1. This subject had the largest negative value of the scaled score residual for AGE and the largest positive scaled score residual for DRUG in Figure 8.2. The second largest value of the likelihood displacement statistic in Figure 8.3 corresponds to a 43-year-old subject in Figure 8.1. The subject had the largest positive value of the scaled score residual for AGE and the largest negative scaled score residual for DRUG in Figure 8.2. To assess the effect that these two subjects had on the coefficients, we refit the model without them. The result was that the coefficients for AGE, DRUG and the constant decreased from their values in Table 8.2 by 4.5, 1.2 and 3.7 percent, respectively. Although the diagnostic statistics for these two subjects are large relative to the other values, the data were not extreme enough to exert an important change in the model coefficients.

The values of the diagnostic statistics depend on complicated relationships between the covariates, survival time, and censoring status. For this reason, it is not possible to provide specific cutpoint values which, when exceeded, indicate that an important change in a coefficient will occur. We recommend that one always carefully examine the data for all subjects identified as being extreme in any diagnostic statistic and refit the model deleting these subjects. Clinical criteria should always be used to make decisions about the continuing role of such subjects in the analysis.

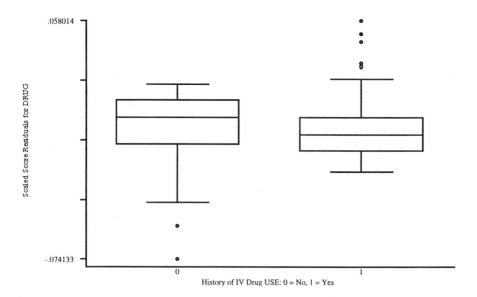

Figure 8.2 Box plot of the scaled score residuals for DRUG computed from the model in Table 8.2.

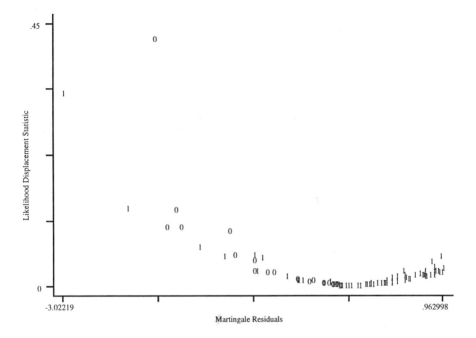

Figure 8.3 Graph of the likelihood displacement or cook distance statistic versus the martingale residuals from the model in Table 8.2 with the censoring variable used as the plotting symbol.

In Chapter 6, we discussed methods for assessing the assumption of proportional hazards. With a fully parametric model, this analysis step is replaced with one that determines whether the data support the particular parametric form of the hazard function. The most frequently employed method uses the model-based estimate of the cumulative hazard function to form the Cox–Snell residuals that were briefly discussed in Section 6.2. Cox and Snell (1968) noted that the values of the estimated cumulative hazard function may be thought of as observations from a censored sample from an exponential distribution with parameter equal to one. The diagnostic plot compares the model-based cumulative hazard to one obtained from a nonparametric (Kaplan–Meier or Nelson–Aalen) estimator. The nonparametric estimator uses the model-based estimates of the cumulative hazard at each observed time as the time variable and the censoring indicator from the original survival time variable as the censoring variable. If the parametric model is correct, this plot should follow a line through the origin with slope equal to one.

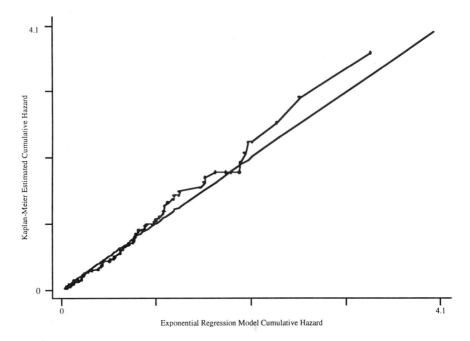

Figure 8.4 Graph of the Kaplan–Meier estimate of the cumulative hazard ver- sus the exponential regression model estimate of the cumulative hazard from the model in Table 8.2. The solid line is the referent line with slope = 1.0 and inter- cept = 0.

The estimator of the Cox–Snell residuals from an exponential re- gression model is obtained by exponentiating the additive residuals on the log-time scale. Specifically,

$$\hat{H}\left(t_i, \mathbf{x}_i, \hat{\boldsymbol{\beta}}\right) = \exp\left(y_i - \mathbf{x}_i'\hat{\boldsymbol{\beta}}\right) = t_i e^{-\mathbf{x}_i'\hat{\boldsymbol{\beta}}}, \quad i = 1, 2, \ldots, n \ . \qquad (8.24)$$

The estimates for the current fitted model are obtained using the coeffi- cients in Table 8.2. The pairs $\left(\hat{H}\left(t_i, \mathbf{x}_i, \hat{\boldsymbol{\beta}}\right), c_i\right)$, $i = 1, 2, \ldots, n$, are used to compute the Nelson–Aalen cumulative hazard estimator in (2.30) or the Kaplan–Meier estimator shown in (2.1). That is, the values of $\hat{H}\left(t_i, \mathbf{x}_i, \hat{\boldsymbol{\beta}}\right)$ define time and the values of c_i define the events. We de- note this estimator as $\hat{H}_{\text{km},i}$. Figure 8.4 presents a plot of the pairs, $\left[\hat{H}\left(t_i, \mathbf{x}_i, \hat{\boldsymbol{\beta}}\right), \hat{H}_{\text{km},i}\right]$, $i = 1, 2, \ldots, 100$, along with the referent straight line.

In this figure, the plotted pairs of points initially fall on the referent line and then falls mostly above the referent line. This provides some evidence that an assumption of a constant baseline hazard function based on the fitted exponential regression model may not be appropriate for these data.

We complete our assessment of fit with the Grønnesby-Borgan test described in Section 6.5. The risk score in this setting is based on the coefficients in Table 8.2, and it refers to risk on the time scale rather than on the hazard scale. As before, we create 10 nearly equal sized groups using cutpoints that do not split subjects with the same risk score. We can use either the score, likelihood ratio or Wald test for the significance of adding 9 design variables to the model. The likelihood ratio test has a value of $G = 1.97$, which, with 9 degrees-of-freedom, yields a p-value of 0.99. A summary of the observed and estimated expected number of events within each decile is presented in Table 8.3. The individual p-values are not significant, and the overall test supports model fit.

At this point we have two choices: We can take the view that the departure from model fit seen in Figure 8.3 is not significant and proceed with model interpretation, or we can see whether an alternative parametric model provides a better fit than the exponential regression model. We comment on the first choice before proceeding with the details of the second in the next section.

Table 8.3 Observed Number of Events, Estimated Number of Events, z-Score and Two-Tailed p-Value within Each Decile of Risk Based on the Model in Table 8.2

Decile of Risk	Observed Number of Events	Estimated Number of Events	z	p-Value
1	7	7.47	−0.17	0.87
2	10	11.55	−0.46	0.65
3	9	9.02	−0.01	0.99
4	7	5.78	0.51	0.61
5	8	8.18	−0.06	0.95
6	8	5.95	0.84	0.40
7	10	8.93	0.36	0.72
8	9	10.49	−0.60	0.55
9	6	5.56	0.19	0.85
10	6	6.56	−0.12	0.83
Total	80	80		

The methods for providing an interpretation of a fitted exponential accelerated failure time regression model follow the same procedure-sused in Section 6.6 for the proportional hazards model. The only difference is that exponentiated coefficients provide estimates of the multiplicative effect on the time rather than estimating the hazard rate. As in Section 6.6, the key step is to specify the values of the covariates yielding a clinically interesting comparison. The time ratio is the exponentiation of the difference in the linear part of the model. That is, the time ratio comparing two levels of the covariates in the model is, in general,

$$\hat{TR}(x = x_1, x = x_0) = \exp\left[(x_1 - x_0)'\hat{\beta}\right]. \tag{8.25}$$

For example, from the model in Table 8.2, the estimated time ratio comparing presence and absence of a history of IV drug use, controlling for age, is

$$\hat{TR}(DRUG = 1, DRUG = 0, AGE = a) = \exp\left[(1 - 0) \times -1.01\right] = 0.36.$$

We obtain the endpoints of a 95 percent confidence interval by exponentiating the endpoints of the Wald-based confidence interval for the coefficient for DRUG shown in Table 8.2. The endpoints of this interval are $(0.24, 0.57)$. The interpretation is that, after controlling for age, the survival times for subjects with a history of IV drug use are estimated to be 36 percent of those for subjects without a history of IV drug use. The confidence interval suggests that the estimate could be between 24 and 57 percent. Another way to interpret the point estimate is to say that the survival times for subjects with a history of IV drug use is estimated to be 64 percent shorter than for subjects without a history, and they could be between 74 and 43 percent shorter.

The point estimate of the effect of a 10-year increase in age, controlling for history of IV drug use, is

$$\hat{TR}(AGE = a + 10, \ AGE = a, DRUG = d)$$
$$= \exp\left\{\left[(a + 10) - a\right] \times -0.092\right\} = 0.40 \ .$$

The endpoints of the 95 percent confidence interval are obtained by multiplying the endpoints of the confidence interval for the coefficient for AGE in Table 8.2 by 10 and then exponentiating, yielding

$$\exp(10 \times -0.124) = 0.29 \quad \text{and} \quad \exp(10 \times -0.06) = 0.55.$$

The interpretation of these estimates is that the effect of a 10-year in-crease in age is to reduce survival time by an estimated 60 percent, and the reduction could be between 45 and 71 percent.

Before we conclude that the model in Table 8.2 is our final model, we examine other possible parametric models, such as the Weibull.

8.3 THE WEIBULL REGRESSION MODEL

The basic form of a Weibull accelerated failure time regression model was presented in (8.2). The main difference between it and the expo-nential regression model discussed in the previous section is that the pa-rameter σ in the distribution of the "error" term of the accelerated failure time form of the model can be different from 1.0. The inclusion of this parameter in the model leads to a slightly more complicated haz-ard function and related regression model parameters. For this reason, we begin our discussion of the Weibull model in the single covariate setting to compare and contrast it with the exponential model.

The hazard function for the single covariate Weibull regression model is

$$h(t, x, \boldsymbol{\beta}, \lambda) = \frac{\lambda t^{\lambda-1}}{\left(e^{\beta_0 + \beta_1 x}\right)^{\lambda}}, \tag{8.26}$$

where, for convenience, we use $\lambda = 1/\sigma$. This hazard function may be re-expressed in proportional hazards or accelerated failure time form. The proportional hazards form of the function is obtained as follows:

$$h(t, x, \boldsymbol{\beta}, \lambda) = \lambda t^{\lambda-1} e^{-\lambda(\beta_0 + \beta_1 x)}$$

$$= \lambda t^{\lambda-1} e^{-\lambda\beta_0} e^{-\lambda\beta_1 x}$$

so

$$h(t, x, \boldsymbol{\beta}, \lambda) = \lambda\gamma t^{\lambda-1} e^{-\lambda\beta_1 x}$$

$$= h_0(t) e^{\theta_1 x}, \tag{8.27}$$

where $\gamma = \exp(-\beta_0/\sigma) = \exp(\theta_0)$, $\theta_1 = -\beta_1/\sigma$ and the baseline hazard function is $h_0(t) = \lambda\gamma t^{\lambda-1}$. Although the parameter σ is a variance-like parameter on the log-time scale, $\lambda = 1/\sigma$ is commonly called the *shape parameter*. In the remainder of this chapter, we will refer to σ as the shape parameter. The parameter γ is called a *scale parameter*. The expression for the hazard function in (8.27) leads to a hazard ratio interpretation of the parameter θ_1.

The accelerated failure time form of the hazard function is obtained by re-expressing (8.26) as follows:

$$h(t, x, \boldsymbol{\beta}, \lambda) = \lambda t^{\lambda-1} e^{-\lambda(\beta_0 + \beta_1 x)}$$

$$= \lambda \gamma \left(te^{-\beta_1 x}\right)^{\lambda-1} e^{-\beta_1 x}. \qquad (8.28)$$

The rationale for presenting both forms of the hazard function is that different software packages use different parameterizations when fitting a model. For example, SAS and BMDP use the accelerated failure time model parameterization in (8.28) and report estimates of the $\boldsymbol{\beta}$ form of the coefficients and σ. STATA, on the other hand, offers the option of providing estimates of either parameterization. If the accelerated failure time or log-time form is chosen, estimates of $\boldsymbol{\beta}$ and σ are provided. If the log-hazard form is chosen, estimates of $\boldsymbol{\theta}$ and σ are provided such that the relationship between the two sets of coefficients is $\boldsymbol{\theta} = -\boldsymbol{\beta}/\sigma$. In the remainder of this section, we use the accelerated failure time formulation of the model.

The survivorship function corresponding to the accelerated failure time form of the hazard function in (8.28) is

$$S(t, x, \boldsymbol{\beta}, \sigma) = \exp\left\{-t^\lambda \exp\left[(-1/\sigma)(\beta_0 + \beta_1 x)\right]\right\}. \qquad (8.29)$$

We obtain the equation for the median survival time by setting the survivorship function equal to 0.5 and solving for time yielding

$$t_{50}(x, \boldsymbol{\beta}, \sigma) = \left[-\ln(0.5)\right]^\sigma e^{\beta_0 + \beta_1 x}. \qquad (8.30)$$

For example, if the covariate is dichotomous and coded 0/1, the time ratio at the median survival time is

$$\text{TR}(x = 1, x = 0) = \frac{t_{50}(x = 1, \boldsymbol{\beta}, \sigma)}{t_{50}(x = 0, \boldsymbol{\beta}, \sigma)} = \frac{\left[-\ln(0.5)\right]^\sigma e^{\beta_0 + \beta_1}}{\left[-\ln(0.5)\right]^\sigma e^{\beta_0}} = e^{\beta_1}. \qquad (8.31)$$

A similar result would be obtained for any percentile of survival time. Thus, we see that the interpretation of the $\boldsymbol{\beta}$ form of the coefficients is the same as in the exponential regression model.

Rather than providing the details of parameter estimation for both the univariable and multivariable models as was done with the exponential regression model, we present only the general multivariable model. The equation for the log-likelihood function for a sample possibly containing right-censored data is obtained from (1.6) using

$$z_i = \frac{y_i - \mathbf{x}_i'\boldsymbol{\beta}}{\sigma},$$

and $f(z)$ is replaced with $f(z)/\sigma$. This yields the log-likelihood function

$$L(\boldsymbol{\beta},\sigma) = \sum_{i=1}^{n} c_i\left(-\ln(\sigma) + z_i\right) - e^{z_i}. \qquad (8.32)$$

The score equation for the jth regression coefficient is obtained by taking the derivative of (8.32) with respect to β_j and setting it equal to zero, yielding

$$\frac{\partial L(\boldsymbol{\beta},\sigma)}{\partial \beta_j} = \sum_{i=1}^{n} \frac{-x_{ij}}{\sigma}\left(c_i - e^{z_i}\right) = 0, \quad j = 0,1,2,...,p. \qquad (8.33)$$

The score equation for the shape parameter, σ, is

$$\frac{\partial L(\boldsymbol{\beta},\sigma)}{\partial \sigma} = \frac{-m}{\sigma} + \sum_{i=1}^{n} \frac{-z_i}{\sigma}\left(c_i - e^{z_i}\right) = 0. \qquad (8.34)$$

Some software packages, for example, STATA, parameterize the log-likelihood function in terms of $-\ln(\sigma)$ since this makes the score equation easier to solve from a computational point of view. Regardless of what form is used, all packages report an estimate of σ. The solutions to (8.33) and (8.34) are denoted as $\hat{\boldsymbol{\beta}}$ and $\hat{\sigma}$, respectively.

As is the case in any application of maximum likelihood, the estimator of the covariance matrix of the parameter estimator is obtained from the observed information matrix. The individual elements of this matrix to be evaluated are

$$-\frac{\partial^2 L(\boldsymbol{\beta},\sigma)}{\partial \beta_j \, \partial \beta_k} = \frac{1}{\sigma^2} \sum_{i=1}^{n} x_{ij} x_{ik} e^{z_i} \, ,$$

$$-\frac{\partial^2 L(\boldsymbol{\beta},\sigma)}{\partial \beta_j \, \partial \sigma} = \frac{1}{\sigma^2} \sum_{i=1}^{n} x_{ij} z_i e^{z_i} \, ,$$

and

$$-\frac{\partial^2 L(\boldsymbol{\beta},\sigma)}{\partial \sigma \, \partial \sigma} = \frac{m}{\sigma^2} + \frac{1}{\sigma^2} \sum_{i=1}^{n} z_i^2 e^{z_i} \, .$$

When evaluated at the solution to the likelihood equations, the information matrix may be expressed as

$$\mathbf{I}(\hat{\boldsymbol{\beta}},\hat{\sigma}) = \frac{1}{\hat{\sigma}^2} \begin{bmatrix} \mathbf{X}'\hat{\mathbf{V}}\mathbf{X} & \mathbf{X}'\hat{\mathbf{V}}\hat{\mathbf{z}} \\ \hat{\mathbf{z}}'\hat{\mathbf{V}}\mathbf{X} & \hat{\mathbf{z}}'\hat{\mathbf{V}}\hat{\mathbf{z}}+m \end{bmatrix}, \tag{8.35}$$

where $\mathbf{X}$ is an n by $p+1$ matrix containing the values of the covariates, $\hat{\mathbf{V}} = \text{diag}(e^{\hat{z}_i})$, an n by n diagonal matrix, and $\hat{\mathbf{z}}' = (\hat{z}_1, \hat{z}_2, ..., \hat{z}_n)$, with

$$\hat{z}_i = \frac{y_i - \mathbf{x}_i'\hat{\boldsymbol{\beta}}}{\hat{\sigma}}.$$

The estimator of the covariance matrix of the estimators of the parameters is

$$\hat{\text{Var}}(\hat{\boldsymbol{\beta}},\hat{\sigma}) = \left[\mathbf{I}(\hat{\boldsymbol{\beta}},\hat{\sigma}) \right]^{-1}. \tag{8.36}$$

The details of the model building process for the Weibull regression model are the same as those presented in the previous section for the exponential regression model. The martingale residuals used in the Grambsch, Therneau and Fleming (1995) plots for checking the scale of continuous covariates and for model assessment are

$$\hat{M}_i = c_i - \exp(\hat{z}_i)$$

$$= c_i - t_i^{\hat{\lambda}} \exp(-\hat{\lambda} \mathbf{x}_i' \hat{\boldsymbol{\beta}}), \tag{8.37}$$

where $\hat{\lambda} = 1/\hat{\sigma}$.

The model development process yields the same model for the Weibull model as was obtained for the exponential model shown in Table 8.2. Demonstrating this is left as an exercise. Table 8.4 presents the results of this Weibull fit.

A related model development issue is whether the Weibull model, with its additional parameter, offers an improvement over the simpler exponential model for a given set of covariates. One step in this evaluation is a test of the hypothesis that $\sigma = 1$. This hypothesis may be tested using a Wald test or the confidence interval formed from its estimate and associated standard error. Shown at the bottom of Table 8.4 is a Wald test and confidence interval for the log form of the parameter σ, as estimated in STATA, and similar statistics for the parameter estimate itself. The confidence interval for σ was formed from the endpoints of the interval for $-\ln(\sigma)$. The Wald test for σ is not reported because, based on properties of other similarly bounded parameters such as the odds ratio, one would expect tests and estimates based on the log form to have better statistical properties. The confidence interval for $-\ln(\sigma)$ does not contain zero and the p-value for the Wald test is 0.042. This indicates that the Weibull regression model presented in Table 8.4 may provide a better fit than the exponential model presented in Table 8.2.

Other packages report only the parameter estimate of σ and its estimated standard error. In SAS, the test of the hypothesis that $\sigma = 1$ is performed when one fits the exponential model and it is based on the score test for the addition of σ to the model.

As noted above, the significance of the Wald test for the shape parameter indicates that the Weibull model may provide a better fit to the data than the exponential regression model. To confirm this, we need to use the same diagnostic statistics, plots and tests that are used to assess

Table 8.4   Estimated Coefficients, Standard Errors, z-Scores, Two-Tailed p-Values and 95% Confidence Intervals for the Preliminary Final Weibull Regression Model for the HMO-HIV+ Study ($n = 100$)

Variable	Coeff.	Std. Err.	z	P>|z|	95% CIE
AGE	−0.091	0.014	−6.67	<0.001	−0.117, −0.064
DRUG	−1.049	0.189	−5.55	<0.001	−1.420, −0.679
Constant	6.148	0.511	12.04	<0.001	5.147, 7.149
-ln(Sigma)	0.175	0.086	2.03	0.042	0.006, 0.344
Sigma	0.839	0.072			0.709, 0.994

Log likelihood = −128.502.

the fit of the exponential regression model. The value of the score residuals for the ith subject on the jth regression coefficient is obtained by evaluating the individual terms in (8.33) at the estimator, namely

$$\hat{L}_{ij} = -\frac{x_{ij}}{\hat{\sigma}}\left(c_i - e^{\hat{z}_i}\right). \tag{8.38}$$

The score residuals for the shape parameter are obtained from (8.34) in a similar manner, and are

$$\hat{L}_{ip+1} = -\frac{c_i}{\hat{\sigma}} - \frac{\hat{z}_i}{\hat{\sigma}}\left(c_i - e^{\hat{z}_i}\right). \tag{8.39}$$

If we parameterize the model in terms of $-\ln(\sigma)$, the score residuals are

$$\hat{L}_{ip+1} = c_i + \hat{z}_i\left(c_i - e^{\hat{z}_i}\right). \tag{8.40}$$

The vector of $p+2$ score residuals is

$$\hat{\mathbf{L}}_i' = \left(\hat{L}_{i0}, \hat{L}_{i1}, \ldots, \hat{L}_{ip}, \hat{L}_{ip+1}\right).$$

The scaled score residuals that provide estimates of the effect that each subject has on individual parameter estimates can be obtained by extending (8.21) to the current model using the expression for $\widehat{\text{Var}}\left(\hat{\boldsymbol{\beta}},\hat{\sigma}\right)$ in (8.36), namely

$$\Delta\left(\hat{\boldsymbol{\beta}},\hat{\sigma}\right)_i = \widehat{\text{Var}}\left(\hat{\boldsymbol{\beta}},\hat{\sigma}\right)\hat{\mathbf{L}}_i. \tag{8.41}$$

The likelihood displacement or Cook's-distance-type measure of overall effect is, by extension of (8.22),

$$ld_i = \mathbf{L}_i'\widehat{\text{Var}}\left(\hat{\boldsymbol{\beta}},\hat{\sigma}\right)\hat{\mathbf{L}}_i. \tag{8.42}$$

As in the previous section, we plot the individual scaled score residuals for AGE and DRUG against the covariate values, shown in Figures 8.5a and 8.5b. The scaled score residuals for the $-\ln(\sigma)$ parametrization of the shape parameter are plotted against the estimated martingale residuals in (8.37), shown in Figure 8.5c. The likelihood displacement or Cook's distance measure is plotted against the estimated martingale residuals in (8.37), shown in Figure 8.5d.

The plot for AGE in Figure 8.5a is nearly identical to the one shown in Figure 8.1 for the fitted exponential regression model. The plot has the same basic hourglass shape, fanning out from its narrowest point at approximately the mean age. The plot shows two subjects, one in the bottom left corner and one in the upper right corner, that could have an influential effect on the coefficient for AGE.

The plot in Figure 8.5b is quite similar to Figure 8.2. We see one subject with no previous history and one with a history of IV drug use that could have an influence on the coefficient for DRUG.

In Figure 8.5c there are two subjects that could have an influence on the magnitude of the shape paprameter. These correspond to the subjects with the smallest and largest values of the estimated martingale residuals.

The plot in Figure 8.5d is nearly identical to Figure 8.3. The plot identifies two subjects with large valus of this statistic. These same two subjects may be seen in the bottom left corner of Figure 8.5c and are among the most influential for the shape parameter.

Analysis of the individual diagnostic statistics reveals that the two extreme points correspond to the 20- and 43-year-old subjects identified using the diagnostic statistics for the fitted exponential regression model in the previous section. When we refit the Weibull regression model deleting these subjects, the maximum change in any coefficient is less than 4 percent when compared to the estimates from the full data model in Table 8.4. There is a 30.5 percent change in the estimate of the shape parameter. At this point we continue the analysis with all subjects included keeping in mind the influence on the shape parameter of the two subjects.

The next step in the assessment of model fit is to plot the Kaplan-Meier estimator against its Cox–Snell residuals, the estimated cumulative hazard function, in the same manner as shown in Figure 8.4 for the exponential regression model. The estimator of the cumulative hazard function for the Weibull regression model is

$$\hat{H}\left(t_i, \mathbf{x}_i, \hat{\boldsymbol{\beta}}, \hat{\sigma}\right) = \exp\left[\left(y_i - \mathbf{x}_i'\hat{\boldsymbol{\beta}}\right)\big/\hat{\sigma}\right]$$

$$= \exp(\hat{z}_i)$$

so

$$\hat{H}\left(t_i, \mathbf{x}_i, \hat{\boldsymbol{\beta}}, \hat{\sigma}\right) = \left(t_i e^{-\mathbf{x}_i'\hat{\boldsymbol{\beta}}}\right)^{\hat{\lambda}},$$

for $i = 1, 2, \ldots, n$. Alternatively, one may calculate the values from the martingale residuals as

$$\hat{H}\left(t_i, \mathbf{x}_i, \hat{\boldsymbol{\beta}}, \hat{\sigma}\right) = c_i - \hat{M}_i.$$

Figure 8.6 presents the Cox–Snell residual diagnostic plot. When we compare the plot in Figure 8.4 to the one in Figure 8.6, we see that the estimated cumulative hazard for the Weibull regression model falls closer to the referent line, indicating better adherence to the parametric assumptions. The likelihood ratio form of the Grønnesby–Borgan test has a value of $G = 5.93$ which, with 9 degrees-of-freedom, yields a p-value of 0.746, which also supports overall model fit.

We do not provide the associated table of observed and expected numbers of events, as it supports the conclusions about model fit. However, we note that the actual value for the Grønnesby–Borgan test that

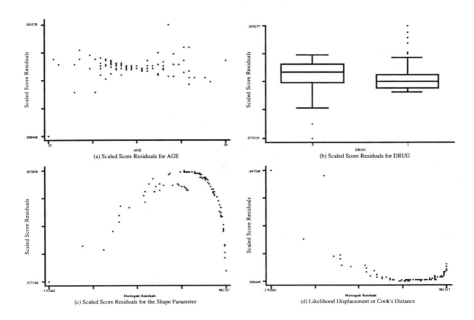

Figure 8.5 Plots of the scaled score residuals for (a) AGE, (b) DRUG, (c) the Shape parameter, $-\ln(\sigma)$, and (d) the likelihood displacement or Cook's distance based on the fitted weibull regression model in Table 8.4.

one obtains in any example will depend on the particular cutpoints used. In settings where there are multiple tied values that result in multiple choices for how to form the 10 groups, as is the case in the current example, many different values for the test are possible. However, if the model fits, then any choice of cutpoints should yield a test statistic that supports model fit. As yet, the dependence of the Grønnesby–Borgan test on choice of cutpoints has not been studied.

At this point, it appears that both the exponential and Weibull models fit the data reasonably well. However, on the basis of the plot in Figure 8.6 as well as the significance of the shape parameter, the Weibull model may be preferable. Estimates of time ratios for the Weibull model are obtained by exponentiating the product of a coefficient and the stated differences in the covariate values [see (8.31)]. The process is the same as that used for the exponential regression model. Comparing the estimates of the coefficients for AGE and DRUG in Tables 8.2 and 8.4, we find that the two models yield basically the same estimates of

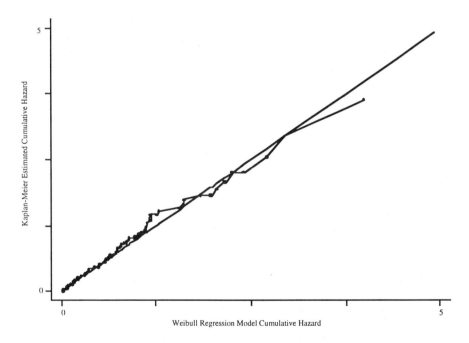

Figure 8.6 Graph of the Kaplan–Meier estimate of the cumulative hazard against the Weibull regression model estimate of the cumulative hazard based on the model in Table 8.4. The solid line is the referent line with slope = 1.0 and intercept = 0.

effect. We leave as an exercise the actual calculation of the time ratios with confidence intervals for the model in Table 8.4.

In this particular example, it is constructive to examine the estimated baseline hazard functions for the fitted exponential and Weibull regression models. The estimator of the baseline hazard for the exponential model is obtained from (8.8) and is

$$h_{0e}\left(t, \hat{\boldsymbol{\beta}}\right) = \exp\left(-\hat{\beta}_0\right). \tag{8.43}$$

The estimator of the baseline hazard for the Weibull model is obtained from (8.27) and is

$$h_{0w}\left(t, \hat{\boldsymbol{\beta}}, \hat{\sigma}\right) = \frac{1}{\hat{\sigma}} \exp\left(-\beta_0/\hat{\sigma}\right) t^{\left((1/\hat{\sigma})-1\right)}. \tag{8.44}$$

The text symbols "e" and "w" in the subscripts of (8.43) and (8.44) refer to exponential and Weibull, respectively. As in the case of the proportional hazards model, for the value of the baseline hazard function to be biologically meaningful, one must define the covariate value of zero to be meaningful. To do this, we refit the models in Tables 8.2 and 8.4, centering AGE at its mean. Covariates equal to zero correspond to a subject of average age and no history of IV drug use. Plots of the two estimated baseline hazard functions are shown in Figure 8.7.

Figure 8.7 illustrates the difference in the shapes of the two estimated baseline hazard functions. The estimated baseline hazard function of the exponential model, by definition, has a constant value of 0.06. The Weibull model, on the other hand, has a baseline hazard that begins at a value lower than the exponential baseline hazard, rises sharply, and then starts to level off after about 30 months. The average baseline hazard for the Weibull is also 0.06, however, so the exponential has the correct average baseline rate but the Weibull model is much more specific. For these data, the pattern of a progressive increase in risk over time is quite reasonable. In this example, when all aspects of the two models are considered, we favor the Weibull model. It fits the data better and has a clinically plausible hazard function.

This comparative analysis is only possible because both the exponential and Weibull models are also proportional hazards models. This type of model comparison could not be done using the accelerated failure time hazard functions, because the function for the Weibull model in (8.28) depends on both covariate values and coefficients in such a

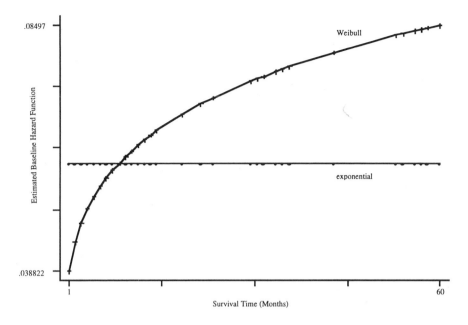

Figure 8.7 Plots of the estimated baseline hazard function for the exponential regression model in Table 8.2 and the Weibull regression model in Table 8.3, with age centered at its mean.

way as to prevent a baseline hazard function being independent of covariates.

Modeling survival time data via accelerated failure time models is different from normal errors linear regression modeling in one important aspect: we have the choice of several possible "error" distributions. So far, we have considered two of them. In the next section we consider another popular and useful model: the log-logistic accelerated failure time model.

8.4 THE LOG-LOGISTIC REGRESSION MODEL

The single covariate log-logistic accelerated failure time model may be expressed as

$$\ln(T) = \beta_0 + \beta_1 x + \sigma\varepsilon, \tag{8.45}$$

where the error term ε follows the standard logistic distribution, which we will discuss in more detail later in this section. The appealing feature of the log-logistic model is that the slope coefficient can be expressed in such a way that it can be interpreted as an odds-ratio.

In order to develop the odds-ratio interpretation, we begin by expressing the survivorship function for the model in (8.45) as

$$S(t,x,\beta,\sigma) = \left[1 + \exp(z)\right]^{-1}, \qquad (8.46)$$

where z, as before, is the standardized log-time outcome variable, i.e., $z = (y - \beta_0 - \beta_1 x)/\sigma$ and $y = \ln(t)$. The odds of a survival time of at least t is

$$\frac{S(t,x,\beta,\sigma)}{1 - S(t,x,\beta,\sigma)} = \exp(-z). \qquad (8.47)$$

As an example, assume that the covariate is dichotomous and coded 0/1. The odds-ratio at time t formed from the ratio of the odds in (8.47) evaluated at $x = 1$ and $x = 0$ is

$$OR(t, x=1, x=0) = \frac{\exp\left[\dfrac{-(y - \beta_0 - \beta_1 \times 1)}{\sigma}\right]}{\exp\left[\dfrac{-(y - \beta_0 - \beta_1 \times 0)}{\sigma}\right]} = \exp(\beta_1/\sigma). \quad (8.48)$$

Note that the ratio in (8.48) is independent of time. If the odds-ratio in (8.48) was 2.0, the interpretation would be that the odds of survival beyond time t among subjects with $x = 1$ is twice that of subjects with $x = 0$, and this holds for all t.

An alternative interpretation is obtained when we express the median survival time as a function of the regression coefficients. Setting the survivorship function in (8.46) equal to 0.5 and solving, we obtain an equation for the median survival time of

$$t_{50}(x,\beta,\sigma) = \exp(\beta_0 + \beta_1 x), \qquad (8.49)$$

and the time ratio at the median is

$$\text{TR}\left(t_{50}, x=1, x=0\right) = \exp\left(\beta_1\right). \qquad (8.50)$$

As expected with an accelerated failure time model, the exponentiated coefficient provides the acceleration factor on the time scale. In particular, if the time ratio in (8.50) is 2.0, the median survival time in the group with $x=1$ is twice that of the group with $x=0$. Since the percentiles of the survival time distribution in (8.46) are of the form

$$t_p\left(x, \beta, \sigma\right) = \left[(1-p)/p\right]^\sigma \exp\left(\beta_0 + \beta_1 x\right),$$

the result in (8.50) holds at all values of time. Cox and Oakes (1984) show that the log-logistic model is the only accelerated failure time model with the proportional odds property in (8.47).

As was the case with the exponential and Weibull models, maximum likelihood is the method usually employed to fit a log-logistic model to a set of data subject to right censoring. It follows from results for the standard logistic distribution [see Evans, Hastings and Peacock (1993) or Klein and Moeschberger (1997)] that the contribution of a noncensored time to the likelihood is

$$\left(1/\sigma\right)\exp(z)\Big/\left[1+\exp(z)\right]^2$$

and that of a censored time is

$$\left[1+\exp(z)\right]^{-1},$$

where, for a multivariable model, $z = (y - \mathbf{x}'\boldsymbol{\beta})/\sigma$. It follows that the log-likelihood function for a sample of n independent observations of time, covariates and censoring indicator, denoted $\left(t_i, \mathbf{x}_i, c_i\right), i = 1, 2, \ldots, n$, is

$$L(\boldsymbol{\beta}, \sigma) = \sum_{i=1}^{n} c_i \left\{ -\ln(\sigma) + z_i - 2\ln\left[1 + \exp(z_i)\right] \right\} - \left(1 - c_i\right) \ln\left[1 + \exp(z_i)\right].$$

$$(8.51)$$

The score equation for the jth regression coefficient is obtained by taking the derivative of the log-likelihood in (8.51) with respect to β_j, and the score equation for σ is obtained in a similar manner. These equations were presented in detail for the exponential and Weibull models, as they provided the basis for diagnostic statistics to assess the effect that individual observations had on parameter estimates and to assess model fit. Since these same diagnostic statistics have not yet been developed for the log-logistic model, we do not present the score equations or the elements in the matrix of second derivatives.

The model development issues are the same for the log-logistic model as for the other models. Unfortunately, most of the methods used in the previous sections and chapters for the proportional hazards type models have not been extended for use with the log-logistic model. We conclude this section with an example, fitting a model from the HMO-HIV+ study.

A number of packages have the capability of fitting the log-logistic model. The results shown in Table 8.5 were obtained from SAS, and the estimated coefficients are reasonably similar to those obtained from the other models. The estimate of the effect on the time scale for having a history of IV drug use is

$$\hat{TR} = \exp(-0.891) = 0.41.$$

The interpretation is that survival times of subjects with a history of IV drug use are estimated to be 41 percent of those without a history. The endpoints of a 95 percent confidence interval obtained by exponentiating the endpoints of the Wald-based confidence interval for the coefficient in Table 8.5 are 0.27 and 0.62. The confidence interval estimate suggests that the survival times for those with a history could be between 27 and 62 percent of those without a history of IV drug use. Point and interval estimates for the effect of age are calculated and interpreted in a manner similar to previous sections.

Table 8.5 Estimated Coefficients, Standard Errors, z-Scores, Two-Tailed p-Values and 95% Confidence Intervals for the Log-Logistic Regression Model in the HMO-HIV+ Study ($n = 100$)

Variable	Coeff.	Std. Err.	z	$P > \lvert z \rvert$	95% CIE
AGE	−0.087	0.016	−5.61	<0.001	−0.117, −0.057
DRUG	−0.891	0.214	−4.17	<0.001	−1.310, −0.472
Constant	5.540	0.575	9.39	<0.001	4.390, 6.690
Sigma	0.588	0.054			

Log likelihood = −129.106.

One diagnostic assessment that is possible with the log-logistic model is the plot of the Cox–Snell residuals. The estimator of these residuals, the log-cumulative hazard, is obtained from the estimator of the multivariable form of the survivorship function in (8.46), and for the ith subject it is

$$\hat{H}\left(t_i, \hat{\boldsymbol{\beta}}, \hat{\sigma}\right) = \ln\left[1 + \exp(\hat{z}_i)\right], \qquad (8.52)$$

where $\hat{z}_i = \left(y_i - \mathbf{x}_i'\hat{\boldsymbol{\beta}}\right)\!\big/\hat{\sigma}$ and $\hat{\boldsymbol{\beta}}$ and $\hat{\sigma}$ denote the estimators maximizing (8.51). Figure 8.8 presents the plot of the Kaplan–Meier estimator against these residuals.

For the most part, the plotted points follow the referent line. There is slightly more departure from the line at larger values of the cumulative hazard than was seen in the corresponding plot for the Weibull model shown in Figure 8.6. Based on this plot, it would appear that the log-logistic model provides a reasonable fit to the data.

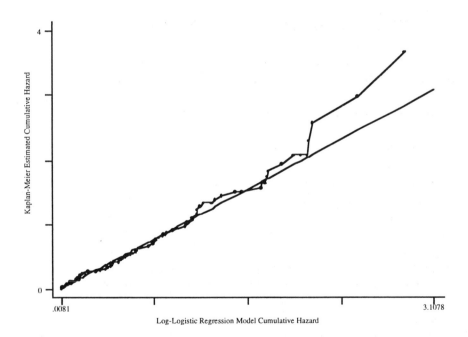

Figure 8.8 Graph of the Kaplan–Meier estimate against the log-logistic regression model estimate of the cumulative hazard based on the fitted model in Table 8.5. The solid line is the referent line with slope = 1.0 and intercept = 0.

The three models fit to the data from the HMO-HIV+ study are not directly comparable. Akaikie (1974) proposed an information criterion (AIC) statistic to compare different models and/or models with different numbers of parameters. For each model the value is computed as

$$\text{AIC} = -2 \times (\log\text{-likelihood}) + 2(p + 1 + s), \quad (8.45)$$

where p denotes the number of covariates in the model not including the constant term, $s = 0$ for the exponential model and $s = 1$ for the Weibull and log-logistic models. The values of AIC are 266.79 for the exponential model in Table 8.2, 265.00 for the Weibull model in Table 8.4 and 266.21 for log-logistic model in Table 8.5. These three values are quite similar and reflect what was seen in the analysis. There are not large differences in the fit or the inferences from the three models. Each offers an accelerated failure time interpretation, and the estimates of effect are not that different. The Weibull model, with the smallest value of AIC, seems to be the best fitting of the three.

8.5 OTHER PARAMETRIC REGRESSION MODELS

There is an extensive literature on parametric models to analyze survival time data. A number of texts that present a significant amount of material on parametric analysis of survival time data are listed at the end of Section 8.1. Delving much further into this topic would take us beyond the scope of this text, and we conclude this chapter with a few observations and comments on parametric survival time models.

The principal benefit of any parametric model is the specificity it provides for modeling the data, especially the hazard function's description of the underlying aging process. Parametric models offer a wide variety of possible shapes for the hazard function, ranging from a simple constant hazard to complex "bathtub" shaped functions that might be appropriate for modeling human life over a long time span. In general, these models are most effectively used when the investigator has considerable knowledge of the aging process in the subjects being studied. Clinical plausibility is a vital aspect of any choice of hazard function.

One should not use a complicated parametric model without prior knowledge that the hazard function is plausible for the problem at hand. In the vast majority of current applications of survival time data analysis that follow humans over time, this prior knowledge is not available.

Therefore, use of the semi-parametric models discussed in the previous chapters provides a safe and proven method for analysis. This isn't to say that parametric models should be avoided, but rather that they should be used with caution and with a keen eye as to the plausibility of the parametric form of the hazard.

EXERCISES

The first three problems involve simple two variable exponential and Weibull regression models fit to data from the UIS. The goal of these problems is to estimate and compare time and hazard ratios in a simple setting. For this reason, important model building details, such as checking the scale of age, are skipped.

1. Fit an exponential regression model containing AGE and TREAT to the UIS data.

(a) Using the fitted model compute point and 95 percent confidence interval estimates of the time ratio for treatment and for a 10-year increase in age. Interpret these estimates within the context of the UIS.

(b) Using the fitted model compute point and 95 percent confidence interval estimates of the hazard ratio for treatment and for a 10-year change in age. Interpret these estimates within the context of the UIS.

(c) Compare the time ratio and hazard ratio estimates computed in problems 1(a) and 1(b). In particular, which estimate, time or hazard ratio, do you think would be more easily understood by non-statistically trained study staff?

(d) Why can the fitted exponential regression model be used to compute both time and hazard ratio estimates?

2. Repeat problem 1 using the Weibull regression model.

3. Which model, the exponential regression model fit in problem 1 or the Weibull regression model fit in problem 2, is the better fitting model? Justify your response using plots and statistical tests.

4. Verify that the appropriate scale for the covariate AGE in the exponential regression model fit to the data from the HMO-HIV+ study (results shown in Table 8.2) is linear.

5. Using the results from the fitted Weibull model shown in Table 8.4, compute point and 95 percent confidence limits for the time ratio for history of IV drug use and a 10-year increase in age.

6. Fit an exponential regression model to the UIS data using the covariates in the main effects proportional hazards model shown in Table 5.5.

(a) Assess the scale of AGE, BECKTOTA and NDRUGTX. Are the results the same or different from those obtained for the proportional hazards model in Chapter 5?

(b) Using the correctly scaled model from problem 6.(a), assess the need for interactions in the exponential regression model. Are the interactions the same or different from those selected in Chapter 5 for the proportional hazards model?

(c) Compute the diagnostic statistics to assess the model obtained in problem 6(b). Explore the effect on the estimates of the coefficients of any influential subjects identified through use of the diagnostic statistics.

(d) Assess the overall fit of the model from problem 6(c) via the Borgan–Grønnesby test.

(e) Assess the adherence of the fitted model to the exponential errors assumption via the plot of the Cox–Snell residuals.

(f) Describe the estimates of effect of the covariates in the final exponential regression model using time ratios, with 95 percent confidence intervals.

7. Repeat problem 6 using the Weibull regression model. Is the fit of the Weibull regression superior to the exponential model?

8. Fit a log-logistic regression model containing the covariates in the final Weibull regression model from problem 6. Repeat problem 6(f) for the log-logistic model. Is the fit of the log-logistic regression superior to the Weibull or exponential regression models?

CHAPTER 9

Other Models and Topics

9.1 INTRODUCTION

Modeling survival time when there is a single event of interest that terminates observation is the central theme of the previous eight chapters. This situation, in which each subject may experience the event of interest only once, describes the majority of applied settings when follow-up time is observed. There are situations, however, when the event of interest may occur more than one time for each subject. For example, a subject's cancer may be treated and then recur some time later, and this process may be repeated several times. The participants in the UIS could be followed and intervals of time when they remained "drug free" could be recorded. Recording the interval of time between heart attacks in a cohort of subjects at high risk for this event would be another example. Data of this type are referred to as *recurrent event* data. We discuss several approaches to modeling recurrent event data in Section 9.2.

Another extension of the standard modeling situation occurs when subgroups of responses are correlated due to study design considerations. This lack of independence of response can occur in any setting in which survival time is influenced by unmeasured factors that are the same within groups of subjects and are thought to have significant group to group variability. When these factors are present in the usual normal errors linear model setting, they are called *random effects*. Survival analysis models incorporating such factors are called *frailty models*. These models have also been suggested for use in the recurrent event setting. We discuss frailty survival time models in Section 9.3.

Often, investigators employing statistical methods have less data than they would like to have. Sample sizes tend to be limited by cost and time constraints. Alternatively, there are other situations when follow-up time is available in such a large cohort that it is impractical to use all the data. When this happens, nested case-control studies have been suggested as a way to model a covariate's effect at a considerably reduced cost. We discuss this approach to modeling survival time data in Section 9.4.

In previous chapters we used regression models in which the covariate's effect on the hazard rate was multiplicative. Models of this type yield hazard rate ratio estimates of effect that are meaningful and easily interpreted. We have largely ignored additive models, as they are not as easy to fit or interpret. We discuss a few additive models in Section 9.5.

9.2 RECURRENT EVENT MODELS

The idea that an event can occur multiple times in the course of the follow-up of a subject is a conceptually easy extension of the single event model. We encounter things that break, are repaired, and then break again all the time. A car is a good example. Another example is cancer. Following treatment, the cancer may go into remission but, at a later point in time, may recur. The process of treatment and recurrence may be repeated multiple times in a subject. The defining characteristic of a recurrent event is that we observe the same event in a single subject multiple times during the follow-up period.

Modeling of recurrent event data is not typically addressed in a text of this level. One of the few summaries available is a technical report by Therneau (1995). Clayton (1994) discusses recurrent events and compares them to generalized linear models, such as Poisson and logistic regression.

A number of proportional hazards-type models have been proposed for use with recurrent event data. The easiest way to explain the similarities and differences in the models is by describing how we handle the data for two hypothetical subjects using each model. Suppose we have n independent subjects in our study. One of these subjects experienced the event of interest at 9, 13 and 28 months of follow-up, and was followed for another 3 months with no additional events before the study ended. The second subject experienced the event at 10 and 15 months, and follow-up ended at the second event. We also assume that a sufficient number of subjects had four recurrent events to allow modeling this number of recurrent events.

The simplest modeling approach is to use the counting process formulation described in detail in Andersen, Borgan, Gill and Keiding (1993). In this formulation, follow-up time is broken into segments defined by the events. The data for our hypothetical subjects under this model are shown in the panel labeled "Counting Process" in Table 9.1. We describe the data using time intervals, event indicators and strata. The purpose of the stratum variable will become clear when we describe the other modeling approaches. In the counting process model, the events are assumed to be independent, and a subject contributes to the risk set for an event as long as the subject is under observation at the time the event occurs. The first hypothetical subject will be in the risk set for any event occurring between 0 and 31 months. Under an assumption of no tied event times, this subject contributes the event defining the risk set at 9, 13 and 28 months. The second subject will be in the risk sets for any events occurring between 0 and 15 months. This subject contributes the event defining the risk set at 10 and 15 months. The data for the first subject could be described as data for four different subjects: the first begins follow-up at time 0 and has the event at 9 months, the second has delayed entry at 9 months and is followed until 13 months when the event occurs, the third has delayed entry at 13 months and is followed until 28 months when the event occurs and the

Table 9.1 Data Layout under Four Recurrent Event Models for Two Hypothetical Subjects

Model	Subject 1			Subject 2		
	Time Interval	Event	Stratum	Time Interval	Event	Stratum
Counting Process	(0, 9]	1	1	(0, 10]	1	1
	(9, 13]	1	1	(10, 15]	1	1
	(13, 28]	1	1			
	(28, 31]	0	1			
Conditional A	(0, 9]	1	1	(0, 10]	1	1
	(9, 13]	1	2	(10, 15]	1	2
	(13, 28]	1	3			
	(28, 31]	0	4			
Conditional B	(0, 9]	1	1	(0, 10]	1	1
	(0, 4]	1	2	(0, 5]	1	2
	(0, 15]	1	3			
	(0 , 3]	0	4			
Marginal	(0, 9]	1	1	(0, 10]	1	1
	(0, 13]	1	2	(0, 15]	1	2
	(0, 28]	1	3	(0, 15]	0	3
	(0, 31]	0	4	(0, 15]	0	4

fourth has delayed entry at 28 months and is followed until 31 months and is censored at that time. Data for our second hypothetical subject may be described in a similar manner.

We see from the way the data are constructed that the model treats the events as being independent and does not differentiate a first event from a second or third, and so on. This model can handle time-varying covariates of any type.

In theory, we could model the data in this setting using any hazard function, but we will use the proportional hazards function. It follows that the partial likelihood is identical to the standard proportional hazards partial likelihood [see (3.17)]. If we denote the number of recurrent events for the ith subject as m_i then the total number of events modeled is $m = \Sigma m_i$. The correlation due to observing multiple events within the same subject is accounted for by adjusting the estimates of the standard errors using a method that we describe after discussing the data layout for each model.

The counting process model is particularly easy to fit, as many software packages allow follow-up time to be described by a time defining the beginning of an interval and a time defining the end of an interval where the subject was either lost to follow-up or experienced the event. For example SAS, STATA and S-PLUS allow this type of data input.

Two conditional models for recurrent event data have been suggested in Prentice, Williams and Peterson (1981). The models are conditional in the sense that a subject is assumed not to be at risk for a subsequent event until a prior event has occurred. For example, hypothetical subject 1 is assumed not to be at risk for a third event until the second event occurred and is not at risk for a fourth event until the third event occurs. A stratum variable is used to keep track of the event number.

The difference between the two conditional models is the time scale used. One model uses time defined by the beginning of the study while the second uses time since the previous event. The data layouts for the two models are labeled "Conditional A" and "Conditional B," respectively, in Table 9.1. The follow-up time for the two hypothetical subjects under the A model is handled via the same type of time intervals and event indicators that were used in the counting process model. The stratum variable indicates the specific event number the subject is at risk of having. In the B model, follow-up time begins at zero for each event and ends at the length of time until the next event.

Under the assumption that all the covariates are fixed at the beginning of the study, the proportional hazards function for the sth event under conditional model A is

$$h_s(t,\mathbf{x},\boldsymbol{\beta}_s)=h_{0s}(t)\exp(\mathbf{x}'\boldsymbol{\beta}_s),\tag{9.1}$$

and under conditional model B it is

$$h_s(t,\mathbf{x},\boldsymbol{\beta}_s)=h_{0s}(t-t_{s-1})\exp(\mathbf{x}'\boldsymbol{\beta}_s),\tag{9.2}$$

where t_{s-1} denotes the time at which the previous event occurred. Parameter estimates for either model may be obtained by using the stratified partial likelihood in (7.2) and (7.3) with data as shown in Table 9.1. Event-specific parameter estimates are obtained by including stratum by covariate interactions in the model. Time-varying covariates can also be used. The risk sets used in the partial likelihood are composed of subjects at risk for a specific event at time t. These sets will be completely different under the two models, since each uses a different time scale. Thus, we expect the two models to yield different values for the estimate of the effect of the same covariate. We consider this point in more detail in the example below.

Wei, Lin and Weissfeld (1989) proposed a marginal event-specific model for the analysis of recurrent event data. The model is marginal in the sense that each event is considered as a separate process. By definition, time for each event starts at the beginning of follow-up for each subject. All subjects are considered to be at risk for all events, regardless of how many events they actually had. The data layout for the two hypothetical subjects is shown in the panel labeled "Marginal" in Table 9.1. Recall that in this study we are modeling up to four recurrent events. Our hypothetical subject 1 had three events, and the first three intervals record the total length of time to each event, the event indicator variable is equal to one, and the stratum variable records the specific event. Note that this subject has an additional fourth interval that records the total length of follow-up, has event indicator variable equal to zero and stratum variable equal to four. This interval records the "marginal" time the subject was at risk for the fourth event. The data for our second hypothetical subject follow this same pattern. This subject had two events, and the first two intervals denote the total follow-up time until each event. The third interval records the marginal follow-up time to the third event, and, since the event was not observed, the event indicator variable is equal to zero. The marginal follow-up time for the fourth event is the same and is repeated with stratum variable equal to four. All subjects in the study contribute follow-up times to all possible recurrent events, whether they experienced that particular recurrence or not.

Under the assumption that all covariates are fixed, the proportional hazards function for the sth event is of the same form as the one in (9.1), and parameter estimates can be obtained using the stratified partial likelihood in (7.2) and (7.3). Stratum-specific parameter estimates are obtained by including stratum by covariate interactions in the model. This model will also accommodate time-varying covariates.

From a purely computational point of view, once the data for each subject has been put into the specific form shown in Table 9.1, we can use existing software to fit any of the models. The actual model or models we use will depend on which event process we wish to describe. In any setting, the four models may yield parameter estimates of a similar value for the same covariate, yet their interpretation is quite different because entirely different processes are being modeled. We return to this point in an example.

Before considering an example, we discuss a method for adjusting the estimates of the variance of the coefficients to account for the correlation among the observations on an individual subject. Lin and Wei (1989) proposed an extension of White's (1980, 1982) robust variance estimator to the proportional hazards model setting. The extension is similar to the "information sandwich" estimator proposed by Liang and Zeger (1986a, 1986b) for use with correlated data in generalized linear models. Lin and Wei proposed that the estimator be based on the vector of changes in the estimate when the ith subject is deleted. This is the influence diagnostic defined in (6.23). The Lin and Wei robust estimator for the usual single-event proportional hazards model is

$$\hat{R}(\hat{\beta}) = \hat{Var}(\hat{\beta})[\hat{L}'\hat{L}]\hat{Var}(\hat{\beta}), \qquad (9.3)$$

where $\hat{Var}(\hat{\beta})$ is the information-matrix-based estimator in (3.33), and $\hat{L}$ is the n by p matrix whose rows contain the vector of score residuals, $\hat{L}_i$, defined in (6.17).

In the recurrent event setting, subjects may contribute more than one score residual. The actual number will depend on which of the four models is fit. The robust estimator is still based on (9.3), but it uses a value of $\hat{L}_i$ obtained by summing over all the score residuals contributed by the ith subject. This estimator is available as an option in many software packages, including SAS, S-PLUS and STATA. We do not discuss the robust estimator further, but refer the reader to Lin and Wei (1989) for details. In the examples, we will provide both the robust and information-matrix-based estimators for comparative purposes.

The model building details involved in fitting any of the four models are the same as those discussed in detail in Chapter 5, namely checking the scale of continuous covariates, dealing with issues of adjustment or confounding, and assessing the inclusion of interactions. Since all four models are fit using conventional software for a one-event proportional hazards model, the methods are identical to those discussed and illustrated in Chapter 5. Model fit can be assessed using the same residual and influence measures discussed in Chapter 6.

The May and Hosmer (1998) approach to calculating the Grønnesby-Borgan goodness-of-fit test may be used with the counting process model, since this model treats each event as if it were a first event. The other three models make an explicit differentiation between events. The extension of the goodness-of-fit test to the setting of multiple events has not been studied.

To illustrate the use of recurrent event models, we use data from a study in which successive endpoints occur, but the event defining the endpoints is a landmark in the course of the treatment of psoriasis. The data are from a recent study of the influence of a mindfulness-based stress reduction intervention in the treatment of psoriasis. Kabat–Zinn et al. (1998) report the details of the study's design and implementation and the results of an analysis aimed at a general medical audience. Patients in the study had severe psoriasis and were beginning a treatment program prescribed by their physician. The treatment (LIGHT) was either phototherapy (LIGHT = 0) or photochemotherapy (LIGHT = 1). A portion of the intervention consisted of listening to a tape of soothing music while undergoing treatment. Patients within each light group were randomized to one of two tape groups (TAPE). One tape group (TAPE = 1) received instruction on the use of mindfulness-based stress reduction techniques and employed these during the treatment sessions. The other tape group (TAPE = 0) did not receive any aspect of the mindfulness-based intervention.

Four possible successive endpoints were observed. The first endpoint was the number of days until a first "response" to treatment was noted by study personnel. The second endpoint was the number of days until a specific major lesion showed evidence of treatment, called the turning point. The third endpoint was the number of days until this key lesion was half-cleared (called the half-way point). The fourth and final endpoint was the number of days until the key lesion cleared completely (called the clearing point).

Our goal in this example is not to provide a definitive analysis of the Kabat–Zinn study, but rather to illustrate the use of the recurrent event models. One covariate that affected the response to treatment was the number of years the patient had psoriasis (YRSPSOR), and it is included in all of our models.

Table 9.2 presents the results of fitting the counting process, conditional A, conditional B and marginal models to the psoriasis data. Results from partial likelihood ratio tests (not shown) indicate that the stratum by covariate interaction terms, required by the conditional A, conditional B and the marginal models to obtain event-specific coefficient estimates, are not significant. The results of checking the scale of years with psoriasis support treating this covariate as linear in the log-hazard function. In addition, the interaction between LIGHT and TAPE is not significant in any of the four models. The results in Table 9.2 are obtained by defining event type as a stratification variable. The z-scores and two-tailed p-values in Table 9.2 are calculated using the robust estimate of the variance.

Since the primary purpose of the example is to compare the four models, we focus this discussion on the estimated coefficients for TAPE. The coefficient for TAPE is significant in all four models, but it varies in magnitude. Even though the rank order of the coefficients is the same for all three variables in Table 9.2, it is not possible to conclude that the observed order holds in all cases. Broad categorical generalizations about the comparative magnitude of the robust and information matrix estimators of the standard errors are not possible either, so we consider the models one at a time.

The counting process model uses time defined from the beginning of the study, treats all events as if they were the same type of event, and assumes all events are independent. In the psoriasis study, the follow-up times for the 32 subjects, when expanded into time intervals using the method shown in Table 9.1, created 110 intervals of which 96 ended in an event. The counting process model uses a partial likelihood formed from the observed survival times of the 96 events. Anyone under observation at the particular time is in the risk set for that event. For example, if a subject experienced his/her second event at 34 days, the risk set would contain all subjects who were still being followed (under treatment in our example) at 34 days regardless of how many events they may have already experienced. For example, the risk set could

**Table 9.2 Coefficient Estimates, Standard Errors, Robust
z-Scores and Two-Tailed p-Values for Four Recurrent
Event Models Fit to the Psoriasis Data (n = 32)**

Model	Var.	Coeff.	Std. Err.	Robust Std. Err.	Robust z	$P>|z|$
	TAPE	0.390	0.216	0.126	3.09	0.002
CP	LIGHT	0.401	0.212	0.123	3.26	0.001
	YRSPSOR	0.009	0.012	0.006	1.59	0.112
	TAPE	0.838	0.247	0.254	3.30	0.001
C-A	LIGHT	1.118	0.270	0.272	4.10	<0.001
	YRSPSOR	0.021	0.013	0.014	1.52	0.129
	TAPE	0.616	0.232	0.249	2.48	0.013
C-B	LIGHT	0.811	0.232	0.227	3.57	<0.001
	YRSPSOR	0.016	0.013	0.013	1.26	0.208
	TAPE	1.030	0.243	0.325	3.17	0.002
M	LIGHT	1.540	0.254	0.312	4.94	<0.001
	YRSPSOR	0.028	0.014	0.019	1.47	0.142

CP: Counting Process Model, C-A: Conditional A Model,
C-B: Conditional B Model and M: Marginal Model.

contain a subject who had not yet had his/her first event as well as a
subject who had experienced 3 events. This property of not taking into
account the order or type of event makes the counting process model
simple but somewhat unrealistic in many settings. It is probably not a
useful model for the psoriasis data, in which there is a clear time sequence in the events. Having said this, we will interpret the coefficient
for TAPE as if the model did make sense to use. We obtain estimates of
hazard ratios, as before, by exponentiating estimated coefficients. The
estimated hazard ratio for the effect of the mindfulness intervention is
1.48. This can be interpreted to mean that subjects having the mindfulness therapy are estimated to be experiencing events at a rate that is 48%
higher than subjects not having this treatment adjunct. Estimated hazard ratios could be computed and interpreted in a similar manner for
the other covariates. As is the case with all models, estimated hazard
ratios for any continuous covariate should be based on a meaningful
change in the covariate. In summary, the counting process model is
perhaps the simplest of the four models to fit and interpret. Its simplicity is both its major strength and weakness when applied to recurrent
event data.

The two conditional models take the specific order of events into account, but define time in different ways. Model A uses time defined from the beginning of the study, while model B "resets the clock" after an event is observed, and the risk sets are different for the two models. For example, consider the time to the second event for the two subjects in Table 9.1. Under model B, both subjects are in the risk set for the second event from time 0 to 4 months. Under model A, the first subject enters the risk set at 9 months and leaves at 13 months, while the second subject does not enter the risk set until 10 months and leaves at 15 months. Not only are the times of entry and removal from the risk sets different under the two models, but the lengths of time the subjects are in the risk sets together are different.

It is not possible to make comparative statements about the magnitude of the coefficients from the two models. Model A is the logical choice if one is interested in modeling the full time course of the recurrent event process. Model B should be used when the goal of the analysis is to model the gap time between events. The logical choice for the psoriasis data is model A, since the goal is to document the full time course of the therapy.

Estimated hazard ratios are computed and interpreted in the standard manner. Under model A, the estimated hazard ratio for the effect of the addition of the mindfulness therapy is 1.85. This is interpreted to mean that the rate of attainment of the endpoints is 85 percent higher in the group receiving the mindfulness adjunct to therapy. Estimated hazard ratios could be computed and interpreted in a similar manner for the other variables in the model. As was recommended for the counting process model for recurrent events, confidence interval estimates and Wald-type tests should use the robust standard error. In summary, the two conditional approaches provide models that allow one to take into account both the occurrence of multiple events and the time order of the events.

The marginal model takes a different approach to the multiple event process. Under this model, the total time to each of the possible recurrent events is modeled. Each subject contributes to the risk set for each event as long as he/she is still being followed at the time defining the risk set. For example, consider the time to the third event for the two subjects in Table 9.1. The first subject experienced a third event at 28 months and is in the risk set for the third event from time 0 to 28 months. The second subject experienced two events, yet under the marginal model is considered to be at risk for the third (and fourth for that matter) from time 0 to 15 months. In a sense, the marginal model looks at each event separately and models all of the available data for that event.

Estimated hazard ratios and confidence intervals are computed in the usual manner. The estimated hazard ratio for the effect of the mindfulness therapy is 2.80. The interpretation is similar to the other models, but the rate of attainment of the endpoints is 180 percent greater in the group receiving the mindfulness adjunct to therapy. Hazard ratios for the other covariates in the model could be computed and interpreted in a similar manner.

The estimated models presented in Table 9.2 are somewhat simpler than one might encounter in other examples or settings since none of the stratum by covariate interaction terms were significant. The effect of each of the covariates could be described by a single coefficient. In settings where the interactions are significant, stratum- or event-specific estimates of the effect must be computed. Depending on how the design variables for the interaction are created, the estimate of the effect of the covariates may involve coefficients for main effects as well as interaction coefficients. The method for estimating and interpreting a hazard ratio in this case is the same as that discussed and illustrated with the model from the UIS in Section 6.6.

As with any fitted statistical model, it is important to assess its adequacy and fit before using the estimated coefficients for inferential purposes. Furthermore, it is important to evaluate how well the fitted proportional hazards model adheres to the proportional hazards assumption. Since each of the models is fit using standard proportional hazards software, the methods for carrying out this step are the same as those described in Chapter 6 for the single event setting.

Modeling of recurrent event data is no more difficult than modeling single event data in that the same software may be used with data augmented in the manner shown in Table 9.1. The challenge, however, lies in deciding which of the four models to use. This choice should be based on careful consideration of the goals of the analysis, so that the interpretation of the coefficients is appropriate to the reseach question.

9.3 FRAILTY MODELS

Up to this point, all of the statistical models we have used to describe the distribution of survival time have assumed that the hazard function is completely specified given the baseline hazard function and the values of the covariates (i.e., there are no other factors influencing survival).

For example, in a study comparing two forms of treatment for a particular type of cancer, we might, in addition to treatment, include age and gender in the model and assume that the proportional hazards model is correct. In this case, all subjects of the same treatment, age and gender are assumed to have the same underlying distribution of survival time. This is not to say that all observed subjects of the same treatment, age and gender will have the same observed survival time. Rather, we assume that, if there is no censoring, the observed survival times are independent observations from a distribution with the same parameters. In some studies, particularly those involving human subjects, there may be factors other than the measured covariates that significantly affect the distribution of survival time. This condition is often referred to as heterogeneity of the subjects. Among the early papers on this subject is the work by Vaupel, Manton and Stallard (1979) who used the concept of frailty to describe differences in survival time among apparently similar individuals. Considerable work has been done in this area, but much of it is beyond the mathematical level assumed for this text. In this section, therefore, we present an introductory overview of frailty models. Hougaard (1995) presents an excellent overview of the models proposed for use in this area. Aalen (1994) also provides a relatively non-technical summary with some examples, with a focus on fully parametric models. Klein and Moeschberger (1997) present methods based on incorporating frailty in a proportional hazards model and discuss some of the technical details. Nielsen, Gill, Andersen and Sørensen (1992) present a more theoretical treatment based on the counting process approach. Andersen, Borgan, Gill and Keiding (1993) also discuss frailty models from the counting process perspective and illustrate the use of these models with examples.

The basic idea of a frailty model is to incorporate an unmeasured "random" effect in the hazard function to account for heterogeneity in the subjects. When the observed data consist of triples $(t_i, \mathbf{x}_i, c_i)$, $i = 1, 2, \ldots, n$ denoting the observed follow-up times, the vector of p covariates, and a right censoring indicator variable, the hazard function at time t for the ith subject is, under the proportional hazards model,

$$h(t, \mathbf{x}_i, \beta) = h_0(t) \exp(\mathbf{x}_i' \boldsymbol{\beta}). \tag{9.4}$$

This idea extends to models with time-varying covariates, with the usual change in notation.

A frailty model includes, in the hazard function, the value of an additional unmeasured covariate, the frailty, denoted z_i, yielding a hazard function

$$h_f(t, \mathbf{x}_i, \boldsymbol{\beta}) = z_i h(t, \mathbf{x}_i, \boldsymbol{\beta}). \qquad (9.5)$$

We use the subscript f in (9.5) to represent a hazard function that has been modified by the inclusion of a frailty. An important statistical assumption is that the frailty is independent of any censoring that may take place. Much of the work in this area [see Hougaard (1995), Aalen (1994) and Klein and Moeschberger (1997)] has dealt with the choice of statistical distribution for the frailty. Since the hazard cannot be negative, distributions must have positive values. This and other technical issues have led to the use of the gamma distribution. In particular, the most frequently used model assumes that the frailties represent a sample from a gamma distribution with mean equal to one and variance parameter θ. If the value of the frailty in (9.5) is greater than one, the subject has a larger than average hazard and is said to be more "frail." On the other hand, if the value of the frailty is less than one, the subject is less "frail" than an average subject. Aalen (1994) points out that there are advantages to using a fully parametric model, such as the Weibull regression model, with a frailty. Not only is estimation easier, but it is possible to describe explicitly the effect that frailties have on hazard ratios over time. In particular, due to the fact the "most" frail individuals tend to fail early in the follow-up, the average hazard ratio tends to decrease over time [see Aalen (1994) for an example and Hougaard (1995) for additional discussion of this point]. Since the major thrust of this text is modeling with the proportional hazards model, we do not consider parametric models with frailties in any more detail.

Frailty models have often been used when groups of subjects have responses that are likely to be dependent in some general way. For example, in an animal carcinogenicity study the responses of members of the same litter are not likely to be independent. Another example occurs when multiple events have been observed on the same subject, as discussed in the previous section. Liang, Self, Bandeen–Roche and Zeger (1995) discuss the use of frailty models with multivariate failure time data. In these settings, if the value of the frailty is assumed to be constant within groups, the models are called *shared frailty models*. Recently, the shared frailty model has been extended by Pickels et al. (1994) and Yashin, Vaupel and Jachine (1995) to allow different but correlated frailties among observations within a group. In the remainder of this section, we focus on adding a frailty to the hazard for a single subject as described in (9.5).

The idea of using an unmeasured covariate or random effect to account for heterogeneity or dependence of responses among groups of subjects is not new or unique to survival time models. This idea has been proposed for use in many generalized linear models. [Clayton (1994) and Neuhaus (1992) provide broad overviews of the topic and Collett (1991, Chapter 8) discusses random effects logistic regression models.]

The major problem faced by the practitioner wishing to use a gamma frailty model is the lack of readily available software. Some privately developed software has been made publicly available. Klein and Moeschberger (1997) provide SAS macros that fit the gamma frailty proportional hazards model at their web site. Jenkins (1997) has developed a set of programs for use in STATA that fit, with and without a gamma frailty, the interval-censored or discrete-time proportional hazards model discussed in Chapter 7. However, until these individually developed routines have been thoroughly tested and incorporated into a major software package, we would not recommend their use to anyone who does not have the programming and statistical skills necessary to debug problems that are likely to be encountered.

The method typically used to fit the continuous-time gamma frailty model in (9.5) is an application of the Estimation-Maximization (EM) algorithm.[1] One may find a detailed discussion of the implementation of the EM algorithm in Klein and Moeschberger (1997). They suggest that a simpler implementation proposed by Nielsen, Gill, Andersen and Sørensen (1992) can also be used. We discuss this simpler approach, a method also discussed and illustrated in Andersen, Borgan, Gill and Keiding (1993).

The EM algorithm is used to estimate the regression parameters for each of a fixed set of values of the variance parameter of the gamma frailty distribution. The solution is the one yielding the largest value of a "profile" log-likelihood. The specific steps are as follows:

---

[1] The EM algorithm has been used in many settings since it was first described by Dempster, Laird and Rubin (1977). This algorithm and its application in medical settings are discussed in a series of papers in *Statistical Methods in Medical Research* **6**(1) 1997.

Step 1: Fit the proportional hazards model containing the covariates of interest. Following the fit, obtain the estimate of the baseline cumulative hazard function for each subject, $\hat{H}_0(t_i)$, discussed in Section 3.5. Use this to obtain the estimate of the cumulative hazard for each subject,

$$\hat{H}\left(t_i, \mathbf{x}_i, \hat{\boldsymbol{\beta}}\right) = \hat{H}_0(t_i) \exp\left(\mathbf{x}_i' \hat{\boldsymbol{\beta}}\right).$$

Step 2: Create a set of possible values for the variance parameter of the gamma frailties, θ. Typically, this set is constructed by beginning with a small value, for example, 0.25, and incrementing by 0.25 until reaching some maximum value, for example, 4 or 5. For each of the values of θ in this set, steps 3, 4 and 5 are followed.

Step 3: The estimation step (E) consists of computing for each subject an estimate of the value of their frailty as

$$\hat{z}_i = \frac{1 + \theta \times c_i}{1 + \theta \times \hat{H}\left(t_i, \mathbf{x}_i, \hat{\boldsymbol{\beta}}\right)}. \tag{9.6}$$

Step 4: The maximization step (M) consists of fitting the proportional hazards model with the same covariates, but including $\hat{z}_i$ in the hazard function, as shown in (9.5). Alternatively, one may include $\ln(\hat{z}_i)$ as a model covariate with a coefficient fixed and equal to 1.0. Including a term in a model with a fixed coefficient equal to 1.0 is common in applications of Poisson regression, and sometimes logistic regression, where it is called an *offset*. The use of an offset variable is not as common in applications of the proportional hazards model, and few software packages allow the user to specify one in the associated routines. This is the major stumbling block to application of the EM algorithm in the gamma frailty setting. One must customize software to fit a proportional hazards model with hazard function

$$h_f\left(t, \hat{z}_i, \mathbf{x}_i, \boldsymbol{\beta}\right) = h_0(t)\hat{z}_i \exp\left(\mathbf{x}_i' \boldsymbol{\beta}\right), \tag{9.7}$$

where $\hat{z}_i$ is given in (9.6). In addition, one must customize software to modify the estimated baseline cumulative hazard to include the estimated frailty term. Thus, one must evaluate for each subject the estimate of the baseline hazard function

$$\hat{h}_{f_0}(t_i) = \frac{c_j}{\sum\limits_{l \in R(t_j)} \hat{z}_i \exp\left(\mathbf{x}_i'\hat{\boldsymbol{\beta}}\right)}, \tag{9.8}$$

the cumulative baseline hazard function

$$\hat{H}_{f_0}(t_i) = \sum\limits_{t_j \leq t_i} \hat{h}_{f0}(t_j) \tag{9.9}$$

and, finally, the cumulative hazard containing the frailty, namely

$$\hat{H}_f\left(t_i, \mathbf{x}_i, \hat{\boldsymbol{\beta}}\right) = \hat{H}_{f_0}(t_i) \exp\left(\mathbf{x}_i'\hat{\boldsymbol{\beta}}\right). \tag{9.10}$$

The E- and M-steps are repeated until convergence is achieved. In the example below, the criterion for convergence was that the change in both the log-likelihood for fitting the model in the M-step and the sum of the estimated frailties in the E-step was less than 0.0001 between successive iterations of the two steps. Note that in the second and subsequent applications of the E-step, one uses the cumulative hazard containing the frailty, (9.10), computed in the previous M-step.

Step 5: Evaluate the profile log-likelihood using the specified value of θ and the results from the fit in the M-step at convergence. The equation for the profile log-likelihood may be found in Klein and Moeschberger [1997, equation (13.3.2)]. There is a small mistake in Klein and Moeschberger's equation that has been corrected in the errata section of the web site for their book. The equation shown below is algebraically simplified and is the correct form for the model described in (9.5). One should consult Klein and Moeschberger (1997) for the correct form when fitting a shared frailty model. Specifically, one evaluates

$$L(\theta, \hat{\boldsymbol{\beta}}) = \sum\limits_{i=1}^{n} c_i\left\{\mathbf{x}_i'\hat{\boldsymbol{\beta}} + \ln\left[h_{f0}(t_i)\right]\right\} - \sum\limits_{i=1}^{n} \left(\frac{1}{\theta} + c_i\right)\ln\left[1 + \theta\hat{H}_f\left(t_i, \mathbf{x}_i, \hat{\boldsymbol{\beta}}\right)\right]. \tag{9.11}$$

Note that the value of the first term in (9.11) is, in fact, the log-likelihood for the model fit in the M-step.

The entire procedure, steps 1 to 5, is performed for all chosen values of θ. The maximum likelihood estimate of θ is the value maximizing (9.11). Nielsen, Gill, Andersen and Sørensen (1992) describe an easily employed method for using the calculated values of (9.11) to obtain an empirical estimate of the MLE, $\hat{\theta}$. One selects the value of θ yielding the maximum value of (9.11) and fits a quadratic equation in θ to the values of (9.11) over points in a neighborhood of the maximum. We describe this procedure in greater detail in the example.

Significance tests for model coefficients may be performed in two ways. The more complicated approach is to use a Wald test with estimated standard errors from a covariance matrix obtained from the negative of the inverse of the matrix of second derivatives of the profile log-likelihood. Klein and Moeschberger (1997) provide equations for the elements of this matrix. These equations have recently been modified by Andersen, Klein, Knudsen and Tabanera y Placios (1997) to correct for an underestimation in the variance of the estimate of the MLE of the gamma variance parameter. A simpler approach suggested by Nielsen, Gill, Andersen and Sørensen (1992) is to delete the covariate from the model, refit it using the same value of $\hat{\theta}$, and do a likelihood ratio test based on twice the difference in the respective values of (9.11). We illustrate the second method in the example.

A significance test for the variance parameter, θ, can also be performed in one of two ways. Commenges and Andersen (1995) proposed a score test that is described in some detail in Klein and Moeschberger (1997). A more easily performed test is to use the likelihood ratio test suggested by Nielsen, Gill, Andersen and Sørensen (1992). Under the hypothesis that $\theta = 0$, the value of the profile log-likelihood in (9.11) is the partial log-likelihood from fitting the proportional hazards model minus the number of noncensored follow-up times. That is,

$$L\big(\theta = 0, \hat{\boldsymbol{\beta}}\big) = \sum_{i=1}^{n} c_i \big\{ \mathbf{x}_i' \hat{\boldsymbol{\beta}} + \ln\big[h_0(t_i)\big] \big\} - \sum_{i=1}^{n} c_i. \qquad (9.12)$$

The lead term in (9.12) is simply the partial log-likelihood from the model fit in step 1 of the EM procedure, and the second term is the number of events. The likelihood ratio test statistic is

$$G = 2\big[L\big(\hat{\theta}, \hat{\boldsymbol{\beta}}\big) - L\big(0, \hat{\boldsymbol{\beta}}\big)\big], \qquad (9.13)$$

with a p-value computed using a chi-square distribution with one degree-of-freedom.

We use a model from the UIS containing AGE and BECKTOTA to illustrate the process of fitting a proportional hazards model with a frailty. In this data set 591 subjects had complete data on AGE and BECKTOTA. We fit the frailty model using the stated five-step procedure, with values of $\theta = 0.25, 0.50, 0.75, \ldots, 3.75$. This procedure yielded a maximum profile log-likelihood at $\theta = 0.75$. Following the method outlined in Nielsen, Gill, Andersen and Sørensen (1992), we fit two additional models, one using $\theta = (0.5 + 0.75)/2 = 0.625$ and the other using $\theta = (0.75 + 1.0)/2 = 0.875$. The maximum observed value of the profile log-likelihood was still at $\theta = 0.75$. We repeated this procedure and fit two more models using $\theta = (0.625 + 0.75)/2 = 0.6875$ and $\theta = (0.75 + 0.875)/2 = 0.8125$. The observed maximum still occurred at 0.75. Table 9.3 presents the values of θ and the associated values of the profile log-likelihood at the various stages of this process.

In order to refine our estimate of the gamma variance parameter, we fit a quadratic equation to the values of the profile log-likelihood using the seven values from rows 2 through 8 of Table 9.3. (We actually used y = (profile log-likelihood + 3210) to eliminate possible numerical problems in the computations.) This yielded the equation

$$y = -2.067 + 0.9519 \times \theta - 0.6435 \times \theta^2 .$$

The maximum likelihood estimate of θ was chosen to be the value maximizing this fitted quadratic equation, namely

$$\hat{\theta} = (0.9519/2 \times 0.6435) = 0.74. \tag{9.14}$$

We ran the M-step using this value of the gamma variance parameter. The resulting value of the profile log-likelihood is

$$L(\hat{\theta} = 0.74, \hat{\beta}) = -3211.7145.$$

Table 9.4 presents the results of fitting this model as well as the model with $\theta = 0$. As noted above, the standard errors provided when fitting the model in the M-step are not correct and thus are not included in Table 9.4. The elements of the information matrix provided in Andersen, Klein, Knudsen and Tabanera y Placios (1997) can be used to obtain estimates of the variances. The reported significance tests in Table 9.4 are based on likelihood ratio tests.

Table 9.3 Values of the Profile Log-likelihood at Specified Values of the Gamma Variance Parameter θ

θ	Profile Log-likelihood
0.25	−3211.8799
0.5	−3211.7522
0.625	−3211.7226
0.6875	−3211.7160
0.75	−3211.7146
0.8125	−3211.7183
0.875	−3211.7268
1.0	−3211.7575
1.25	−3211.8672
1.5	−3212.0305
1.75	−3212.2358
2.0	−3212.4734
2.25	−3212.7356
2.5	−3213.0165
2.75	−3213.3112
3.0	−3213.6161
3.25	−3213.9281
3.5	−3214.2451
3.75	−3214.5653

The likelihood ratio test for the inclusion of the gamma variance parameter is not significant, with a p-value of 0.535. This indicates that there may not be unmeasured heterogeneity in the responses of the subjects. This confirms the adequacy of the proportional hazards assumption for this model.

The interpretation of coefficients from a model containing a frailty is not as straightforward as that for a model without a frailty. In particular, the exponentiation of a coefficient yields a hazard ratio only when the frailty is held constant. In the model presented in Table 9.4, the frailty is not shared but is at the individual subject level.

Using a 10-point increase in the Beck score as an example, the quantity

$$\exp(10 \times 0.01781) = 1.19$$

is the estimated hazard ratio comparing two hypothetical subjects with the same frailty. The complicated nature of these comparisons is one of

Table 9.4 Results of Fitting the Proportional Hazards Models Without and With a Gamma Frailty Using the UIS Data ($n = 591$)

Variable	Without Gamma Frailty			With Gamma Frailty		
	Coeff.	G	p-Value	Coeff.	G	p-Value
AGE	−0.0131	3.18	0.075	−0.0196	3.40	0.065
BECKTOTA	0.0105	4.84	0.028	0.0178	6.52	0.011
θ	0.0			0.74	0.38	0.535
Profile log-likelihood	−3212.0976			−3211.7145		

the motivating factors for Aalen's (1994) favoring the use of a fully parametric model. In the parametric model, the actual values of the un-observed frailties can be "averaged" out of the model in the way they are handled in random effects logistic regression. Andersen, Klein, Knudsen and Tabanera y Placios (1997) also suggest that one consider the use of parametric models in place of the proportional hazards models. Further research will determine whether use of parametric models will be supported or whether methods for handling frailties in the proportional hazards model can be improved.

A few additional comments on fitting frailty models are in order. To provide a set of benchmark values that others can use when developing their own implementation of the EM algorithm, we have presented in Tables 9.3 and 9.4 more detail than is shown in other texts and papers. Also, we caution first time users of an EM algorithm that convergence can be quite slow. Some of the models fit in Table 9.3 required more than 60 iterations to converge. In combination with the iterative calculations necessary to fit the proportional hazards model, this leads to a set of statistical fitting methods that requires considerable patience on the part of the user, even with a fast personal computer. This limitation will diminish over time as computers get faster, but one should not expect to fit frailty models quite as quickly as is possible with the usual proportional hazards model.

9.4 NESTED CASE-CONTROL STUDIES

In some situations, such as large epidemiologic studies with time-varying covariates, the data may be so extensive that it would be impractical to fit survival time models using the entire cohort of subjects. In other cases, new research may indicate that data on an additional covariate should have been collected, but the process of collecting new information on all subjects would be too extensive to be feasible. In studies of this type, one may choose to use what is known as a *nested case-control study*.[2]

In a nested case-control study, one may estimate the regression parameters in the proportional hazards model with risk sets that contain a reduced number of subjects. Briefly, the data for a nested study is obtained by selecting a sample (without replacement) of a fixed number of subjects who did not fail from the risk set at each observed survival time. These subjects become the controls for the case whose survival time defines the risk set. This design is discussed from an epidemiologic perspective in Clayton and Hills (1993). Construction of the partial likelihood was first considered by Oakes (1981). A series of papers by Borgan and Langholz (1993), Langholz and Borgan (1995) and Borgan, Goldstein and Langholz (1995) review the construction of the partial likelihood, derive score equations for estimation, and present methods for inference, as well as examples.

Methods for implementing the design assume that there are no tied survival times, so one must decide how to handle ties that might exist in the data. Borgan and Langholz (1993) suggest that tied survival times be broken randomly and that the random order be used as if it was the real order in the survival times. We illustrate one method of breaking ties in the example.

The number of control subjects selected will have an impact on the efficiency of the nested case-control analysis relative to an analysis based on the complete risk set at each survival time. Clayton and Hills (1993) point out that the ratio of the standard errors of coefficient estimates from the two methods of analysis is approximately

$$\frac{\text{SE}\left(\hat{\beta} \,|\, \text{case-control}\right)}{\text{SE}\left(\hat{\beta} \,|\, \text{full cohort}\right)} \approx \sqrt{1 + \frac{1}{n_c}},$$

where n_c is the number of controls selected from each risk set. This expression indicates that little would be gained from using more than five controls.

Once the number of controls to be selected is determined, the sampling of controls from each risk set must be done independently of all

---

[2] Those readers who are unfamiliar with the case-control study may wish to consult Rothman and Greenland (1998).

other study factors. Eligible subjects may be selected as controls in more than one risk set. In order to simplify the notation somewhat, we denote the $n_c + 1$ subjects, consisting of the n_c controls and one case, in the nested case-control study risk set at survival time t_i as $\tilde{R}(t_i)$. If we assume there are no time-varying covariates among the p covariates in the model, the partial likelihood is

$$l_{ncc}(\boldsymbol{\beta}) = \prod_{i=1}^{n_c} \frac{e^{x_i'\boldsymbol{\beta}}}{\sum_{j \in \tilde{R}(t_i)} e^{x_j'\boldsymbol{\beta}}}.$$ (9.15)

The subscript "ncc" in (9.15) is used to differentiate this partial likelihood from the one in (3.17) for the full cohort. The distinction between the two partial likelihoods goes well beyond a subscript, since the model in (9.15) cannot be fit using a proportional hazards regression program. The risk sets for the partial likelihood for the full cohort analysis are nested. That is, subjects in a risk set at time t are contained in all risk sets at time less than t, but that is not true of the risk sets in (9.15). The likelihood in (9.15) is identical to that of a matched case-control study, with one case and n_c controls per matched set, with survival times defining the matched sets [see Hosmer and Lemeshow (1989, Chapter 7)]. Thus, the regression coefficients from the nested case-control study may be obtained by using any matched logistic regression routine that can handle multiple controls per case. From a practical point of view, the only impediment to using a nested case-control design is the actual selection of the sample of controls from the risk set. Clayton and Hills (1997) have eased the burden by providing a command for use in STATA that will execute the sampling and create an output data set containing all the information needed to fit the model in (9.15) using the conditional logistic regression program. Users of other packages are encouraged to check their favorite package's web site and/or make an inquiry to its user group list server.

Time-varying covariates can be included in the partial likelihood in (9.15) simply by letting the covariate values depend on time, for example, by using $\mathbf{x}(t_i)$ in place of $\mathbf{x}$. What this means in practice is that fewer calculations of the values of the time-varying covariates are necessary because one only needs the values of the covariates at the times that define the case-control sets, and only for those subjects in the set.

As an example, we reanalyze the data from the UIS using a nested case-control study with five controls per observed survival time, which in this study was time to return to drug use. The sampling was performed using the Clayton and Hills (1997) routines for STATA. The software automatically breaks tied survival times randomly, but we performed this step manually to have better control of the analysis. Since the unit of time is months, the procedure we used to break the ties was to subtract from each observed survival time a randomly generated number between 0 and 1. This process broke all the tied survival times but left ties among the censored observations intact.

The model we intend to fit is the final model in Table 6.6. Among the 575 subjects with complete data there were 464 observed times of return to drug use, and application of the sampling procedure produced a new data set with $2784 = 6 \times 464$ data records. Key data for the first 3 sets of one case and 5 sampled controls is shown in Table 9.5. Selecting 5 controls per case was possible since the minimum number in any risk set was 8. In order to have the sampling be fully independent, survival times with fewer than n_c nonfailing subjects would not be used.

In the UIS data set there were two subjects with an observed time of

Table 9.5 Key Variables from Creation of the Nested Case-Control Study from the UIS, for Three Risk Sets

Set	Case-Control Indicator	Study ID	Case Time	Follow-up Time
1	0	372	3.016	546
1	0	434	3.016	192.650
1	0	396	3.016	185.615
1	0	160	3.016	45.143
1	0	281	3.016	99.233
1	1	334	3.016	3.016
2	0	186	3.161	93.279
2	0	203	3.161	119.144
2	0	211	3.161	25.898
2	0	386	3.161	72.991
2	0	625	3.161	17.108
2	1	143	3.161	3.161
3	0	521	5.009	549
3	0	261	5.009	179.251
3	0	353	5.009	569
3	0	394	5.009	20.971
3	0	487	5.009	433.268
3	1	454	5.009	5.009

4 months. The procedure used to break the ties created a new time of 3.016 months for subject 334 and 3.161 months for subject 143. The study ID numbers for the subjects selected are shown in the third column of Table 9.5, and the survival times for cases, with ties broken, are shown in the fourth column. Follow-up time for each subject is shown in the last column. We note that these times are both noninteger and integer valued. The integer valued times correspond to subjects whose observed time is censored, while noninteger times correspond to subjects who actually returned to drug use. The follow-up time listed in the table are the recorded times to return to drug use minus a uniform (0,1) random number. Thus subjects who eventually defined case-control sets and subjects whose follow-up time was censored were selected. We also note that in each set, follow-up time is longer for the controls than the case, as expected. An analysis of the controls (not shown) indicated that many subjects appeared in multiple sets. In fact, one subject was in 60 of the 464 sets created by the sampling procedure. The fact that subjects appear in multiple sets is easily explained by the fact that the number in the full cohort risk sets becomes increasingly smaller as follow-up time increases, thus increasing the chance that a subject will be selected as a control in later sets.

Conditional logistic regression was used to fit the model presented in Table 6.6 to the nested case-control data set. The first two columns of Table 9.6 present the results of this fit. The next two columns contain the estimated coefficients and standard errors from the full data set. The last column contains the ratio of estimated standard errors of the coefficients.

The results in Table 9.6 indicate that the coefficient estimates obtained from the nested case-control study are close to those obtained from the full data set. In addition, the ratios of the estimated standard errors are each close to the theoretical ratio of $1.095 = \sqrt{1.2}$. The conclusions would certainly be the same for both analyses. In particular, the methods used in Chapter 6 to provide estimates and confidence intervals for hazard ratios may be used with the results of a nested case-control analysis.

Borgan, Goldstein and Langholz (1995) derive the estimator of the baseline survivorship function. We can use this estimator in a manner identical to that shown in Chapter 6 to provide estimates and graphs of covariate-adjusted survivorship probabilities. The estimator of the baseline survivorship for the nested case-control study is similar in format to the one used for the frailty models in the previous section in that it includes a specific weighting factor for each risk set. The estimator of the

Table 9.6 Estimated Coefficients, Standard Errors and the Ratio of the Estimated Standard Errors Comparing the Fit in the Nested Case-Control Sample to the Full Data Set for the Final Model for the UIS Study

Variable	Nested Case-Control Coeff.	Std. Err.	Full Data Set Coeff.	Std. Err.	Ratio
AGE	−0.038	0.011	−0.041	0.010	1.12
BECKTOTA	0.014	0.005	0.009	0.005	1.07
NDRUGFP1	−0.604	0.141	−0.574	0.125	1.13
NDRUGFP2	−0.222	0.055	−0.215	0.049	1.12
IVHX_3	0.213	0.118	0.228	0.109	1.08
RACE	−0.454	0.146	−0.467	0.135	1.08
TREAT	−0.280	0.103	−0.247	0.094	1.09
SITE	−1.204	0.595	−1.317	0.531	1.12
AGEXSITE	0.030	0.018	0.032	0.016	1.12
RACEXSITE	1.036	0.278	0.850	0.248	1.12

baseline hazard at the ith observed survival time is given by the equation

$$\hat{h}_{0ncc}(t_i) = \frac{1}{\sum\limits_{j \in \tilde{R}(t_i)} w_j e^{x'_j \hat{\beta}}}, \tag{9.16}$$

where the value of the weight is $w_i = n(t_i)/(n_c + 1)$ and $n(t_i)$ denotes the number of subjects in the risk set at time t_i in the original cohort. The estimator of the cumulative baseline hazard function at time t is

$$\hat{H}_{0ncc}(t) = \sum_{t_i \le t} \hat{h}_{0ncc}(t_i), \tag{9.17}$$

and the estimator of the cumulative hazard function for the ith subject at time t is

$$\hat{H}_{ncc}(t, x_i, \hat{\beta}) = \hat{H}_{0ncc}(t) \exp(x'_i \hat{\beta}). \tag{9.18}$$

The estimator of the survivorship function is of the same form as the estimator when the full cohort is analyzed using the proportional hazards model, namely

$$\hat{S}_{\text{ncc}}\left(t, \mathbf{x}_i, \hat{\boldsymbol{\beta}}\right) = \exp\left[-\hat{H}_{\text{ncc}}\left(t, \mathbf{x}_i, \hat{\boldsymbol{\beta}}\right)\right]$$

$$= \left[\hat{S}_{0\text{ncc}}(t)\right]^{\exp\left(\mathbf{x}_i'\hat{\boldsymbol{\beta}}\right)}, \tag{9.19}$$

where the estimator of the baseline survivorship function is

$$\hat{S}_{0\text{ncc}}(t) = \exp\left[-\hat{H}_{0\text{ncc}}(t)\right]. \tag{9.20}$$

Note that we again use the subscript "ncc" in (9.16)–(9.20) to emphasize that these estimators are obtained from a nested case-control study.

As an example of the use of the estimator of the survivorship function, we calculate the modified risk score discussed in Chapter 6, to obtain covariate-adjusted survivorship functions describing the effect of treatment. The estimated risk score for the ith subject in the nested case-control study is obtained using the coefficients in the second column of Table 9.6, i.e., $\hat{r}_i = \mathbf{x}_i'\hat{\boldsymbol{\beta}}$. The estimate of the modified risk score is $\hat{rm}_i = \hat{r}_i - (-0.280)\text{TREAT}$, and the median value of the modified risk score over the 2784 observations is $\hat{rm}_{50} = -1.995$. Thus, the covariate-adjusted survivorship function at time t for the shorter of the two treatments in the UIS is

$$\hat{S}_{\text{ncc}}\left(t, \hat{rm}_{50}\right) = \left[\hat{S}_{0\text{ncc}}(t)\right]^{\exp(-1.995)}, \tag{9.21}$$

and for the longer of the two treatments it is

$$\hat{S}_{\text{ncc}}\left(t, \hat{rm}_{50}\right) = \left[\hat{S}_{0\text{ncc}}(t)\right]^{\exp(-1.995-0.280)}. \tag{9.22}$$

Figure 9.1 presents graphs of the estimated survivorship functions in (9.21) and (9.22), plotted at the 464 observed times of return to drug use in the nested case-control study. The estimated survivorship functions and their graphs may be used to estimate quantiles in the same way as described in detail in Chapter 6.

Due to the fact that a random process was used to select the control subjects, it would be impossible for the reader to reproduce the exact results shown in Table 9.6 and Figure 9.1. Each time a new nested case-control data set is created, the composition of the control group will be different. Results computed using different controls should be similar

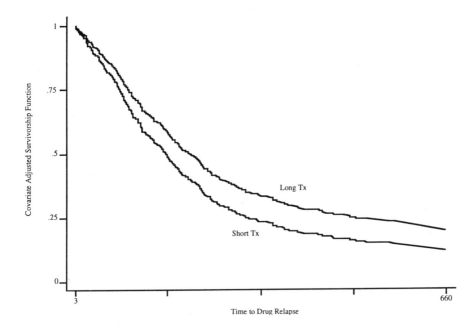

Figure 9.1 Graph of the covariate-adjusted survivorship functions for the longer and shorter treatments computed using the fitted model from the nested case-control study in Table 9.6.

to each other, but they will not be identical.

   In summary, the nested case-control study offers the possibility of fitting data from a follow-up study using information on fewer subjects than the total number of subjects in the complete cohort. If an adequate number of controls are used, the results from the nested study should agree with those that would have been obtained had the entire cohort been used.

9.5 ADDITIVE MODELS

Up to this point we have focused on *multiplicative* regression models for the analysis of survival time. In these models, the combined effect of the covariates is obtained by multiplying their separate effects. For example, in a proportional hazards model containing sex and age, the multiplicative nature of the hazard function is seen if we rearrange the terms as follows:

$$h(t, \mathbf{x}, \boldsymbol{\beta}) = h_0(t)e^{\beta_1 \text{SEX} + \beta_2 \text{AGE}}$$

$$= h_0(t)e^{\beta_1 \text{SEX}}e^{\beta_2 \text{AGE}}$$

$$= h_0(t) \times (\text{effect of sex}) \times (\text{effect of age}). \qquad (9.23)$$

In a similar manner, the fully parametric models presented in Chapter 8 can also be shown to be multiplicative.

As we showed in detail, multiplicative models are extremely useful in practice since either the estimated coefficients themselves or simple functions of them can be used to provide estimates of hazard ratios. In addition, statistical software is readily available and easy to use to fit models, check model assumptions and assess model fit. The widespread use of multiplicative models, particularly the proportional hazards model, in applied settings is, to a large extent, due to these factors.

However, there may be times when a measure of the additive effect of a covariate is preferred over a relative measure. There are several different forms of additive models, including the *additive relative hazard model* mentioned in Section 3.2. If we use a hypothetical model containing sex and age, the additive relative hazard function is

$$h(t, \mathbf{x}, \boldsymbol{\beta}) = h_0(t)(1 + \beta_1 \text{SEX} + \beta_2 \text{AGE})$$

$$= h_0(t) \times (1 + \text{effect of sex} + \text{effect of age}). \qquad (9.24)$$

If we code $\text{SEX} = 1$ for male and $\text{SEX} = 0$ for female, then under the model in (9.24) the hazard ratio for male versus female at $\text{AGE} = a$ is

$$\text{HR}(\text{SEX} = 1, \text{SEX} = 0, \text{AGE} = a) = \frac{(1 + \beta_1 + \beta_2 a)}{(1 + \beta_2 a)}.$$

The coefficient for sex, β_1, is the additive increase or decrease in the hazard ratio due to male gender. One rather obvious problem with this model is that, if inferences are based on hazard ratios, it is impossible, except in a univariate model, to isolate the effect of a single covariate. Under this model, the difference in the hazard rate for males and females, at $\text{AGE} = a$, is

$$h(t, \text{SEX} = 1, \text{AGE} = a, \boldsymbol{\beta}) - h(t, \text{SEX} = 0, \text{AGE} = a, \boldsymbol{\beta}) = h_0(t)\beta_1,$$

which, unfortunately, depends on both the coefficient for sex and the unspecified baseline hazard function.

Two other forms of the additive relative hazard model that have been suggested, expressed in terms of the two variable example, are

$$h(t, \mathbf{x}, \boldsymbol{\beta}) = h_0(t)\left(1 + e^{\beta_1 \text{SEX}} + e^{\beta_2 \text{AGE}}\right) \tag{9.25}$$

and

$$h(t, \mathbf{x}, \boldsymbol{\beta}) = h_0(t)\left(1 + e^{\beta_1 \text{SEX} + \beta_2 \text{AGE}}\right). \tag{9.26}$$

In theory, the additive relative hazard models may be easily expanded to include interactions between covariates, nonlinear terms for continuous covariates, time-varying covariates and stratification variables. Thus, fitting the models should be no more difficult than fitting the proportional hazards model. In fact, both the BMDP and EGRET software packages offer the option to fit the additive relative hazard models shown in (9.24) and (9.25). However, it has been our experience that these software packages have trouble fitting additive relative hazard models containing more than just a few dichotomous covariates. This is due to the fact that the range of allowable values for the coefficients, those yielding a positive hazard function, is tightly constrained by the additive form of the model, and the iterative methods used by these programs have trouble finding a solution to the likelihood equations. Despite the possible clinical appeal of additive relative hazard models, they are not terribly practical, which may be why they have not been used more frequently in applied research. In due time, advances in statistical computing may make it practical to fit these models.

Two fully additive models have received attention in the statistical literature. One of these is a semi-parametric model [see Breslow and Day (1987)] where, for the hypothetical two-variable model, the hazard function is

$$h(t, \mathbf{x}, \boldsymbol{\beta}) = h_0(t) + \beta_1 \text{SEX} + \beta_2 \text{AGE}$$

$$= h_0(t) + \text{effect of sex} + \text{effect of age}. \tag{9.27}$$

Under this model, the difference in the hazard rate for males and females of a common age is

$$h(t, \text{SEX} = 1, \text{AGE} = a, \boldsymbol{\beta}) - h(t, \text{SEX} = 0, \text{AGE} = a, \boldsymbol{\beta}) = \beta_1.$$

The regression coefficient, β_1, is the additive change from the baseline hazard function for males, holding age fixed. Similarly, the quantity β_2 is the additive change in hazard rate for a 1-year increase in age, holding sex fixed at either male or female. The model in (9.27) can easily be expanded to include nonlinear terms for continuous covariates, interactions, time-varying covariates and stratification variables.

Lin and Ying (1994) derive an estimator of the regression coefficients from models of the form shown in (9.27) that does not require iterative calculations. They prove that their estimator has a large-sample normal distribution, with mean equal to the true coefficients, and they derive an estimator of the covariance matrix of the estimator of the coefficients. One can use these estimators, with the standard normal distribution or chi-square distribution, for hypothesis tests and confidence intervals. They also derive an estimator of the baseline cumulative hazard function that can, when combined with the estimates of the coefficients and a vector of covariate values, be used to estimate the covariate-adjusted survivorship function. One potential problem with their estimator of the coefficients is that it can yield an estimate of the baseline cumulative hazard function that is a nonmonotonic increasing function of time. In addition, the estimate can be negative, leading to estimated survival probabilities greater than one. They suggest that one work around these problems by using an empirical procedure that forces the estimator of the baseline cumulative hazard to be a monotonic increasing function of time. Lin and Ying (1994) illustrate the use of the semiparametric model with a few examples. Unfortunately, software to fit this model is, at this time, not readily available in a major statistical software package. Assuming we have a model with p fixed covariates, the Lin and Ying estimator is

$$\hat{\boldsymbol{\beta}} = \left\{ \sum_{i=1}^{n} \sum_{t_j \geq t_i} \left[\mathbf{x}_i - \overline{\mathbf{x}}(t_j) \right]^{\otimes 2} \right\}^{-1} \left\{ \sum_{i=1}^{n} c_i \left[\mathbf{x}_i - \overline{\mathbf{x}}(t_i) \right] \right\}, \qquad (9.28)$$

where the expression $\mathbf{x}^{\otimes 2}$ denotes the matrix operation $\mathbf{x}\mathbf{x}'$, which is a p by p matrix, and

$$\overline{\mathbf{x}}(t_i) = \frac{\sum\limits_{j \in R(t_i)} \mathbf{x}_j}{n_i}$$

is the arithmetic mean of the covariates among those at risk at time t_i.

Aalen (1989) has developed a more general additive model. He discusses issues of estimation, testing and assessment of model fit in two applied papers [Aalen (1989 and 1993)]. His model is fully additive and nonparametric and values of the regression coefficients are allowed to vary over time. We present his model in more detail than the other additive models as it provides the opportunity to not only fit an additive model, but the results of the fit can be used to provide graphical descriptions that supplement fits of other models, such as the proportional hazards model.

The Aalen model, for $p+1$ fixed covariates, $\mathbf{x}' = (1, x_1, x_2, \ldots x_p)$, has hazard function at time t equal to

$$h(t, \mathbf{x}, \boldsymbol{\beta}(t)) = \beta_0(t) + \beta_1(t)x_1 + \beta_2(t)x_2 + \cdots + \beta_p(t)x_p. \qquad (9.29)$$

The coefficients in this model provide the change at time t, from the baseline hazard rate $\beta_0(t)$, for a one-unit change in the respective covariate. Unlike any other model considered in this text, the Aalen model allows the effect of the covariate to change continuously over time. The cumulative hazard function obtained from the hazard function in (9.29) is

$$H(t, \mathbf{x}, \mathbf{B}(t)) = \int_0^t h(u, \mathbf{x}, \boldsymbol{\beta}(u)) \, du \,,$$

so

$$H(t, \mathbf{x}, \mathbf{B}(t)) = \sum_{k=0}^p x_k \int_0^t \beta_k(u) \, du$$

$$= \sum_{k=0}^p x_k B_k(t) \,, \qquad (9.30)$$

where $x_0 = 1$ and $B_k(t)$ is called the *cumulative regression coefficient* for the kth covariate. It follows from (9.30) that the baseline cumulative hazard function is $B_0(t)$. Klein and Moeschberger (1997) consider the Aalen model and illustrate its use in settings in which the model contains only a few discrete covariates. Zahl and Tretli (1997) illustrate the use of the Aalen model with an analysis of data on survival of breast cancer patients in Norway, and Borgan and Langholz (1997) extend the model for use in nested case-control studies. McKeague and Sasieni (1994)

consider a model that combines the time-varying coefficients of the
Aalen model in (9.29) with the time-fixed coefficients in the model as
illustrated in (9.27).

Aalen (1989) notes that the cumulative regression coefficients are
easier to estimate than the regression coefficients themselves and presents an easily computed estimator. Aalen (1993) discusses methods for
estimation of the individual regression coefficients, but these involve
smoothing techniques that are beyond the mathematical scope of this
text.

We concentrate on the estimates of the cumulative regression coefficients and how they can be used to provide insight as to the effect
covariates have over time. Assume that we have n independent observations of time (with no tied survival times), p fixed covariates and a right
censoring indicator variable, independent of time, denoted by the usual
triplet (t_i, x_i, c_i). Aalen's estimator of the cumulative regression coefficients is a least-squares-like estimator and is most easily presented using
matrices and vectors. We let $\mathbf{X}_j$ denote an n by $p+1$ matrix, where the
ith row contains the data for the ith subject, $\mathbf{x}_i'$, if the ith subject is in the
risk set at time t_j, otherwise the ith row is all 0's. We let $\mathbf{y}_j$ denote a 1
by n vector, where the jth element is 1 if the jth subject's observed time,
t_j, is a survival time (i.e., $c_j = 1$), otherwise all the values in the vector
are 0. The estimator of the cumulative regression coefficient at time t is

$$\hat{\mathbf{B}}(t) = \sum_{t_j \le t} \left(\mathbf{X}_j' \mathbf{X}_j \right)^{-1} \mathbf{X}_j' \mathbf{y}_j . \tag{9.31}$$

We note that the value of the estimator changes only at observed survival
times and is constant between observed survival times. Huffer and
McKeague (1991) discuss weighted versions of the estimator in (9.31)
but they are much more complicated to implement. The increment in
the estimator is computed only when the matrix $\left(\mathbf{X}_j' \mathbf{X}_j \right)$ can be inverted.
The matrix is singular, and therefore cannot be inverted, when there are
fewer than $p+1$ subjects in the risk set. Other data configurations can
also yield a singular matrix. For example, if the model contains a single
dichotomous covariate and all subjects who remain at risk have the same
value for the covariate, the matrix will be singular. Aalen's estimator of
the covariance matrix of the estimator of the cumulative regression coefficients at time t is the following $p+1$ by $p+1$ matrix,

$$\hat{\text{Var}}\big[\hat{\mathbf{B}}(t)\big] = \sum_{t_j \le t} \left(\mathbf{X}'_j\mathbf{X}_j\right)^{-1}\mathbf{X}'_j\mathbf{I}_j\mathbf{X}_j\left(\mathbf{X}'_j\mathbf{X}_j\right)^{-1}, \tag{9.32}$$

where $\mathbf{I}_j$ is an n by n diagonal matrix with $\mathbf{y}_j$ on the main diagonal of the matrix. It follows from (9.30) and (9.31) that the estimator of the cumulative hazard function for the ith subject at time t is

$$\hat{H}\big(t, \mathbf{x}_i, \hat{\mathbf{B}}(t)\big) = \sum_{k=0}^{p} x_{ik}\hat{B}_k(t), \tag{9.33}$$

and an estimator of the covariate-adjusted survivorship function is

$$\hat{S}\big(t, \mathbf{x}_i, \hat{\mathbf{B}}(t)\big) = \exp\big[-\hat{H}\big(t, \mathbf{x}_i, \hat{\mathbf{B}}(t)\big)\big]. \tag{9.34}$$

We note, as does Aalen (1989), that it is possible for an estimate of the cumulative hazard in (9.33) to be negative and to yield a value for (9.34) greater than 1.0. This is most likely to occur for small values of time, and one way to avoid this problem is to use zero as the lower bound for the estimator in (9.33). One benefit of fitting the Aalen additive model is to provide graphical evidence of the effect of a covariate over time, rather than to provide an additive covariate-adjusted survivorship function.

The graphical presentation most often used with the Aalen model is a plot of $\hat{B}_k(t)$ versus t, along with the upper and lower endpoints of a pointwise confidence interval. For a 95 percent interval, one would plot

$$\hat{B}_k(t) \pm 1.96 \hat{\text{SE}}\big[\hat{B}_k(t)\big],$$

where $\hat{\text{SE}}\big[\hat{B}_k(t)\big]$ is the estimator of the standard error of $\hat{B}_k(t)$, obtained as the square root of kth diagonal element of the variance estimator in (9.32).

At present there are two sources for software to fit the Aalen model. Aalen and Fekjær provide macros for use with S-PLUS on a web site.[3] Klein and Moeschberger (1997) provide macros for use with SAS at the web site for their book. Calculations for the example presented in this section were performed using Aalen and Fekjær's S-PLUS routines.

---

[3] The web site for the S-PLUS macros is http://www.med.uio.no/imb/stat/addreg/.

In order to illustrate the use of the Aalen additive model, we fit it to some of the data from the UIS that were modeled using the proportional hazards model in Chapters 5 and 6. The proportional hazards model fit to these data contained main effects for age, Beck score, number of previous drug treatments, history of recent IV drug use, race, treatment and intervention site. Recall that we modeled age and Beck score as linear in the hazard function and the number of previous drug treatments with two non-linear terms (see Table 6.6 for the final fitted proportional hazards model). We used Aalen's method [Aalen (1989)] of subtracting a computer-generated uniform random number to break ties in observed survival time. We did not modify tied censored times, as they do not contribute to the estimators in (9.31) and (9.32). In addition, we centered age, Beck score and the number of previous drug treatments at their respective means. The baseline cumulative regression coefficient estimates the cumulative hazard for a subject of average age, average Beck score, no previous history of IV drug use, white race and randomized to the short interventions at site A. We refer to this subject as the "baseline subject" when discussing the results of fitting the Aalen model. Since the goal of this analysis is to look for possible time-dependent effects in the covariates, we fit a model containing only the seven main effect terms. We present plots of the cumulative regression coefficients along with pointwise 95 percent confidence limits in Figure 9.2. The maximum observed survival time when the matrix $\left(\mathbf{X}'_j \mathbf{X}_j\right)$ was nonsingular was 569 days. However, we restrict the plots in Figure 9.2 to the interval 0 to 500 days, because there are only 6 uncensored values of time to return to drug use greater than 500 days, resulting in large estimated standard errors and wide confidence interval estimates. The same imprecision in the estimator of the right tail of the survivorship function was seen in Chapter 2.

Before we discuss the eight plots in Figure 9.2, we describe what the plots of the cumulative regression coefficients are expected to look like under different types of covariate effects. If a regression coefficient in (9.29) is constant over time, it follows from (9.30) that the plot of the estimated cumulative regression coefficient should look like a straight line through the origin, with slope equal to the value of the coefficient. That is, if the coefficient for x_k is $\beta_k(t) = \beta_k$ in (9.29), then $B_k(t) = t\beta_k$ in (9.30).

Deviation from a straight line in any time interval in the plot provides empirical evidence for a time-varying effect in the covariate. One form of time-varying effect is for a covariate to have a constant effect in an initial interval of time and then to have no effect at later intervals. In this situation, it follows from (9.30) that the graph of the estimated cumulative regression coefficient would be linear in the initial interval of time and would remain constant over the remaining observed time interval. For example, suppose the "true" coefficient for covariate x_k in (9.29) is 0.02 in the interval from 0 to 200 days and is zero for time greater than 200 days. We expect the graph of the estimated cumulative regression coefficient to look like the straight line $B_k(t) = 0.02t$ in the interval $[0, 200]$. Since the coefficient is equal to zero after 200 days, there is no change in the cumulative regression coefficient and we expect the graph to look constant with value $B_k(t) = 4.0$ for $t > 200$.

Another form of time-varying effect is for a covariate to have a constant effect in an initial interval of time and then to have the effect weaken over time. In this case we expect the graph of the estimated cumulative regression coefficient to be linear in the initial interval when the effect is constant. The shape of the plot over the remainder of the follow-up time depends on how the effect changes. For example, if the effect is still constant, but less than it was in the initial interval, then the plot will still be linear but with an attenuated slope. In the previous example, if the coefficient is 0.01 for $t > 200$, we expect the plot to look like the line $B_k(t) = 4.0 + 0.01t$ for $t > 200$. On the other hand, if the coefficient is –0.02 for $t > 200$, we expect the plot to look like $B_k(t) = 4.0 - 0.02t$ for $t > 200$.

If a covariate has no effect and its coefficient in (9.28) is equal to zero over the entire observed time interval, we expect the plot to cross back and forth over the zero line, a line with both slope and intercept equal to 0, with its pointwise confidence limits falling on either side of the zero line. After discussing the plots in Figure 9.2, we discuss a test of the hypothesis that a coefficient is equal to zero for the observed range of time over which the model can be fit.

In general, the shape of a plot provides evidence of the values of the individual regression coefficients in (9.29). It is incorrect to assume that the observed shape implies that the same type of parametric relationship with time holds in other models, such as the proportional hazards model, although the form of the plot can provide guidance on how to model a covariate [see Mau (1986)]. Henderson and Milner (1991) discuss this point and suggest including in the plot an estimate of what the cumulative regression coefficient is expected to look like if the proportional hazards model is correct. As yet, this embellishment has not been added to available software fitting the Aalen model.

Figure 9.2 contains eight separate plots, one for the baseline cumulative hazard model, and one for each term in the fitted model. Each of the eight subfigures contains the plot of an estimated cumulative regression coefficient, along with its upper and lower 95 percent pointwise confidence limits.

Figure 9.2a presents the graph of the estimated baseline cumulative hazard function. We note that the function increases sharply in a nearly linear fashion over the first 350 days, suggesting that the hazard for the baseline subject described above is approximately constant. There is little or no further increase beyond 350 days, which could be due to the relatively few uncensored times of return to drug use larger than 350 days.

Figure 9.2b presents the graph for AGE. The estimated cumulative regression coefficient decreases nearly linearly over the entire 500-day interval. There is a slight upwards bump in the plot between 300 and 375 days, but the plot continues to decrease linearly after 375 days. This bump is likely an artifact of chance variability. Overall, the plot suggests that there is a decrease in the hazard rate with increasing age that remains in effect over the entire time period.

Figure 9.2c presents the graph for BECKTOTA. The plot is nearly linear with a positive slope for the first 175–200 days, at which point it decreases slowly toward zero. We also note that after about 200 days the zero line is contained within the band for the lower 95 percent confidence limit. This pattern suggests that increasing values of the Beck score initially increase the hazard rate of return to drug use and then have no effect. In other words, the initial level of depression, as measured by the Beck score, has no effect after about 200 days. This suggests that we might try modeling Beck score as a time-dependent covariate, which could be done by adding an interaction term between Beck score and a time-dependent dichotomous covariate that takes on the value 1 after 200 days.

Figure 9.2d presents the graph for NDRUGTX. The plot is nearly linear, with a positive slope over the entire 500 days. This plot suggests that the effect of the number of previous drug treatments does not change over time, and that the number of treatments increases the hazard over the entire time period. We note that in Chapter 5, when we considered the scale of this covariate, the method of fractional polynomials and the smoothed residual plots yielded a highly nonlinear but clinically plausible transformation of this covariate (see Table 5.8). The next step might be to fit the Aalen model containing the two nonlinear covariates formed from the number of previous drug treatments. We

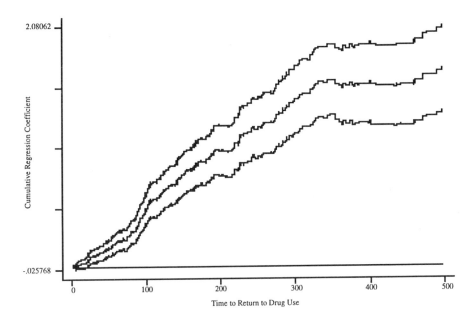

Figure 9.2 (a) Baseline cumulative hazard function.

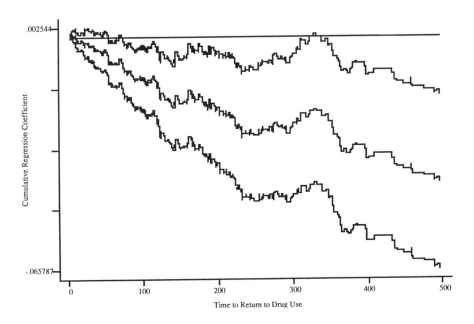

Figure 9.2 (b) Cumulative regression coefficient for AGE.

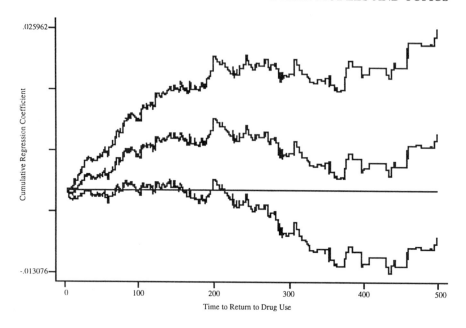

Figure 9.2 (c) Cumulative regression coefficient for BECKTOTA.

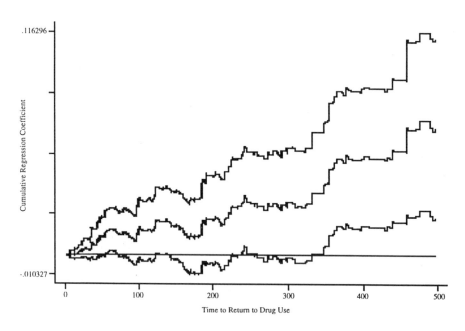

Figure 9.2 (d) Cumulative regression coefficient for NDRUGTX.

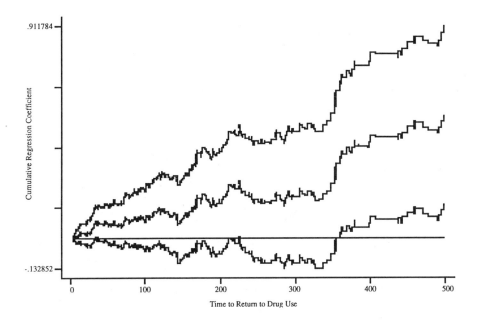

Figure 9.2 (e) Cumulative regression coefficient for IVHX_3.

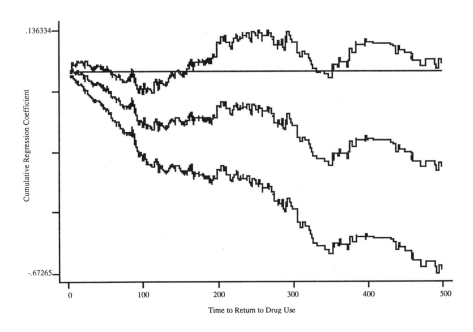

Figure 9.2 (f) Cumulative regression coefficient for RACE.

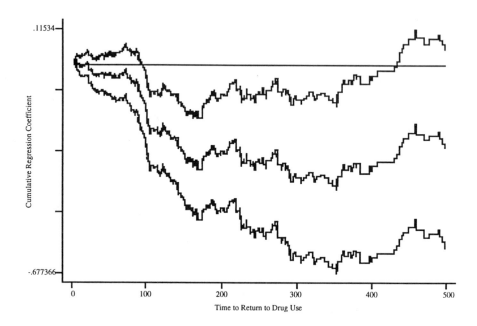

Figure 9.2 (g) Cumulative regression coefficient for TREAT.

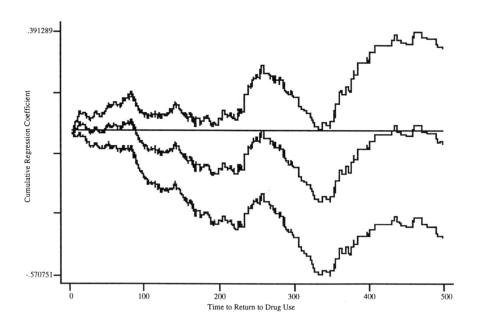

Figure 9.2 (h) Cumulative regression coefficient for SITE.

leave fitting this model as an exercise.

Figure 9.2e presents the graph for IVHX_3. The plot is nearly linear with a slight positive slope over the first 350 days. However, in this time period the zero line is contained within the lower 95 percent confidence band, which suggests that the covariate may not provide a significant additive increase to the hazard rate during the first 350 days of follow-up. We note that the plot is linear with a much steeper slope after 350 days. This suggests that recent history of IV drug use may have a late or delayed effect. We might try to model history of recent IV drug use in the proportional hazards model with an interaction between it and a time-dependent dichotomous covariate of the type suggested above for Beck score. Another way to explore the possible late effect of history of recent IV drug use is to begin the calculations for the plot at 350 days. In this plot, the confidence bands would be narrower than those in Figure 9.2e and thus lend potentially more precision to the graph above 350 days.

Figure 9.2f presents the graph for RACE. The plot decreases linearly for the first 100 days of follow-up, so it appears that, in this time interval, non-white race is associated with a constant and significant additive decrease in the hazard rate. After about 125–150 days, it appears that the covariate no longer has any effect, as the upper confidence band contains the zero line. This plot suggests that non-white race has only an early effect on the hazard rate, which could be explored further within the context of the proportional hazards model with an interaction of the type suggested for Beck score and history of recent IV drug use.

Figure 9.2g presents the graph for TREAT. Recall that this covariate is an indicator of randomization to the longer of the two interventions. As discussed in Section 1.3, the short intervention is of 3 months planned duration, and the long one is of 6 months planned duration. Examining the plot we see that during the first 90 days, when all subjects have the possibility to be under treatment, the covariate has no effect. This conclusion is based on the observation that the confidence bands contain the zero line in this interval. For the next 90 days the plot decreases sharply and nearly linearly, and the confidence bands no longer include the zero line. This suggests that assignment to the long treatment provides a significant additive decrease in the hazard rate. We conclude that subjects in the long intervention group are less likely to return to drug use over the observed time period. Interestingly, during the period from 180 to about 360 days of follow-up after the end of the long treatment, the plot continues to decrease linearly but with a less steep slope than in the interval from 90 to 180 days. The fact that the upper confidence band does not contain the zero line is noteworthy. One possible explanation is that there is a late benefit due to randomization to the long treatment. After about 360 days of follow-up, the plot slowly begins to decrease linearly. Interpretation here is difficult due to low statistical power as there are few uncensored times larger than 360 days. As we suggested in the discussion of Figure 9.2e, a possible way to explore this late effect is to calculate a new plot beginning at 360 days. The overall picture that emerges from Figure 9.2g is that the long intervention may continue to have a significant additive decrease in the hazard rate for about 6 months post-treatment. The time-varying effects of treatment could be explored within the context of the proportional hazards model by forming dichotomous time-varying covariates for the three intervals and including their interaction with treatment in the model.

Figure 9.2h presents the graph for SITE. The plot shows no consistent trend in any time interval, and the zero line is contained within the 95 percent confidence bands. Thus, there appears to be no significant additive increase or decrease in the hazard rate associated with site.

Aalen (1989) presents and illustrates a method for testing the hypotheses that the coefficients in the model are equal to zero. Tests can be made for individual coefficients as well as for the overall significance of the model. Test statistics are formed from the components of the vector

$$\hat{\mathbf{U}} = \sum_{t_j} \mathbf{K}_j \left(\mathbf{X}_j' \mathbf{X}_j \right)^{-1} \mathbf{X}_j' \mathbf{y}_j . \tag{9.35}$$

The summation is over all noncensored times when the matrix $\left(\mathbf{X}_j' \mathbf{X}_j \right)$ is non-singular, and $\mathbf{K}_j$ is a $(p+1) \times (p+1)$ diagonal matrix of weights. Aalen (1989) considers two possible choices for weights. One choice is the number in the risk set at t_j, i.e., $\mathbf{K}_j = \mathrm{diag}(n_j)$. He suggests that, since the estimator in (9.31) looks like a least squares estimator from linear regression, one uses as weights the inverse of a least-squares-like variance, namely the inverse of the diagonal elements of $\left(\mathbf{X}_j' \mathbf{X}_j \right)^{-1}$, that is

$$\mathbf{K}_j = \left\{ \mathrm{diag}\left[\left(\mathbf{X}_j' \mathbf{X}_j \right)^{-1} \right] \right\}^{-1} . \tag{9.36}$$

The variance estimator of $\mathbf{U}$ in (9.35) is obtained from the variance estimator in (9.32) and is

$$\widehat{\mathrm{Var}}\left(\hat{\mathbf{U}}\right) = \sum_{t_j} \mathbf{K}_j \left(\mathbf{X}_j' \mathbf{X}_j\right)^{-1} \mathbf{X}_j' \mathbf{I}_j \mathbf{X}_j \left(\mathbf{X}_j' \mathbf{X}_j\right)^{-1} \mathbf{K}_j. \qquad (9.37)$$

Tests of individual coefficients use the individual elements of the vector $\mathbf{U}$ in (9.35) and are scaled by the estimator of their standard error obtained as the square root of the appropriate element from the diagonal of the matrix in (9.37). Aalen remarks that this ratio has approximately the standard normal distribution when the hypothesis is true and the sample is sufficiently large. He reports that limited simulations indicate greater power from the least-squares-like weights and suggests that these be used in practice. These weights have been implemented in the software.

Table 9.7 reports the results of applying the test to each of the seven main effect variables. The results from the fit of the proportional hazards model shown in Table 5.5 are repeated here for comparison. The results of the significance tests for the regression coefficients in the Aalen model agree with those from the proportional hazards model. If a covariate has a significant effect on survival time, both models identify this fact. A disadvantage of the Aalen model is that it does not yield a set of coefficients that can be used to provide point and interval estimates of covariate effects. Instead, the model provides estimates of the additive effect at each observed survival time.

Table 9.7 Estimated Coefficients, Standard Errors, z-Scores, Two-Tailed p-Value and for the Main Effects Proportional Hazards Model and the z-Scores and Two-Tailed p-values from the Aalen Model for the UIS ($n = 575$)

Variable	Proportional Hazards Model				Aalen Model	
	Coeff.	Std. Err.	z	$P>|z|$	z	$P>|z|$
AGE	−0.026	0.008	−3.25	0.001	−3.47	<0.001
BECKTOTA	0.008	0.005	1.70	0.090	1.61	0.107
NDRUGTX	0.029	0.008	3.54	<0.001	2.93	0.003
IVHX_3	0.256	0.106	2.41	0.016	2.34	0.019
RACE	−0.224	0.115	−1.95	0.051	−2.12	0.034
TREAT	−0.232	0.093	−2.48	0.013	−2.59	0.010
SITE	−0.087	0.108	−0.80	0.422	−0.65	0.516

In summary, the plots of the cumulative regression coefficients obtained from a fit of the Aalen model to the UIS data have identified a number of potentially interesting time-dependent effects for the model covariates. Since the Aalen model is an additive model, the postulated time-dependent relationships may not be statistically significant when added to the proportional hazards model shown in Table 9.7.

The next step, which we leave as an exercise, is to expand the main effects model to include these time-varying covariates. One would then examine the extent to which they are significant, thus allowing one to sharpen inferences about the effect of model covariates on time to return to drug use.

In general, the additive models discussed in this section should be used with some caution. While there may be times when it is more meaningful clinically to express survival experience and covariate effects in terms of an additive increase or decrease in the hazard rate, one must be sure that, if there are any iterative calculations, the program, in fact, converges. In all cases, the fitted model must yield clinically plausible estimates of effect.

EXERCISES

1. Consider the following hypothetical study of two treatment modalities to reduce the occurrence of muscle soreness among middle-aged men beginning weight training. Study participants were 400 middle-aged men who joined a health club for the specific purpose of weight training. Subjects were randomized into one of two instructional programs designed to prevent muscle soreness. The control treatment consisted of the standard written brochures and instructions used by the health club to explain proper technique, including suggestions for frequency and duration of training. The new method included 1 hour with a personal trainer as well as the brochures. Subjects were followed and the dates on which muscle soreness limited the prescribed workout were recorded. The dates were converted into the number of days between soreness episodes.

The data may be found on the statistical data set web site at the University of Massachusetts/Amherst and the John Wiley web site discussed in Section 1.3 and the Preface. The data are in a file called RECUR.DAT. The variables are: ID (1–400), AGE (years), TREAT (0 = NEW, 1 = CONTROL), TIME0 (day of the previous episode), TIME1 (day of new episode), CENSOR (1 = muscle soreness episode occurred at TIME1, 0 = subject left the study or the study ended at TIME1) and EVENT (1–4 muscle soreness episode). The maximum number of episodes observed was 4.

Every study subject had one episode, 386 had two, 324 had three, and 186 had four. Thus the data file has 1296 records and is in the form shown for the counting process recurrent event model shown in Table 9.1. The data for this hypothetical study were generated to have sufficient power to detect particular differences in the models. A careful analysis will uncover these differences.

(a) Fit the counting process recurrent event model to these data obtaining both the information matrix-based estimates and the robust estimates of the standard errors of the estimated coefficients for AGE and TREAT.

(b) Repeat problem 1(a) fitting the conditional A recurrent event model. Does the effect of the covariates depend on the episode number?

(c) Repeat problem 1(a) fitting the conditional B recurrent event model. Does the effect of the covariates depend on the episode number?

(d) Repeat problem 1(a) fitting the marginal recurrent event model. Does the effect of the covariates depend on the episode number?

(e) Prepare a table of estimated hazard ratios, along with 95 percent confidence intervals, comparing the new method to the control method, corresponding to a 10-year change in age, for each of the four models fit. Compare and contrast the point and interval estimates under the four models. Compare and contrast the interpretation of the four sets of point and interval estimates.

2. One of the classic papers on recurrent event models is Wei, Lin and Weissfeld (1989) (WLW) who propose the marginal method and illustrated its use on recurrence times to bladder cancer. Therneau (1995) presents a detailed discussion of fitting the recurrent events models to the WLW data. These data are available in a library of data sets maintained by the Statistics Department at Carnegie Mellon University. The internet address for the WLW data, Table 2 of Wei, Lin and Weissfeld (1989), is http://lib.stat.cmu.edu/datasets/tumor. Download these data and fit the four recurrent event models using the possibility of four events. Prepare a table of estimated hazard ratios, along with 95 percent confidence intervals. Compare and contrast the interpretation of the four sets of estimates and intervals.

3. In the WHAS a reasonable proportional hazards model, when the analysis is restricted to grouped cohort one (YRGRP =1), contains age and left heart failure complications (CHF). Use the methods for including a subject-specific unobserved frailty to explore possible unaccounted for heterogeneity among the subjects in grouped cohort one.

4. In order to provide a data set for applying methods for the analysis of a nested case-control study, we created a small data set from the main WHAS data (this is not a subset of the WHAS data described in Section 1.3). Before performing the case-control sampling we broke ties in survival times by subtracting a uniform (0,1) random variable from each value of LENFOL corresponding to a death, and censored values were not changed. The modified version of LENFOL is denoted as T. The sampling procedure selected five controls for each case. These data are in the file WHASNCC.DAT that may be found on the web sites containing the data sets from this text. The variables are (see Table 1.4 for a description) SET, CASE, T, LENFOL, FSTAT, AGE, SEX, CHF, MIORD, YRGRP, LENSTAY and NR. The variable describing the set of one case and five controls is SET where CASE = 1 for the indexed death time and NR is the number of subjects in the risk set in the original cohort.

(a) Use the methods for the analysis of nested case-control studies to fit the proportional hazards model to these data.

(b) Following the fit of the model in problem 4(a) prepare a table of estimated hazard ratios with corresponding 95 percent confidence intervals.

(c) Graph the estimated covariate-adjusted survivorship functions comparing the survival experience of those with and without left heart complications, that is, CHF = 0 vs. CHF = 1.

5. Review the discussion in Section 9.5 of the plots of the estimated cumulative regression coefficients obtained from fitting Aalen's additive model to the UIS data.

(a) Fit a proportional hazards model using the time-varying covariates suggested in the plots. Assess the need to include additional interactions. Compare this new "final" model to the previously obtained final model shown in Tables 5.11 and 6.6. Which model provides the better description of the effect of the covariates? Consider this question from both statistical and clinical perspectives.

(b) Use plots from the estimated cumulative regression coefficients from a fit of the Aalen additive model to explore the possibility of a time-varying effect in NDRUGFP1 and NDRUGFP2.

6. Fit the Aalen model to the WHAS data using a model containing AGE, SEX, CHF and MIORD. Use plots from the estimated cumulative regression coefficients from a fit of the Aalen additive model to explore the possibility of a time-varying effect in AGE, SEX, CHF and MIORD. Fit the proportional hazards model containing AGE, SEX, CHF and MIORD. Fit the proportional hazards model including any time-varying covariate effects suggested by the plots of the estimated cumulative regression coefficients from the fit of the Aalen model. Are the time-varying effects significant and does their inclusion improve the model from a statistical and clinical perspective?

APPENDIX 1

The Delta Method

A problem faced by statisticians when developing an estimator of a parameter is deriving an expression for an estimator of the variance of the estimator. Both estimators are needed for confidence interval estimation and/or hypothesis testing.

A procedure commonly called the *delta method* has been used by statisticians for many years to obtain an estimator of the variance when the estimator is not a simple sum of observations. The basic idea is to use a method from calculus called a *Taylor series expansion* to derive a linear function that approximates the more complicated function. We refer the reader to any introductory calculus text for a discussion of the Taylor series expansion.

To apply the delta method, the function must be one that can be approximated by a Taylor series and, in general, this means that it is a "smooth" function, with no "corners." Consider such a function of a random variable X denoted as $f(X)$. To apply the delta method we use the first two terms of a Taylor series expansion about the mean of the variable to approximate the value of the function as

$$f(X) \cong f(\mu) + (X - \mu)f'(\mu) \qquad \text{(A.1)}$$

where

$$f'(\mu) = \frac{\partial f(X)}{\partial X}\Big|_{X=\mu}$$

is the derivative of the function with respect to X evaluated at the mean of X. It follows from (A.1) that the variance of the function is approximately

$$\text{Var}\big[f(X)\big] \cong \text{Var}(X - \mu) \times \big[f'(\mu)\big]^2$$

$$\cong \sigma^2 \times \big[f'(\mu)\big]^2, \tag{A.2}$$

where σ^2 is the variance of X. The delta method estimator of the variance of the function is obtained when we use the estimators of μ and σ^2 in (A.2) as follows:

$$\widehat{\text{Var}}\big[f(x)\big] \cong \hat{\sigma}^2 \times \big[f'(\hat{\mu})\big]^2. \tag{A.3}$$

As an example, consider the function $\ln(X)$. The expansion from (A.1) is

$$\ln(X) \cong \ln(\mu) + (X - \mu)\frac{1}{\mu}. \tag{A.4}$$

The delta method estimator of the variance from (A.3) is

$$\widehat{\text{Var}}\big[\ln(X)\big] \cong \hat{\sigma}^2 \frac{1}{\hat{\mu}^2},$$

where $\hat{\sigma}^2$ and $\hat{\mu}$ denote estimators of σ^2 and μ.

As a second example, we provide the details for the development of the delta method estimator of the variance of the log of the Kaplan–Meier estimator of the survivorship function shown in (2.3) and that of the Kaplan–Meier estimator itself in (2.5). The estimator as shown in (2.1) is

$$\hat{S}(t) = \prod_{t_{(i)} \leq t} \frac{n_i - d_i}{n_i},$$

and its log is

$$\ln\big[\hat{S}(t)\big] = \sum_{t_{(i)} \leq t} \ln\left(\frac{n_i - d_i}{n_i}\right)$$

$$= \sum_{t_{(i)} \leq t} \ln(\hat{p}_i),$$

where $\hat{p}_i = (n_i - d_i)/n_i$. The first key assumption in the development of the variance estimator is that the observations of survival among the n_i subjects at risk are independent Bernoulli with constant probability, p_i. Under this assumption the estimator of the constant probability is $\hat{p}_i$ with variance estimator $\hat{p}_i(1 - \hat{p}_i)/n_i$. The Taylor series expansion for the log function in (A.4) yields

$$\ln(\hat{p}_i) \cong \ln(p_i) + (\hat{p}_i - p_i)\frac{1}{\hat{p}_i}$$

and from (A.3) the delta method variance estimator is

$$\widehat{\text{Var}}[\ln(\hat{p}_i)] \cong \frac{1}{\hat{p}_i^2}\frac{\hat{p}_i(1 - \hat{p}_i)}{n_i}$$

$$\cong \frac{d_i}{n_i(n_i - d_i)} .$$

The second key assumption is that observations in different risk sets are independent. Thus the delta method estimator of the variance of the log of the Kaplan–Meier estimator is, as shown in (2.3),

$$\widehat{\text{Var}}\left\{\ln\left[\hat{S}(t)\right]\right\} \cong \sum_{t_{(i)} \le t} \widehat{\text{Var}}[\ln(\hat{p}_i)]$$

$$\cong \sum_{t_{(i)} \le t} \frac{d_i}{n_i(n_i - d_i)} .$$

The estimator of the variance of the Kaplan–Meier estimator comes from a second application of the delta method. In this application the function is

$$f(X) = \exp(X),$$

e.g., $\hat{S}(t) = \exp\left\{\ln\left[\hat{S}(t)\right]\right\}$. It follows from (A.1) that the series expansion is

$$\exp(X) \cong \exp(\mu) + (X - \mu)\exp(\mu)$$

and, from (A.2) the approximate variance is

$$\hat{\text{Var}}[\exp(X)] \cong \sigma^2 [\exp(\mu)]^2 . \tag{A.5}$$

Application of the approximation in (A.5) yields the Greenwood estimator in (2.5), namely

$$\hat{\text{Var}}[\hat{S}(t)] \cong [\hat{S}(t)]^2 \sum_{t_{(i)} \leq t} \frac{d_i}{n_i(n_i - d_i)} .$$

The confidence interval estimator for the Kaplan–Meier estimator discussed in Chapter 2 is based on the log-log survivorship function, that is, $\ln\{-\ln[\hat{S}(t)]\}$. The variance estimator of this function requires a second application of the expansion of the log function. In this case $X = \ln[\hat{S}(t)]$ and application of the approximation to the variance of the log of a random variable yields the estimator shown in (2.6), namely,

$$\hat{\text{Var}}\{\ln[-\ln(\hat{S}(t))]\} \cong \frac{1}{[\ln(\hat{S}(t))]^2} \sum_{t_{(i)} \leq t} \frac{d_i}{n_i(n_i - d_i)} .$$

The results presented in this appendix provide a brief introduction to the use of the delta method within the specific context of deriving an estimator of the variance of the Kaplan–Meier estimator, or functions of it. The technique is quite general and has been used in a variety of settings [see Agresti (1990) for applications in general categorical data models].

APPENDIX 2

An Introduction to the Counting Process Approach to Survival Analysis

We refer to the counting process approach to the analysis of survival time throughout the text. This method has been the source of many new developments in the field since it was first used by Aalen [(1975) and (1978)]. Two recent texts, Fleming and Harrington (1991) and Andersen, Borgan, Gill and Keiding (1993), document the mathematical details of this powerful method in a thorough manner. Fleming and Harrington (1991) focus primarily on the analysis of survival time while Andersen et al. (1993) consider analysis of survival time as well as other, more general statistical problems. We encourage readers of this text to see Andersen et al. (1993, Chapter 1) for an excellent overview of the types of statistical problems that can be formulated as counting processes. Nontechnical mathematical introductions to the approach are provided by Fleming and Harrington (1991, Chapter 0) and Andersen et al. (1993, Section II.1). The purpose of this appendix is to introduce a few of the key ideas and constructs used in the counting process approach to the analysis of survival time. For this reason many of the more technical mathematical assumptions and details will not be discussed.

For the time being, suppose we follow a single subject from time of enrollment, $t = 0$, in a study of a particular cancer until death from this cancer. Furthermore, we assume it is a 5-year study and that this subject is enrolled on the first day of the study. Thus the maximum length of follow-up for this subject is 60 months. We denote the survival time random variable as X. A common approach to modeling the possibility of right censoring is to assume that there is a second random variable, independent of X, that records the time until observation terminates due to anything other than the event of interest, for example, death from another cause or loss to follow-up for reasons unrelated to any study factor. We denote this random variable as Z. The actual observed time random variable is $T = \min(X, Z)$ and the available data for a subject consists of T and an indicator variable C whose value is 1 if $T = X$ and 0 if $T = Z$. Thus the variable T records follow-up time and C is the censoring indicator variable.

Three functions of time central to the counting process approach are: the counting process

$$N(t) = I(T \leq t, C = 1),$$

the at risk process

$$Y(t) = I(T \geq t)$$

and the intensity process

$$\lambda(t)dt = Y(t)h(t)dt \; ,$$

where

$$h(t)dt = \Pr(t \leq T < t + dt, C = 1 | T \geq t)$$

is the hazard function for survival time. The function $I(\cdot)$ is the indicator function whose value is 1 if the argument is true and 0 otherwise.

The counting process records, in our example, whether death from cancer occurs at time t. The function "counts" this by jumping from a value of 0 to a value of 1. For example, suppose our hypothetical subject died from cancer after being in the study 32 months, $(T = 32, C = 1)$. The counting process function for this subject is equal to zero until 32 months. At exactly 32 months the function jumps to a value of 1. The function is equal to 1 for the remaining 28 months of follow-up time. If the subject's follow-up time is right censored, $C = 0$, then a death is not counted and the counting process is equal to 0 for all values of t. For example, if our hypothetical subject was removed from the study at 32 months for reasons unrelated to the cancer, $(T = 32, C = 0)$, the counting process for this subject is equal to zero over the 60 months of follow-up.

The at risk process indicates whether the subject is still being followed, at risk for death, at time t. This function jumps from a value of 1 to a value of 0 when follow-up ends due to death or censoring. For a hypothetical subject with follow-up time of 32 months, the at risk process is equal to 1 from the beginning of follow-up until 32 months. The function jumps/drops to a value of zero just after 32 months as the subject is no longer at risk for the remaining 28 months of the study.

The intensity process may be viewed as being like an "expected number of deaths" at time t. This follows from the fact that the function is of the form " $n \times p$ " (i.e., the expected number of events in a binomial distribution). The at risk process corresponds to " n " and the hazard function to " p ".

The process of following the hypothetical subject from time zero to time t may be thought of as an accumulation of many conditional independent steps, much like the argument used to construct the Kaplan–Meier estimator in Chapter 2. The total expected number of deaths up to time t is obtained from the intensity process in the same manner as the cumulative hazard is obtained from the hazard function, namely by integrating the intensity process over time to obtain

$$\Lambda(t) = \int_0^t \lambda(u) \, du$$

$$= \int_0^t Y(u)h(u) \, du,$$

and this function is called the cumulative intensity process.

Thus one may think of the counting process as the total number of observed events and the cumulative intensity process as the total number of expected events up to time t. The difference between these two quantities is a residual-like quantity called the *counting process martingale*, namely

$$M(t) = N(t) - \Lambda(t).$$

This function is the basis for the martingale residuals that play a central role in model evaluation methods; see Chapter 6. Another way to express the relationship between the counting, intensity and martingale processes is via a linear-like model

$$N(t) = \Lambda(t) + M(t).$$

When expressed in this way we see that the counting process, the observed part of the model, is the sum of a systematic component, the cumulative intensity process and a residual, the martingale process. In our hypothetical example death or censoring can occur only one time and at the actual follow-up time, T, the value of the martingale is

$$M(T) = \begin{cases} 1 - \Lambda(T) \text{ if } C = 1, \\ 0 - \Lambda(T) \text{ if } C = 0. \end{cases}$$

It is well beyond the scope of this appendix to explain what makes a process a martingale and what gives $M(t)$ this quality. We refer the interested reader to the texts cited above for these technical details. It suffices for the purposes of this appendix and text to think of $M(t)$ as being similar to a residual.

Now suppose we have observations of follow-up time and censoring indicator variable on n subjects in our hypothetical cancer study. We assume that observations of time are independent and identically distributed. We denote the actual observed times and right censoring indicator variables in the usual way as (t_i, c_i), $i = 1, 2, \ldots, n$. In this setting, a basic result from counting process theory is that the estimator of the cumulative intensity process for the ith subject at time t is

$$\hat{\Lambda}_i(t) = Y_i(t)\hat{H}(t),$$

where

$$Y_i(t) = I(t_i \geq t),$$

$$\hat{H}(t) = \sum_{t_j \leq t} \frac{c_j}{n_j}$$

is the Nelson–Aalen estimator of the cumulative hazard at t and

$$n_j = \sum_{i=1}^{n} Y_i(t_j)$$

is the number at risk at time t_j. The estimator of the martingale residual for the ith subject at his/her follow-up time is

$$\hat{M}(t_i) = c_i - \hat{\Lambda}(t_i)$$

$$= c_i - Y_i(t_i)\hat{H}(t_i)$$

$$= c_i - \hat{H}(t_i)$$

since $Y_i(t_i) = I(t_i \geq t_i) = 1$. We denote this martingale residual as $\hat{M}_i$. We note that, like residuals from most regression models, $\Sigma \hat{M}_i = 0$.

Assume that we have, in addition to follow-up time and censoring indicator variables, observations on p fixed (not time-varying) covariates. Assume that we fit a proportional hazards regression model. The estimator of the cumulative intensity process for the ith subject at time t is

$$\hat{\Lambda}\left(t, \mathbf{x}_i, \hat{\boldsymbol{\beta}}\right) = -Y_i(t)e^{\mathbf{x}_i'\hat{\boldsymbol{\beta}}} \ln\left[\hat{S}_0(t)\right] .$$

Thus the value of the martingale residual for the ith subject at his/her follow-up time is

$$\hat{M}_i = c_i - e^{\mathbf{x}_i'\hat{\boldsymbol{\beta}}} \ln\left[\hat{S}_0(t_i)\right] .$$

The estimated martingale residuals are the basis for many of the diagnostic methods for assessing various aspects of the fitted model; see Chapters 5 and 6.

One, if not the, major theoretical benefit derived from formulating a survival analysis as a counting process is that a number of theorems from martingale theory may be used to prove many of the distributional results cited in this text. For example, this theory may be used to prove that the maximum partial likelihood estimators of the coefficients in a proportional hazards model are asymptotically normally distributed with a covariance matrix that may be estimated by the observed information matrix (Chapter 3). A second example involves the proof that the Kaplan–Meier estimator and functions of it are asymptotically normally distributed (Chapter 2). The list of applications of this theory in survival analysis is quite long. The central theme in all of the applications involves proving that a particular scaled and centered estimator, such as the Kaplan–Meier estimator, $\sqrt{n}\left[\hat{S}(t) - S(t)\right]$, is a martingale.

In summary we feel that it is important for anyone using the regression methods for the analysis of survival time described in this text to have at least a superficial knowledge of the basics of the counting process paradigm.

APPENDIX 3

Percentiles for Computation of the Hall and Wellner Confidence Band

$1-\alpha$	$\hat{a} = n\hat{\sigma}^2(t_{(m)}) \big/ \left[1 + n\hat{\sigma}^2(t_{(m)})\right]$							
	0.1	0.25	0.40	0.50	0.60	0.75	0.90	1.0
0.90	0.599	0.894	1.062	1.133	1.181	1.217	1.224	1.224
0.95	0.682	1.014	1.198	1.273	1.321	1.354	1.358	1.358
0.99	0.851	1.256	1.470	1.552	1.600	1.626	1.628	1.628

$$\hat{\sigma}^2(t_{(m)}) = \sum_{t_{(i)}} \frac{d_i}{n_i(n_i - d_i)}$$

REFERENCES

Aalen, O.O. (1975). Statistical inference for a family of country processes. Ph.D. Thesis, University of California, Berkeley.

Aalen, O.O. (1978). Non parametric inference for a family of counting processes. *Annals of Statistics*, **6**:701–726.

Aalen, O.O. (1989). A linear regression model for the analysis of life times. *Statistics in Medicine*, **8**:907–925.

Aalen, O.O. (1993). Further results on the non-parametric linear regression model in survival analysis. *Statistics in Medicine*, **12**:1509–1588.

Aalen, O.O. (1994). Effects of frailty in survival analysis. *Statistical Methods in Medical Research*, **3**:227–243.

Agresti, A. (1990). *Categorical Data Analysis*. John Wiley & Sons, Inc. New York.

Akaike, H. (1974). A new look at statistical model identification, *IEEE Transactions on Automatic Control*, **19**:716–723.

Alioum, A. and Commenges, D. (1996). A proportional hazards model for arbitrarily censored and truncated data. *Biometrics*, **52**:512–524.

Altshuler, B. (1970). Theory for the measurement of competing risks in animal experiments. *Math Bioscience*, **6**:1–11.

Andersen, P.K. (1992). Repeated assessment of risk factors in survival analysis. *Statistical Methods in Medical Research*, **1**:297–315.

Andersen, P.K., Borgan, Ø., Gill, R.D. and Keiding, N. (1993). *Statistical Models Based on Counting Processes*. Springer–Verlag, New York.

Andersen, P.K., Klein, J.P., Knudsen, K.M., and Tabanera y Palacios, R. (1997). Estimation of variance in Cox's regression model with shared gamma frailties. *Biometrics*, **53**:1475–1484.

Arjas, E. (1988). A graphical method for assessing goodness–of–fit in Cox's proportional hazards model. *Journal of American Statistical Association*, **83**:204–212.

Barlow, W.E. and Prentice, R.L. (1988). Residuals for relative risk regression. *Biometrics*, **75**:65–74.

Bendel, R.B. and Afifi, A.A. (1977). Comparison of stopping rules in forward regression. *Journal of American Statistical Association*, **72**:46–53.

BMDP Statistical Software (1992). BMDP Classic 7.0 for DOS. SPSS, Inc. Chicago.

Borgan, Ø. and Langholz, B. (1993). Nonparametric estimation of relative mortality from nested case-control studies. *Biometrics*, **49**:593–602.

Borgan, Ø. and Langholz, B. (1997). Estimation of excess risk from case–control data using Aalen's linear regression model. *Biometrics*, **53**:690–697.

Borgan, Ø., Goldstein, L. and Langholz, B. (1995). Methods for the analysis of sampled cohort data in the Cox proportional hazards model. *Annals of Statistics*, **23**:1749–1778.

Borgan, Ø. and Leistøl. K. (1990). A note on confidence bands for the survival curve based on transformations. *Scandanavian Journal of Statistics*, **17**:35–41.

Bendel, R.B. and Afifi, A.A. (1977). Comparison of stopping rules in forward regression. *Journal of the American Statistical Association*, **72**:46–53.

Breslow, N.E. (1970). A generalized Kruskal–Wallace test for comparing K samples subject to unequal patterns of censorship. *Biometrika*, **57**:579–594.

Breslow, N.E. (1974). Covariance analysis of censored survival data. *Biometrics*, **30**:89–100.

Breslow, N.E. and Day, N.E. (1980). *Statistical Methods in Cancer Research. Volume I: The Analysis of Case-Control Studies*. Oxford University Press, Oxford, U.K.

Breslow, N.E. and Day N.E. (1987). *Statistical Methods in Cancer Research. Volume II: The Design and Analysis of Cohort Studies*. Oxford University Press, Oxford, U.K.

Brookmeyer, R and Crowley, J.J. (1982). A confidence interval for the median survival time. *Biometrics*, **38**:29–41.

Bryson, M.C. and Johnson, M.E. (1981). The incidence of monotone likelihood in the Cox model. *Technometrics*, **23**:381–384.

Cain, K.C. and Lange, N.T. (1984). Approximate case influence for the proportional hazards regression model with censored data. *Biometrics*, **40**:493–499.

Carstensen, B. (1996). Regression models for interval censored data: Application to HIV infection in Danish homosexual men. *Statistics in Medicine*, **15**:2177–2189.

Chappell, R. (1992). A note on linear rank tests and Gill and Schumacher's test of proportionality. *Biometrika*, **79**:199–201.

Chiriboga, D., Yarzebeski, J., Goldberg, R.J., Gore, J.M. and Alpert, J.S. (1994). Temporal trends (1975–1990) in the incidence and case-fatality rates of primary ventricular fibrillation complicating acute myocardial infarction: A community wide perspective. *Circulation*, **89**:998–1003.

Clayton, D. (1994). Some approaches to the analysis of recurrent event data. *Statistical Methods in Medical Research*, **3**:244–262.

Clayton, D.G. and Hills, M. (1993). *Statistical Models in Epidemiology*, Oxford University Press, Oxford, U.K.

Clayton D.G. and Hills M. (1997). Analysis of follow-up studies with Stata 5.0. *Stata Technical Bulletin Number 40*. Stata Corporation, College Station, TX.

Cleveland, W.S. (1993). *Visualizing Data*, Hobart Press, Summit, NJ.

Collett, D. (1991). *Modelling Binary Data*. Chapman Hall, London, U.K.

Collett, D. (1994). *Modelling Survival Data in Medical Research*. Chapman Hall, London, U.K.

Commenges, D. and Andersen, P.K. (1995). A score test of homogenity for survival data. *Life Time Data Analysis*, **1**:145–160.

Costanza, M.C. and Afifi, A.A. (1979). Comparison of stopping rules in forward stepwise discriminant analysis. *Journal of the American Statistical Association*, **74**:777–785.

Cox, D.R. (1972). Regression models and life tables (with discussion). *Journal of Royal Statistical Society*: Series B, **34**:187–220.

Cox, D.R. and Oakes, D. (1984). *Analysis of Survival Data*. Chapman Hall, London, U.K.

Cox, D.R. and Snell, E.J. (1968). A general definition of residuals with discussion. *Journal of Royal Statistical Society*: Series A, **30**:248–275.

Crowder, M.J., Kimber, A.C., Smith, R.L. and Sweeting, T.J. (1991). *Statistical Analysis of Reliability Data*. Chapman Hall, London, U.K.

Crowley, J. and Hu, M. (1977). Covariance analysis of heart transplant survival data. *Journal of American Statistical Association*, **78**:27–36.

DeGruttola, V. and Lagakos, S.W. (1989). Analysis of doubly–censored survival data with application to AIDS. *Biometrics*, **45**:1–12.

Dempster, A.P., Laird, N.M. and Rubin, D.R. (1977). Maximum likelihood estimation from incomplete data via the EM algorithm (with discussion). *Journal of the Royal Statistical Society*: Series B, **39**:1–38.

Efron, B. (1977). The efficiency of Cox's likelihood function for censored data. *Journal of American Statistical Association*, **72**:557–565.

Elandt–Johnson, R.C. and Johnson, N.L. (1980). *Survival Models and Data Analysis*. John Wiley & Sons, Inc. New York.

Evans, M., Hastings, N. and Peacock, B. (1993). *Statistical Distributions.* 2nd edition. Wiley–Interscience, John Wiley & Sons, Inc. New York.

Farrington, C.P. (1996). Interval censored survival data: A generalized linear modeling approach. *Statistics in Medicine,* **15**:283–292.

Finkelstein, D.M. (1986). A proportional hazards model for interval censored failure time data. *Biometrics,* **42**:845–854.

Fisher, L.L. (1992). Discussion of Session 1: Clinical trials, survival analysis. *Statistics in Medicine,* **11**:1881–1885.

Fleming, T.R. and Harrington, D.P. (1984). Nonparametric estimation of the survival distribution in censored data. *Communication in Statistics: Theory and Methods,* **13**:2469–2486.

Fleming, T.R. and Harrington, D.P. (1991). *Counting Process and Survival Analysis.* John Wiley & Sons, Inc. New York.

Fleming, T.R., Harrington, D.P. and O'Sullivan, M. (1987). Supremium versions of the logrank and generalized Wilcoxon statistics. *Journal of the American Statistical Association,* **82**:312–320.

Gehan, E.A. (1965). A generalized Wilcoxon test for comparing arbitrarily singly–censored samples. *Biometrics,* **52**:203–223.

Gill, R. and Schumacher, M. (1987). A simple test of the proportional hazards assumption. *Biometrika,* **75**:289–300.

Goldberg, R.J., Gore, J.M., Alpert, J.S. and Dalen, J.E. (1986). Recent changes in the attack rates and survival of acute myocardial infarction (1975–1981): The Worcester Heart Attack Study. *Journal of the American Medical Association,* **255**:2774–2779.

Goldberg, R.J., Gore, J.M., Alpert, J.S. and Dalen, J.E. (1988). Incidence and case fatality rates of acute myocardial infarction (1975–1984): The Worcester Heart Attack Study. *American Heart Journal,* **115**:761–767.

Goldberg, R.J., Gore, J.M., Gurwitz, J.H., Alpert, J.S, Brady, P., Stohsnitter, W., Chen, Z. and Dalen, J.E. (1989). The impact of age on the incidence and prognosis of initial myocardial infarction: The Worcester Heart Attack Study. *American Heart Journal*, **117**:543–549.

Goldberg, R.J., Gore, J.M., Alpert, J.S., Osganian, V., de Groot, J., Bade, J., Chen, Z., Frid, D. and Dalen, J. (1991). Cardiogenic shock after acute myocardial infarction: Incidence and mortality from a community wide perspective, 1975–1988. *New England Journal of Medicine*, **325**:1117–1122.

Goldberg, R.J., Gorak, E.J., Yarzebski, J., Hosmer, D.W., Dalen, P., Gore, J.M. and Dalen, J.E. (1993). A community-wide perspective of gender differences and temporal trends in the incidence and survival rates following acute myocardial infarction and out-of-hospital deaths due to coronary heart disease. *Circulation*, **87**:1997–1953.

Grambsch, P.M. and Therneau, T.M. (1994). Proportional hazards tests in diagnostics based on weighted residuals. *Biometrika*, **81**:515–526.

Grambsch, P.M., Therneau, T.M. and Fleming, T.R. (1995). Diagnostic plots to reveal functional form for covariates in multiplicative intensity models. *Biometrics*, **51**:1469–1482.

Gray, R.J. (1992). Flexible methods for analyzing survival data using splines, with applications to breast cancer prognosis. *Journal of the American Statistical Association*, **87**:942–951.

Greenwood, M. (1926). The natural duration of cancer. *Reports on Public Health and Medical Subjects*. **33**:1–26. Her Majesty's Stationery Office, London.

Gross, A.J. and Clark, V.A. (1975). *Survival Distributions: Reliability Applications in the Biomedical Sciences*. John Wiley & Sons, Inc. New York.

Grønnesby, J.K. and Borgan, Ø. (1996). A method for checking regression models in survival analysis based on the risk score. *Lifetime Data Analysis*, **2**:315–328.

Hald, A. (1990). *A History of Probability and Statistics and Their Applications Before 1750*. John Wiley & Sons, Inc. New York.

Hall, W.J. and Wellner, J.A. (1980). Confidence bands for a survival curve from censored data. *Biometrika*, **67**:133–143.

Hall, W.J., Rogers, W.H. and Pregibon, D. (1982). Outliers matter in survival analysis. *Rand Corporation Technical Report P*-6761. Santa Monica, CA.

Harrell, F.E., Lee, K.L. and Mark, D.B. (1996). Tutorial in biostatistics. Multivariable models: issues in developing models, evaluating assumptions and adequacy and measuring and reducing errors. *Statistics in Medicine*, **15**:361–387.

Harrington, D.P. and Fleming, T.R. (1982). A class of rank test procedures for censored survival data. *Biometrika*, **69**:553–566.

Henderson, R. and Milner, A. (1991). Aalen plots under proportional hazards. *Applied Statistics*, **40**:401–409.

Hosmer, D.W. and Lemeshow, S. (1989). *Applied Logistic Regression*. John Wiley & Sons, Inc. New York.

Hougaard, P. (1995). Frailty Models for Survival data. *Lifetime Data Analysis*, **1**:255–274.

Huffler, F.W. and McKeague, I.W. (1991). Weighted least squares estimation for Aalen's additive risk model. *Journal of the American Statistical Association*, **86**:114–129.

Jenkins, S.P. (1997). Discrete time proportional hazards regression. *Stata Technical Bulletin*. **39**:17–32.

Johnson, N.L. and Kotz, S. (1997). *Discrete Multivariate Distributions*. John Wiley & Sons, Inc. New York.

Kabat–Zinn, J., Wheeler, E., Light, T., Skillings, A., Scharf, M.J., Cropley, T.G., Hosmer, D. and Bernhard, J.D. (1998). Influence of a mindfulness meditation-based stress reduction intervention on rates of skin clearing in patients with moderate to severe psoriasis undergoing phototherapy (UVB) and photochemotherapy (PUVA). *Psychosomatic Medicine*, to appear.

Kalbfleisch, J.D. and Prentice, R.L. (1980). *The Statistical Analysis of Failure Time Data*. John Wiley & Sons, Inc. New York.

Kaplan, E.L. and Meier, P. (1958). Nonparametric estimation from incomplete observations. *Journal of the American Statistical Association,* **53**:457–481.

Klein, J.P. and Moeschberger, M.L. (1997). *Survival Analysis Techniques for Censored and Truncated Data.* Springer–Verlag. New York.

Kleinbaum, D.G. (1997) *Survival Analysis: A Self–Learning Text.* Springer–Verlag. New York.

Kleinbaum, D.G., Kupper, L.L., Muller, K.E. and Nizam, A. (1998). *Applied Regression Analysis and Multivariable Methods.* 3rd edition. Duxbury Press. Pacific Grove, CA.

Kuk, A.Y.C. (1984). All subsets regression in a proportional hazards model. *Biometrika,* **71**:587–592.

Langholz, B. and Borgan, Ø. (1995). Counter-matching: A stratified nested case–control sampling method. *Biometrika,* **82**:69–79.

Lawless, J.F. (1982). *Statistical Models and Methods for Lifetime Data.* John Wiley & Sons, Inc. New York.

Lawless, J.F. and Singhal, K. (1978). Efficient screening of non-normal regression models. *Biometrics,* **34**:318–327.

Le, C.T. (1997). *Applied Survival Analysis.* John Wiley & Sons, Inc. New York.

Lee, E.T. (1992). *Statistical Methods for Survival Data Analysis.* John Wiley & Sons, Inc. New York.

Liang, K.Y. and Zeger, S.L. (1986a). Longitudinal data analysis, using generalized linear models. *Biometrika,* **73**:13–22.

Liang, K.Y. and Zeger, S.L. (1986b). Longitudinal data analysis for discrete and continuous models. *Biometrics,* **42**:121–130.

Liang, K.Y., Self, S.G., Bandeen–Roche, K.J. and Zeger, S. (1995). Some recent developments for regression analysis of multivariate failure time data. *Lifetime Data Analysis,* **1**:403–416.

Lin, D.Y. and Wei, L.J. (1989). The robust inference for the Cox proportional hazards model. *Journal of the American Statistical Association*, **84**:1074–1078.

Lin, D.Y. and Ying, Z. (1994). Semiparametric analysis of the additive risk model. *Biometrika*, **81**:61–71.

Lin, D.Y., Wei, L.J. and Ying, Z. (1993). Checking the Cox model with cumulative sums of Martingale based residuals. *Biometrika*, **80**:557–572.

McCullagh, P. and Nelder, J.A. (1989). *Generalized Linear Models*. 2nd edition. Chapman Hall. London, U.K.

McCusker, J., Vickers–Lahti, M., Stoddard, A.M., Hindin, R., Bigelow, C., Garfield, F., Frost, R., Love, C. and Lewis, B.F. (1995). The effectiveness of alternative planned durations of residential drug abuse treatment. *American Journal of Public Health*, **85**:1426–1429.

McCusker, J., Bigelow, C., Frost, R., Garfield, F., Hindin, R., Vickers–Lahti, M. and Lewis, B.F. (1997a). The effects of planned duration of residential drug abuse treatment on recovery and HIV risk behavior. *American Journal of Public Health*, **87**:1637–1644.

McCusker, J., Bigelow, C., Vickers–Lahti, M., Spotts, D., Garfield, F. and Frost, R. (1997b). Planned duration of residential drug abuse treatment: efficacy versus treatment. *Addiction*, **92**:1467–1478.

McKeague, I.W. and Sasieni, P.D. (1994). A partly parametric additive risk model. *Biometrika*, **81**:501–514.

Makuch, R.W. (1982). Survival curve estimation using covariates. *Journal of Chronic Diseases*, **3**:437–443.

Mallows, C. (1973). Some Comments on Cp. *Technometrics*, **15**:661–676.

Mann, N., Schaefer, R.E. and Singparwalla, N.D. (1974). *Methods for Statistical Analysis of Reliability and Life Data*. John Wiley & Sons, Inc. New York.

Mantel, N. (1966). Evaluation of survival data and two new rank order statistics arising in its consideration. *Cancer Chemotherapy Reports*, **50**:163–170.

Marubini, E. and Valsecchi, M.G. (1995). *Analyzing Survival Data from Clinical Trials and Observational Studies.* John Wiley & Sons, Ltd. Chichester, U.K.

Math Type: Mathematical Equation Editor (1997). Design Sciences, Inc. Long Beach, CA 90803.

Mau, J. (1986). On a graphical method for the detection of time-dependent effects of covariants in survival data. *Applied Statistics,* **35**:245–255.

May, S. and Hosmer, D.W. (1998). A simplified method for calculating a goodness-of-fit test for the proportional hazards model. *Lifetime Data Analysis,* **4**:109–120.

Mickey, J. and Greenland, S. (1989). A study of the impact of confounder selection criteria on effect estimation. *American Journal of Epidemiology,* **129**:125–137.

Miller, R.G. (1981). *Survival Analysis.* John Wiley & Sons, Inc. New York.

Nelson, W. (1969). Hazard plotting for incomplete failure data. *Journal of Quality Technology,* **1**:27–52.

Nelson, W. (1972). Theory and application of hazard plotting for censored failure data. *Technometrics,* **14**:945–965.

Nelson, W. (1982). *Applied Life Data Analysis.* John Wiley & Sons, Inc. New York.

Nelson, W. (1990). *Accelerated Testing, Statistical Models, Test Plans, and Data Analysis.* John Wiley & Sons, Inc. New York.

Neuhaus, J. (1992). Statistical methods for longitudinal and clustered designs with binary outcomes. *Statistical Methods in Medical Research,* **1**:249–273.

Nielsen, G.G., Gill, R.D., Andersen, P.K. and Sørensen, T.I.A. (1992). A counting process approach to maximum likelihood estimation in frailty models. *Scandinavian Journal of Statistics,* **19**:25–43.

Ng'andu, N.H. (1997). An empirical comparison of statistical tests for assessing the proportional hazards assumption of Cox's model. *Statistics in Medicine,* **16**:611–626.

Oakes, D. (1981). Survival times: Aspects of partial likelihood (with discussion). *International Statistical Review*, **49**:235–264.

O'Quigley, J. and Pessione, F. (1989). Score tests for homogenity of regression effect in the proportional hazards model. *Biometrics*, **45**:135–144.

Parmar, M.K.B. and Machin, D. (1995). *Survival Analysis: A Practical Approach*. John Wiley & Sons, Ltd. Chichester, U.K.

Peterson, A.V. (1977). Expressing the Kaplan–Meier estimation as a function of empirical subsurvival functions. *Journal of the American Statistical Association*, **72**:854–858.

Peto, R. and Peto, J. (1972). Asymptotically efficient rank invariance test procedures (with discussion). *Journal of the Royal Statistical Association*. Series A, **135**:185–206.

Pettitt, A.N. and Bin Daud, L. (1989). Case–weighted measures of influence for proportional hazards regression. *Applied Statistics*, **38**:51–67.

Pettitt, A.N. and Bin Daud, L. (1990). Investigating time dependence in Cox proportional hazards model. *Applied Statistics*, **39**:313–329.

Pickels, A., Crouchley, R., Simonoff, E.L., Meyer, J., Rutter, M., Hewitt, J. and Silbery, J. (1994). Survival models for developmental genetic data: Age at onset of puberty and antisocial behavior in twins. *Genetic Epidemiology*, **11**:155–170.

Prentice R.L. (1978). Linear rank tests with right censored data. *Biometrika*, **65**:167–179, Correction **70**:304 (1983).

Prentice, R.L. and Gloecker, L.A. (1978). Regression analysis of grouped survival data with application to breast cancer data. *Biometrics*, **34**:57–67.

Prentice, R.L., Williams, J. and Peterson, A.V. (1981). On the regression analysis of multivariate failure time data. *Biometrika*, **68**:373–379.

Quantin, C., Moreau, T., Asselain, B., Maccario, J. and Lellouch, J. (1996). A regression survival model for testing the proportional hazards hypothesis. *Biometrics*, **52**:874–885.

Rothman, K.J. and Greenland, S. (1998). *Modern Epidemiology*. 3rd edition. Lippincott–Raven. Philadelphia, PA.

Royston, P. and Altman, D.G. (1994). Regression using fractional polynomials of continuous covariates: Parsimonious parametric modeling (with discussion). *Applied Statistics*, **43**:429–467.

Ryan, T. (1997). *Modern Regression Methods*. John Wiley & Sons, Inc. New York.

SAS Institute Inc. (1989). SAS/STAT® User's Guide. Version 6, 4th edition, Volumes 1 and 2. SAS Institute, Cary, NC.

Savage, I.R. (1956). Contributions to the theory of rank-order statistics — the two sample case. *Annals of Mathematical Statistics,* **27**:590–615.

Schemper, M. and Stare, J. (1996). Explained variation in survival analysis. *Statistics in Medicine*, **15**:1999–2012.

Schoenfeld, D. (1980). Chi-squared goodness-of-fit tests for the proportional hazards model. *Biometrika*, **67**:145–153.

Schoenfeld, D. (1982). Partial residuals for the proportional hazards regression model. *Biometrika*, **69**:239–241.

S–PLUS–Statistical Sciences (1993). Statistical Analysis in S-PLUS, Version 3. StatSci, a division of MathSoft, Inc. Seattle, WA.

StataCorp (1997). Stata Statistical Software: Release 5.0. Stata Corporation, College Station, TX.

StatXact 3 for Windows (1995). Cytel Software, Cambridge, MA.

Tarone, R.E. and Ware, J. (1977). On distribution free tests of the equality of survival distributions. *Biometrika*, **64**:156–160.

Therneau, T.M. (1995). Extending the Cox model. Technical Report, Department of Biostatistics, Mayo Clinic, Rochester, MN.

Therneau, T.M., Grambsch, P.M. and Fleming, T.R. (1990). Martingale–based residuals for survival models. *Biometrika*, **77**:147–160.

Thomsen, B.L., Keiding, N. and Altman, D.G. (1991). A note on the calculation of expected survival, illustrated by the survival of liver transplant patients. *Statistics in Medicine*, **10**:733–738.

Tsiatis, A.A. (1980). A note on a goodness–of–fit test for the logistic regression model. *Biometrika*, **67**:250–251.

Vaupel, J.W., Manton, K.G. and Stallard, E. (1979). The impact of heterogenity in individual frailty on the dynamics of mortality. *Demography*, **16**:439–454.

Ware, J.H. and DeMets, D.L. (1976). Reanalysis of some baboon descent data. *Biometrics*, **32**:459–463.

Wei, L.J. (1992). The accelerated failure time model: A useful alternative to the Cox regression model in survival analysis. *Statistics in Medicine*, **11**:1871–1879.

Wei, L.J., Lin, D.Y. and Weissfeld (1989). Regression analysis of multivariate incomplete failure time data by modeling marginal distributions. *Journal of the American Statistical Association*, **84**:1065–1073.

White, H. (1980). A heteroskedasticity consistent covariance matrix estimator and a direct test for heteroskedasticity. *Econometrika*, **48**:817–838.

White, H. (1982). Maximum likelihood estimation of misspecified models. *Economterika*, **50**:1–26.

Yashin, A.I., Vaupel, J.W. and Jachine, I.A. (1995). Correlated individual frailty: An advantageous approach to survival analysis of bivariate data. *Mathematical Population Studies*, **5**:145–149.

Yuan, Shiaw–Shyaun (1993). Prediction of length of oxygen use in BPD babies, Masters Thesis, School of Public Health, University of Massachusetts, Amherst, MA.

Zahl, P.H. and Tretli S. (1997). Long-term survival of breast cancer patients in Norway by age and clinical stage. *Statistics in Medicine*, **13**:1435–1450.

Index

WILEY SERIES IN PROBABILITY AND STATISTICS
ESTABLISHED BY WALTER A. SHEWHART AND SAMUEL S. WILKS

Editors
Vic Barnett, Noel A. C. Cressie, Nicholas I. Fisher,
Iain M. Johnstone, J. B. Kadane, David G. Kendall, David W. Scott,
Bernard W. Silverman, Adrian F. M. Smith, Jozef L. Teugels;
Ralph A. Bradley, Emeritus, J. Stuart Hunter, Emeritus

Probability and Statistics Section

*Now available in a lower priced paperback edition in the Wiley Classics Library.

*Now available in a lower priced paperback edition in the Wiley Classics Library.

Applied Probability and Statistics (Continued)

BATES and WATTS · Nonlinear Regression Analysis and Its Applications

BECHHOFER, SANTNER, and GOLDSMAN · Design and Analysis of Experiments for Statistical Selection, Screening, and Multiple Comparisons

BELSLEY · Conditioning Diagnostics: Collinearity and Weak Data in Regression

BELSLEY, KUH, and WELSCH · Regression Diagnostics: Identifying Influential Data and Sources of Collinearity

BHAT · Elements of Applied Stochastic Processes, *Second Edition*

BHATTACHARYA and WAYMIRE · Stochastic Processes with Applications

BIRKES and DODGE · Alternative Methods of Regression

BLOOMFIELD · Fourier Analysis of Time Series: An Introduction

BOLLEN · Structural Equations with Latent Variables

BOULEAU · Numerical Methods for Stochastic Processes

BOX · Bayesian Inference in Statistical Analysis

BOX and DRAPER · Empirical Model-Building and Response Surfaces

BOX and DRAPER · Evolutionary Operation: A Statistical Method for Process Improvement

BUCKLEW · Large Deviation Techniques in Decision, Simulation, and Estimation

BUNKE and BUNKE · Nonlinear Regression, Functional Relations and Robust Methods: Statistical Methods of Model Building

CHATTERJEE and HADI · Sensitivity Analysis in Linear Regression

CHILÈS and DELFINER · Geostatistics: Modeling Spatial Uncertainty

CHOW and LIU · Design and Analysis of Clinical Trials: Concepts and Methodologies

CLARKE and DISNEY · Probability and Random Processes: A First Course with Applications, *Second Edition*

*COCHRAN and COX · Experimental Designs, *Second Edition*

CONOVER · Practical Nonparametric Statistics, *Second Edition*

CORNELL · Experiments with Mixtures, Designs, Models, and the Analysis of Mixture Data, *Second Edition*

*COX · Planning of Experiments

CRESSIE · Statistics for Spatial Data, *Revised Edition*

DANIEL · Applications of Statistics to Industrial Experimentation

DANIEL · Biostatistics: A Foundation for Analysis in the Health Sciences, *Sixth Edition*

DAVID · Order Statistics, *Second Edition*

*DEGROOT, FIENBERG, and KADANE · Statistics and the Law

DODGE · Alternative Methods of Regression

DOWDY and WEARDEN · Statistics for Research, *Second Edition*

DRYDEN and MARDIA · Statistical Shape Analysis

DUNN and CLARK · Applied Statistics: Analysis of Variance and Regression, *Second Edition*

ELANDT-JOHNSON and JOHNSON · Survival Models and Data Analysis

EVANS, PEACOCK, and HASTINGS · Statistical Distributions, *Second Edition*

FLEISS · The Design and Analysis of Clinical Experiments

FLEISS · Statistical Methods for Rates and Proportions, *Second Edition*

FLEMING and HARRINGTON · Counting Processes and Survival Analysis

GALLANT · Nonlinear Statistical Models

GLASSERMAN and YAO · Monotone Structure in Discrete-Event Systems

GNANADESIKAN · Methods for Statistical Data Analysis of Multivariate Observations, *Second Edition*

GOLDSTEIN and LEWIS · Assessment: Problems, Development, and Statistical Issues

GREENWOOD and NIKULIN · A Guide to Chi-Squared Testing

*HAHN · Statistical Models in Engineering

HAHN and MEEKER · Statistical Intervals: A Guide for Practitioners

HAND · Construction and Assessment of Classification Rules

*Now available in a lower priced paperback edition in the Wiley Classics Library.

*Now available in a lower priced paperback edition in the Wiley Classics Library.

Texts and References Section

AGRESTI · An Introduction to Categorical Data Analysis

ANDERSON · An Introduction to Multivariate Statistical Analysis, *Second Edition*

ANDERSON and LOYNES · The Teaching of Practical Statistics

ARMITAGE and COLTON · Encyclopedia of Biostatistics: Volumes 1 to 6 with Index

BARTOSZYNSKI and NIEWIADOMSKA-BUGAJ · Probability and Statistical Inference

BERRY, CHALONER, and GEWEKE · Bayesian Analysis in Statistics and Econometrics: Essays in Honor of Arnold Zellner

BHATTACHARYA and JOHNSON · Statistical Concepts and Methods

BILLINGSLEY · Probability and Measure, *Second Edition*

BOX · R. A. Fisher, the Life of a Scientist

BOX, HUNTER, and HUNTER · Statistics for Experimenters: An Introduction to Design, Data Analysis, and Model Building

BOX and LUCEÑO · Statistical Control by Monitoring and Feedback Adjustment

BROWN and HOLLANDER · Statistics: A Biomedical Introduction

CHATTERJEE and PRICE · Regression Analysis by Example, *Second Edition*

COOK and WEISBERG · An Introduction to Regression Graphics

COX · A Handbook of Introductory Statistical Methods

DILLON and GOLDSTEIN · Multivariate Analysis: Methods and Applications

DODGE and ROMIG · Sampling Inspection Tables, *Second Edition*

DRAPER and SMITH · Applied Regression Analysis, *Third Edition*

DUDEWICZ and MISHRA · Modern Mathematical Statistics

DUNN · Basic Statistics: A Primer for the Biomedical Sciences, *Second Edition*

FISHER and VAN BELLE · Biostatistics: A Methodology for the Health Sciences

FREEMAN and SMITH · Aspects of Uncertainty: A Tribute to D. V. Lindley

GROSS and HARRIS · Fundamentals of Queueing Theory, *Third Edition*

HALD · A History of Probability and Statistics and their Applications Before 1750

HALD · A History of Mathematical Statistics from 1750 to 1930

HELLER · MACSYMA for Statisticians

HOEL · Introduction to Mathematical Statistics, *Fifth Edition*

HOLLANDER and WOLFE · Nonparametric Statistical Methods, *Second Edition*

HOSMER and LEMESHOW · Applied Survival Analysis: Regression Modeling of Time to Event Data

JOHNSON and BALAKRISHNAN · Advances in the Theory and Practice of Statistics: A Volume in Honor of Samuel Kotz

JOHNSON and KOTZ (editors) · Leading Personalities in Statistical Sciences: From the Seventeenth Century to the Present

JUDGE, GRIFFITHS, HILL, LÜTKEPOHL, and LEE · The Theory and Practice of Econometrics, *Second Edition*

KHURI · Advanced Calculus with Applications in Statistics

KOTZ and JOHNSON (editors) · Encyclopedia of Statistical Sciences: Volumes 1 to 9 wtih Index

KOTZ and JOHNSON (editors) · Encyclopedia of Statistical Sciences: Supplement Volume

KOTZ, REED, and BANKS (editors) · Encyclopedia of Statistical Sciences: Update Volume 1

KOTZ, REED, and BANKS (editors) · Encyclopedia of Statistical Sciences: Update Volume 2

LAMPERTI · Probability: A Survey of the Mathematical Theory, *Second Edition*

LARSON · Introduction to Probability Theory and Statistical Inference, *Third Edition*

LE · Applied Categorical Data Analysis

LE · Applied Survival Analysis

MALLOWS · Design, Data, and Analysis by Some Friends of Cuthbert Daniel

MARDIA · The Art of Statistical Science: A Tribute to G. S. Watson

*Now available in a lower priced paperback edition in the Wiley Classics Library.

WILEY SERIES IN PROBABILITY AND STATISTICS

ESTABLISHED BY WALTER A. SHEWHART AND SAMUEL S. WILKS

Editors
Robert M. Groves, Graham Kalton, J. N. K. Rao, Norbert Schwarz, Christopher Skinner

Survey Methodology Section

*Now available in a lower priced paperback edition in the Wiley Classics Library.